To the REED COMMUNITY
with thanks for many good
years.

Tom Wieting
2001

THE MATHEMATICAL THEORY OF CHROMATIC PLANE ORNAMENTS

MONOGRAPHS AND TEXTBOOKS IN PURE AND APPLIED MATHEMATICS

1. *K. Yano,* Integral Formulas in Riemannian Geometry (1970) *(out of print)*
2. *S. Kobayashi,* Hyperbolic Manifolds and Holomorphic Mappings (1970) *(out of print)*
3. *V. S. Vladimirov,* Equations of Mathematical Physics (A. Jeffrey, editor; A. Littlewood, translator) (1970)
4. *B. N. Pshenichnyi,* Necessary Conditions for an Extremum (L. Neustadt, translation editor; K. Makowski, translator) (1971)
5. *L. Narici, E. Beckenstein, and G. Bachman,* Functional Analysis and Valuation Theory (1971)
6. *D. S. Passman,* Infinite Group Rings (1971)
7. *L. Dornhoff,* Group Representation Theory (in two parts). Part A: Ordinary Representation Theory. Part B: Modular Representation Theory (1971, 1972)
8. *W. Boothby and G. L. Weiss (eds.),* Symmetric Spaces: Short Courses Presented at Washington University (1972)
9. *Y. Matsushima,* Differentiable Manifolds (E. T. Kobayashi, translator) (1972)
10. *L. E. Ward, Jr.,* Topology: An Outline for a First Course (1972) *(out of print)*
11. *A. Babakhanian,* Cohomological Methods in Group Theory (1972)
12. *R. Gilmer,* Multiplicative Ideal Theory (1972)
13. *J. Yeh,* Stochastic Processes and the Wiener Integral (1973) *(out of print)*
14. *J. Barros-Neto,* Introduction to the Theory of Distributions (1973) *(out of print)*
15. *R. Larsen,* Functional Analysis: An Introduction (1973)
16. *K. Yano and S. Ishihara,* Tangent and Cotangent Bundles: Differential Geometry (1973)
17. *C. Procesi,* Rings with Polynomial Identities (1973)
18. *R. Hermann,* Geometry, Physics, and Systems (1973)
19. *N. R. Wallach,* Harmonic Analysis on Homogeneous Spaces (1973)
20. *J. Dieudonné,* Introduction to the Theory of Formal Groups (1973)
21. *I. Vaisman,* Cohomology and Differential Forms (1973)
22. *B.-Y. Chen,* Geometry of Submanifolds (1973)
23. *M. Marcus,* Finite Dimensional Multilinear Algebra (in two parts) (1973, 1975)
24. *R. Larsen,* Banach Algebras: An Introduction (1973)
25. *R. O. Kujala and A. L. Vitter (eds),* Value Distribution Theory: Part A; Part B. Deficit and Bezout Estimates by Wilhelm Stoll (1973)
26. *K. B. Stolarsky,* Algebraic Numbers and Diophantine Approximation (1974)
27. *A. R. Magid,* The Separable Galois Theory of Commutative Rings (1974)
28. *B. R. McDonald,* Finite Rings with Identity (1974)
29. *J. Satake,* Linear Algebra (S. Koh, T. Akiba, and S. Ihara, translators) (1975)

30. *J. S. Golan,* Localization of Noncommutative Rings (1975)
31. *G. Klambauer,* Mathematical Analysis (1975)
32. *M. K. Agoston,* Algebraic Topology: A First Course (1976)
33. *K. R. Goodearl,* Ring Theory: Nonsingular Rings and Modules (1976)
34. *L. E. Mansfield,* Linear Algebra with Geometric Applications: Selected Topics (1976)
35. *N. J. Pullman,* Matrix Theory and Its Applications (1976)
36. *B. R. McDonald,* Geometric Algebra Over Local Rings (1976)
37. *C. W. Groetsch,* Generalized Inverses of Linear Operators: Representation and Approximation (1977)
38. *J. E. Kuczkowski and J. L. Gersting,* Abstract Algebra: A First Look (1977)
39. *C. O. Christenson and W. L. Voxman,* Aspects of Topology (1977)
40. *M. Nagata,* Field Theory (1977)
41. *R. L. Long,* Algebraic Number Theory (1977)
42. *W. F. Pfeffer,* Integrals and Measures (1977)
43. *R. L. Wheeden and A. Zygmund,* Measure and Integral: An Introduction to Real Analysis (1977)
44. *J. H. Curtiss,* Introduction to Functions of a Complex Variable (1978)
45. *K. Hrbacek and T. Jech,* Introduction to Set Theory (1978)
46. *W. S. Massey,* Homology and Cohomology Theory (1978)
47. *M. Marcus,* Introduction to Modern Algebra (1978)
48. *E. C. Young,* Vector and Tensor Analysis (1978)
49. *S. B. Nadler, Jr.,* Hyperspaces of Sets (1978)
50. *S. K. Sehgal,* Topics in Group Rings (1978)
51. *A. C. M. van Rooij,* Non-Archimedean Functional Analysis (1978)
52. *L. Corwin and R. Szczarba,* Calculus in Vector Spaces (1979)
53. *C. Sadosky,* Interpolation of Operators and Singular Integrals: An Introduction to Harmonic Analysis (1979)
54. *J. Cronin,* Differential Equations: Introduction and Quantitative Theory (1980)
55. *C. W. Groetsch,* Elements of Applicable Functional Analysis (1980)
56. *I. Vaisman,* Foundations of Three-Dimensional Euclidean Geometry (1980)
57. *H. I. Freedman,* Deterministic Mathematical Models in Population Ecology (1980)
58. *S. B. Chae,* Lebesgue Integration (1980)
59. *C. S. Rees, S. M. Shah, and Č. V. Stanojević,* Theory and Applications of Fourier Analysis (1981)
60. *L. Nachbin,* Introduction to Functional Analysis: Banach Spaces and Differential Calculus (R. M. Aron, translator) (1981)
61. *G. Orzech and M. Orzech,* Plane Algebraic Curves: An Introduction Via Valuations (1981)
62. *R. Johnsonbaugh and W. E. Pfaffenberger,* Foundations of Mathematical Analysis (1981)

63. *W. L. Voxman and R. H. Goetschel,* Advanced Calculus: An Introduction to Modern Analysis (1981)
64. *L. J. Corwin and R. H. Szczarba,* Multivariable Calculus (1982)
65. *V. I. Istrăţescu,* Introduction to Linear Operator Theory (1981)
66. *R. D. Järvinen,* Finite and Infinite Dimensional Linear Spaces: A Comparative Study in Algebraic and Analytic Settings (1981)
67. *J. K. Beem and P. E. Ehrlich,* Global Lorentzian Geometry (1981)
68. *D. L. Armacost,* The Structure of Locally Compact Abelian Groups (1981)
69. *J. W. Brewer and M. K. Smith, eds.,* Emmy Noether: A Tribute to Her and Work (1981)
70. *K. H. Kim,* Boolean Matrix Theory and Applications (1982)
71. *T. W. Wieting,* The Mathematical Theory of Chromatic Plane Ornaments (1982)

Other Volumes in Preparation

THE MATHEMATICAL THEORY OF CHROMATIC PLANE ORNAMENTS

Thomas W. Wieting

Department of Mathematics
Reed College
Portland, Oregon

MARCEL DEKKER, INC. New York and Basel

Library of Congress Cataloging in Publication Data

Wieting, Thomas W.
The mathematical theory of chromatic plane ornaments.

(Monographs and textbooks in pure and applied mathematics ; v. 71)
Bibliography: p.
Includes index.
1. Tiling (Mathematics) 2. Geometry, Plane.
3. Decoration and ornament. I. Title. II. Title:
Chromatic plane ornaments. III. Series.
QA166.8.W53 511'.6 82-1543
ISBN 0-8247-1517-9 AACR2

MARCEL DEKKER, INC.
270 Madison Avenue, New York, New York 10016

Current printing (last digit):
10 9 8 7 6 5 4 3 2 1

PRINTED IN THE UNITED STATES OF AMERICA

This book is dedicated to my father

Frank J. Wieting

(Summer, 1908 - Spring, 1977)

PREFACE

The object of this book is to develop the coloring theory for plane ornaments, that is, for periodic tilings of the euclidean plane. Such tilings derive from the ornamental art of diverse cultures: from Sumerian cone mosaics, from Egyptian tomb paintings, from Chinese window lattices, from Greek border mosaics and vase paintings, from Islamic wall mosaics, from African textiles. In each of these and in many other cases, artisans have designed plane ornaments of marvelous variety and complexity. Nevertheless, every plane ornament respects certain principles of composition and hence must fall into one of a limited number of classes. Similarly, every coloration of such an ornament respects certain rules of distribution and hence must fall into one of a limited number of subclasses. The object of this book, then, is to define the criterion by which chromatic plane ornaments shall be classified and to develop procedures by which the classification may be implemented.

The first chapter of the book summarizes euclidean plane geometry, in terms congenial to the rest of the text. The second chapter presents in detail the fundamental classification of plane ornaments into 17 types, by associating with each plan ornament its symmetry group and by identifying the 17 isomorphism classes of such groups. Finally, the third chapter introduces the problem of classifying chromatic plane ornaments, then systematically develops procedures by which the problem can be reduced to machine computation.

By implementing the procedures developed in the third chapter, we have obtained and reported counts of the classes of chromatic plane ornaments for a range of colors from 2 through 60. For a range of colors from 2 through 8, we have supplied instructions by which examples of the corresponding ornaments can be constructed. For the case of four colors, we have applied those instructions to assemble a folio of illustrations. These concrete results substantially extend and in some cases correct the published data on the subject.

This book was designed as a text for an undergraduate course in geometry at Reed College. The intent of that course is to show the interplay among geometry, linear algebra, and group theory. In general, students attend the course to acquire understanding of the relation between group theory and the analysis of symmetry. However, many attend specifically to begin studies of tiling problems and of crystallography, others, to learn by example how one may reduce a mathematical problem of classification to a network of algorithms, suitable for machine computation. We have directed the text to these objectives, using the theory of chromatic plane ornaments as a case study. For a smooth reading of the book, one should be familiar with the basic concepts of linear algebra and group theory, in particular, with the idea of transformation group.

In writing this book, we have benefited from consultation with several colleagues and friends, notably, with B. Grünbaum, M. Senechal, D. Alvis, R. Mayer, J. Buhler, M. Geber, and E. Dannenberg. We hope that they and many others will enjoy the outcome of their good advice.

Thomas W. Wieting

CONTENTS

Chapter 1
THE EUCLIDEAN PLANE

1.0 INTRODUCTION

The objectives of this chapter are to describe the euclidean plane, in terms of cartesian coordinate mappings and cartesian coordinate transformations, and to define the euclidean group, together with the various decompositions of that group relative to cartesian coordinate mappings. This essentially arithmetic method of presenting euclidean geometry is well adapted to our subsequent study and classification of (chromatic) plane ornaments, in that it provides a suitable geometric foundation for our study while at the same time it holds close at hand the computationally effective methods of cartesian coordinate geometry.

1.1 THE CARTESIAN PLANE

1° We shall denote by Z the ring of integers, and by Q and R the rational and real number fields, respectively.

By the _cartesian plane_, we mean the set R^2 together with the standard linear and inner product structures. Let us adopt the following notational conventions:

member: $$x = \begin{bmatrix} x_r \\ x_s \end{bmatrix},$$

sum: $$x + y = \begin{bmatrix} x_r + y_r \\ x_s + y_s \end{bmatrix},$$

scalar multiple: $$a.x = \begin{bmatrix} ax_r \\ ax_s \end{bmatrix},$$

inner product: $$x \bullet y = x_r y_r + x_s y_s,$$

norm: $$|x| = (x \bullet x)^{1/2},$$

distance: $$d(x,y) = |x - y| .$$

We shall denote by 0 the <u>origin</u> in R^2, and by r and s the members of R^2 comprising the <u>standard basis</u>:

$$0 = \begin{bmatrix} 0 \\ 0 \end{bmatrix}, \qquad r = \begin{bmatrix} 1 \\ 0 \end{bmatrix}, \qquad s = \begin{bmatrix} 0 \\ 1 \end{bmatrix}.$$

Clearly, for each member x of R^2:

$$x = x_r.r + x_s.s . \tag{1}$$

2° One defines an <u>affine transformation</u> on R^2 to be any mapping carrying R^2 to itself, of the form:

$$A(x \longmapsto t + L(x)),$$

where t is any member of R^2 and where L is any invertible linear mapping carrying R^2 to itself. Clearly, both t and L are uniquely determined by A; we shall refer to t

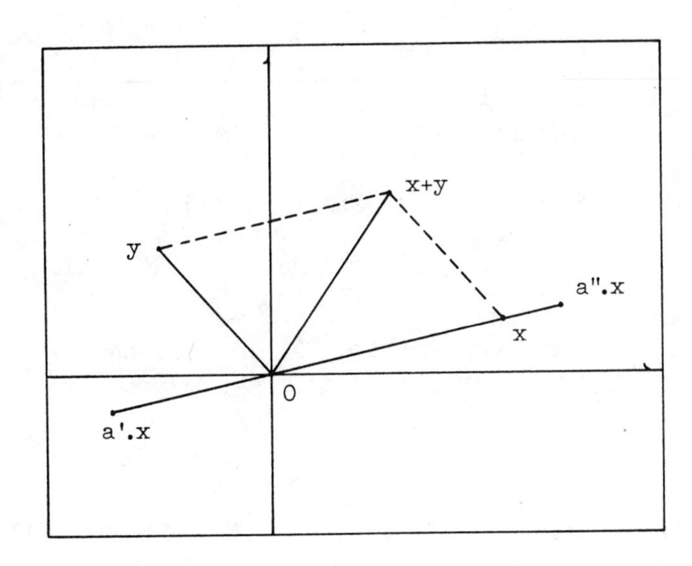

Figure 1

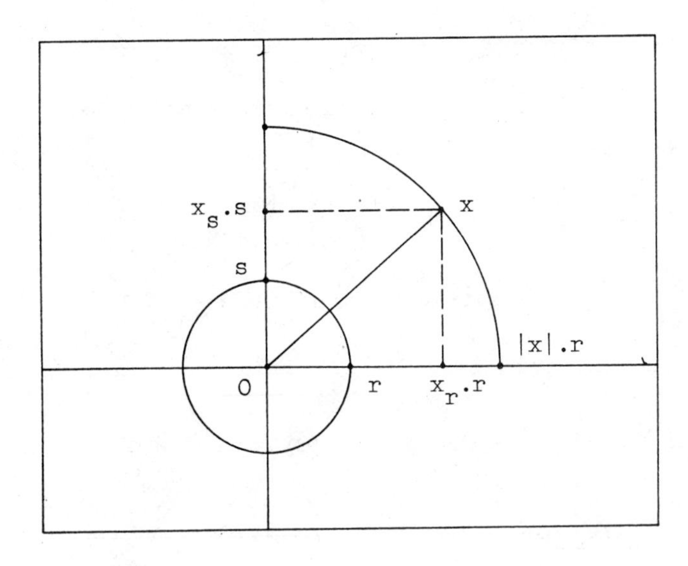

Figure 2

as the translational part and to L as the linear part of A. We shall usually write $[t,L]$ in place of A. The family $\mathscr{A}$ of all affine transformations on R^2 is actually a group of transformations. In fact, we have:

$$[t',L'][t'',L''] = [t' + L'(t''), L'L''], \tag{1}$$

$$[t,L]^{-1} = [-L^{-1}(t), L^{-1}]. \tag{2}$$

When L is the identity mapping I on R^2, then A is the translation

$$(x \longmapsto t + x)$$

determined by t; when $t = 0$, then A coincides with L. We shall denote by $\mathscr{T}$ the family of all translations, and by $\mathscr{L}$ the family of all invertible linear mappings on R^2. Both families are subgroups of $\mathscr{A}$, and $\mathscr{T}$ is abelian. In particular:

$$[t',I][t'',I] = [t' + t'', I], \tag{3}$$

$$[t,I]^{-1} = [-t, I]. \tag{4}$$

One can display the basic relations among $\mathscr{T}$, $\mathscr{A}$, and $\mathscr{L}$ in the sequence

$$R^2 \xrightarrow{q} \mathscr{A} \xrightarrow{p} \mathscr{L},$$

where q is the injective homomorphism

$$(t \longmapsto [t,I])$$

carrying the additive group R^2 to $\mathscr{A}$ and where p is the surjective homomorphism

$$([t,L] \longmapsto L)$$

carrying $\mathscr{A}$ to $\mathscr{L}$. Both the range of q and the kernel of p equal $\mathscr{T}$. Of course, it follows that $\mathscr{T}$ is a normal subgroup of $\mathscr{A}$.

The relation of conjugacy in $\mathscr{A}$ depends upon the computation:

$$[t', L'][t'', L''][t', L']^{-1} = [t' + L'(t'') - (L'L''L'^{-1})(t'), L'L''L'^{-1}]. \quad (5)$$

As a special case:

$$[t', L'][t'', I][t', L']^{-1} = [L'(t''), I]. \quad (6)$$

3° By a <u>cartesian transformation</u> on R^2, we mean any affine transformation for which the linear part is orthogonal. [See the following section.] Since orthogonal linear mappings preserve norms, it is clear that every cartesian transformation is an isometry on R^2. We shall denote by $\mathscr{C}$ the family of all cartesian transformations on R^2. Clearly, $\mathscr{C}$ is a subgroup of $\mathscr{A}$.

4° Let $\mathscr{F}$ be any subgroup of $\mathscr{A}$, and let us denote by T the subgroup $q^{-1}(\mathscr{T} \cap \mathscr{F})$ of R^2 and by $\mathscr{G}$ the subgrou p($\mathscr{F}$) of $\mathscr{L}$. We shall refer to T as the <u>translational</u> part, and to $\mathscr{G}$ as the <u>linear</u> part of $\mathscr{F}$. By restriction, we obtain the sequence

$$T \xrightarrow{q} \mathscr{F} \xrightarrow{p} \mathscr{G}.$$

In this context, we contend that T is <u>invariant</u> under $\mathscr{G}$, which is to say that, for each member L of $\mathscr{G}$, L(T) = T.

<u>Proposition 1</u>. The translational part T of $\mathscr{F}$ is invariant under the linear part $\mathscr{G}$.

Proof. Let L be any member of $\mathscr{G}$, and let t be a member of R^2 such that [t,L] is contained in $\mathscr{F}$. Let u be any member of T. Clearly, $[t,L][u,I][t,L]^{-1}$ is a member of $\mathscr{F}$. By relation (6) in 2°, it follows that [L(u), I] is contained in $\mathscr{F}$, hence that L(u) is a member of T. We infer that $L(T) \subseteq T$. Since the same result holds when L is replaced by L^{-1}, we may conclude that L(T) = T. ///

Now let $\mathscr{G}$ be any subgroup of $\mathscr{L}$, and let T be any subgroup of R^2 which is invariant under $\mathscr{G}$. With reference to the foregoing proposition, we may inquire whether there exist subgroups $\mathscr{F}$ of $\mathscr{A}$ which are compatible with the ordered pair $(T, \mathscr{G})$, in the sense that the translational part of $\mathscr{F}$ is T and the linear part is $\mathscr{G}$.

One such subgroup of $\mathscr{A}$ is the semi-direct product of T and $\mathscr{G}$, defined (and denoted) as follows:

$$T \rtimes \mathscr{G} = \{[t,L] : t \in T, L \in \mathscr{G}\}.$$

In general, there are many other such subgroups of $\mathscr{A}$. However, the problem of describing them all may be very difficult. In due course, we shall see that the classification of plane ornaments depends upon the solution of several special cases of this problem.

The subgroups of $\mathscr{A}$ which are compatible with $(T, \mathscr{G})$ are usually called "extensions" of T by $\mathscr{G}$, and the problem of determining all such subgroups of $\mathscr{A}$ is called a "group extension problem."

5° Let us note in passing that the groups $\mathscr{A}$ and $\mathscr{C}$ are themselves semi-direct products:

$$\mathscr{A} = R^2 \rtimes \mathscr{L},$$

$$\mathscr{C} = R^2 \rtimes \mathscr{O},$$

where $\mathscr{O}$ stands for the group of all orthogonal linear mappings on R^2.

1.2 THE ORTHOGONAL GROUP

1° In this section, we shall develop the geometric structure of the cartesian group, by reviewing the definitions of rotation and reflection and by relating rotations to "angles of rotation" and reflections to "lines of reflection."

Let V be any orthogonal linear mapping on R^2. Such a mapping is characterized by the relation:

$$V(x)\bullet V(y) = x\bullet y, \qquad x, y \in R^2. \tag{1}$$

One can readily show that detV must be either +1 or -1; in the former case, one refers to V as a <u>rotation</u>, in the latter, as a <u>reflection</u>. We shall denote the family of all rotations by $\mathcal{O}^+$, and the family of all reflections by $\mathcal{O}^-$. Clearly, $\mathcal{O}^+$ is a normal subgroup of $\mathcal{O}$.

Let us say that a member u of R^2 is <u>normal</u> provided that $|u| = 1$.

<u>Proposition 2</u>. For each normal member u of R^2, there is precisely one rotation W such that $W(r) = u$ and there is precisely one reflection X such that $X(r) = u$.

Proof. Let us introduce:

$$u^* = \begin{bmatrix} -u_s \\ u_r \end{bmatrix}, \qquad \text{where } u = \begin{bmatrix} u_r \\ u_s \end{bmatrix}.$$

Clearly, u^* is normal, and $u\bullet u^* = 0$.

Let W and X be the linear mappings on R^2 defined by the relations:

$$W(r) = u, \qquad W(s) = u^*;$$

$$X(r) = u, \qquad X(s) = -u^*.$$

Obviously, the matrices for W and X, relative to the standard basis $\{r,s\}$ for R^2, are the following:

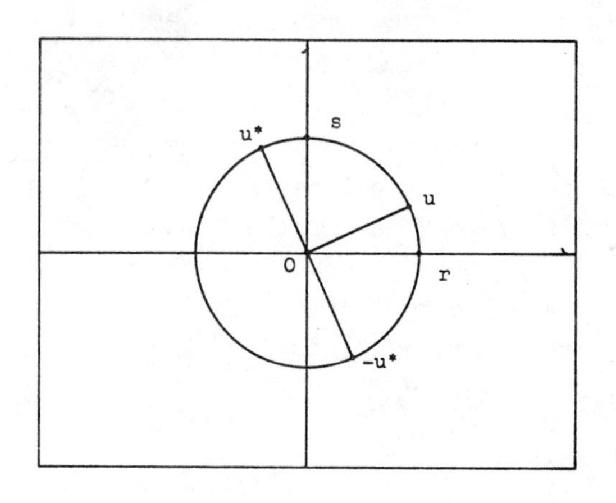

Figure 3

$$W \longleftrightarrow \begin{pmatrix} u_r & -u_s \\ u_s & u_r \end{pmatrix}, \qquad X \longleftrightarrow \begin{pmatrix} u_r & u_s \\ u_s & -u_r \end{pmatrix}.$$

Clearly, W is a rotation and X is a reflection. By definition, both W and X carry r to u.

Now let W' be any rotation for which W'(r) = u. Let W" stand for $W^{-1}W'$. Of course, W" is a rotation, and W"(r) = r. Let v stand for W"(s), so that the matrix for W" relative to {r,s} appears as follows:

$$W'' \longleftrightarrow \begin{pmatrix} 1 & v_r \\ 0 & v_s \end{pmatrix}.$$

Since $r \bullet v = 0$, we have $v_r = 0$. Since $\det W'' = 1$, we have $v_s = 1$. Hence, v = s, and W" must be the identity mapping on R^2. Therefore, W' = W.

Finally, let X' be any reflection for which X'(r) = u. Let X" stand for $X^{-1}X'$. Again, X" is a rotation and X"(r) = r. Hence, X" = I, and therefore X' = X. ///

One can easily generalize the foregoing proposition, by showing that, for any non-zero members v and w of R^2, if $|v| = |w|$ then there is precisely one rotation W such that $W(v) = w$ and there is precisely one reflection X such that $X(v) = w$.

2° Let X be any reflection on R^2. Let u be $X(r)$. Applying Proposition 2, we can write the matrix for X relative to $\{r,s\}$ as follows:

$$X \longleftrightarrow \begin{pmatrix} u_r & u_s \\ u_s & -u_r \end{pmatrix}.$$

By simple calculation, we obtain the following proposition.

Proposition 3. For any reflection X on R^2, $X^2 = I$.

Now let M_X be the subset of R^2 consisting of all members y such that $X(y) = y$. Clearly, M_X is a linear subspace of R^2. We contend that the dimension of M_X is 1.

Let v be any member of R^2. Let v' stand for $(1/2).(v + X(v))$ and v'', for $(1/2).(v - X(v))$. By the foregoing proposition, it is plain that $X(v') = v'$ (so that v' is contained in M_X) and that $X(v'') = -v''$. Moreover, $v = v' + v''$ and $v' \bullet v'' = 0$. Since $X \neq -I$ and $X \neq I$, there must exist members v_1 and v_2 of R^2 such that $v_1' \neq 0$ and $v_2'' \neq 0$. We conclude that $\dim(M_X) = 1$.

One refers to M_X as the axis of reflection for X.

Proposition 4. For any 1-dimensional linear subspace M of R^2, there is precisely one reflection X such that $M_X = M$.

Proof. Let us consider an arbitrary reflection X on R^2. Let y be any member of M_X and let v be any member of R^2. We have:

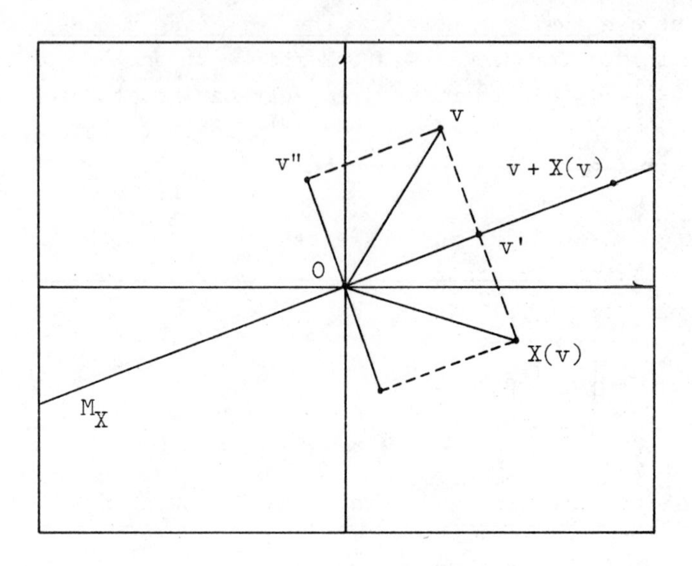

Figure 4

$$X(v)\bullet y = X(v)\bullet X^2(y) \qquad (1)$$
$$= v\bullet X(y)$$
$$= v\bullet y .$$

Now let us take y to be normal: $|y| = 1$. Since v' (which stands for $(1/2).(v + X(v))$) is a member of M_X, there must be some real number s such that $v' = s.y$. Hence, $s = v'\bullet y$, and, by relation (1), $s = v\bullet y$. We conclude that, for each member v of R^2:

$$X(v) = -(v - 2(v\bullet y).y) , \qquad (2)$$

where y is a normal member of M_X.

Relation (2) makes it plain that each reflection on R^2 is uniquely determined by the corresponding axis. Moreover, that same relation provides a method for constructing a reflection with given axis. Thus, let M be any 1-dimensional linear subspace of R^2, and let y be a normal member of M. Let X be the mapping carrying R^2 to itself, defined by relation (2). By straightforward calculation, one can verify that X is a reflection. Since $X(y) = y$, one may conclude that $M_X = M$.

Incidentally, the matrix for X relative to $\{r,s\}$ appears as follows:

$$X \longleftrightarrow \begin{bmatrix} 2y_r y_r - 1 & 2y_r y_s \\ 2y_r y_s & 2y_s y_s - 1 \end{bmatrix},$$

where y is either of the two normal members of M_X. ///

By a routine argument, one can prove the following proposition.

<u>Proposition</u> <u>5</u>. For each reflection X on R^2 and for each invertible linear mapping V on R^2, if VXV^{-1} is a reflection then:

$$M_{VXV^{-1}} = V(M_X).$$

3° Now let us present in detail the structure of the group $\mathcal{O}^+$ of rotations on R^2. In particular, we intend to prove that every finite subgroup of $\mathcal{O}^+$ is cyclic and is uniquely determined by its order, a result which is crucial to the description of geometric and arithmetic classes in Chapter 2. For this purpose, we must develop the relation between rotations and angular measure, in a manner compatible with the prerequisites set for this book.

Let f be a mapping carrying R to R^2. Let f_r and f_s be the components of f, so that, for each real number a:

$$f(a) = \begin{bmatrix} f_r(a) \\ f_s(a) \end{bmatrix}. \qquad (1)$$

Let f^* be the mapping carrying R to R^2 having components $-f_s$ and f_r. Thus, for each real number a:

$$f^*(a) = \begin{bmatrix} -f_s(a) \\ f_r(a) \end{bmatrix} . \qquad (2)$$

We shall say that f is a <u>circular</u> mapping provided that it satisfies the following conditions:

(c1) f is differentiable on R, and $f' = f^*$;

(c2) $f(0) = r$.

These conditions entail that:

(c3) $f'(0) = s$.

<u>Proposition</u> <u>6</u>. There is precisely one circular mapping carrying R to R^2.

Proof. Let us consider the question of uniqueness first. Let f and g be circular mappings, and let h be the function defined on R as follows:

$$\begin{aligned} h(a) &= |f(a) - g(a)|^2 \\ &= (f_r(a) - g_r(a))^2 + (f_s(a) - g_s(a))^2, \end{aligned}$$

where a is any real number. Applying condition (c1) (to both f and g), one can easily show that the derivative of h is constantly 0, so that h is constant. By condition (c2), $h(0) = 0$. It follows that h is constantly 0, and therefore $f = g$.

To establish the existence of a circular mapping, we introduce the power series:

$$f_r(a) = \sum_{n=0}^{\infty} (-1)^n(1/(2n)!)a^{2n}, \qquad (3_r)$$

$$f_s(a) = \sum_{n=0}^{\infty} (-1)^n(1/(2n+1)!)a^{2n+1}. \qquad (3_s)$$

For each of these power series, the radius of convergence is infinite, so both f_r and f_s are defined on R. Obviously, $f_r(0) = 1$ and $f_s(0) = 0$. Applying term by term differentiation, one can readily show that $f_r' = -f_s$ and $f_s' = f_r$. Therefore, the mapping f carrying R to R^2, having components f_r and f_s, is circular. ///

For the present, we shall denote the unique circular mapping by f. In point of fact, the components f_r and f_s of f are the familiar cosine and sine functions of elementary analysis: $f_r = \cos$, $f_s = \sin$. However, we shall adopt the conventional notation only after establishing the basic properties of f.

Let h be the function defined on R as follows:

$$\begin{aligned} h(a) &= |f(a)|^2 \\ &= (f_r(a))^2 + (f_s(a))^2, \end{aligned}$$

where a is any real number. One can easily verify that the derivative of h is constantly 0, and that $h(0) = 1$. Hence, for each real number a:

$$\text{(c4)} \qquad (f_r(a))^2 + (f_s(a))^2 = 1.$$

It follows that the ranges of f_r and f_s are included in the interval $[-1, 1]$.

Let g be the mapping carrying R to R^2, defined as follows:

$$g(a) = \begin{bmatrix} f_r(-a) \\ -f_s(-a) \end{bmatrix},$$

where a is any real number. Clearly, $g' = g^*$ and $g(0) = r$, so g must equal f. Hence, for any real number a:

$$\text{(c5')} \qquad f_r(-a) = f_r(a),$$

$$\text{(c5'')} \qquad f_s(-a) = -f_s(a).$$

That is, f_r is even and f_s is odd.

Let a be any real number. Let g be the mapping carrying R to R^2, having component functions g_r and g_s defined as follows:

$$g_r(b) = f_r(a)f_r(a+b) + f_s(a)f_s(a+b) ,$$

$$g_s(b) = -f_s(a)f_r(a+b) + f_r(a)f_s(a+b) ,$$

where b is any real number. Straightforward calculation yields that $g' = g^*$ and $g(0) = r$. Hence, $g = f$. Algebraic manipulation then yields that:

$$\text{(c6')} \qquad f_r(a+b) = f_r(a)f_r(b) - f_s(a)f_s(b) ,$$

$$\text{(c6")} \qquad f_s(a+b) = f_s(a)f_r(b) + f_r(a)f_s(b) ,$$

where a and b are any real numbers.

Now let Δ be the unit circle in R^2, that is, the set of all normal members of R^2. In virtue of relation (c4), the range of f is included in Δ. For each member u of Δ, let W_u be the rotation on R^2 for which $W_u(r) = u$. By Proposition 2, the mapping $\omega(u \longmapsto W_u)$ carrying Δ to $\mathcal{O}^+$ is bijective. Let ϕ be the mapping ωf carrying R to $\mathcal{O}^+$. Thus, for each real number a:

$$\phi(a) = W_{f(a)} , \qquad (4')$$

$$f(a) = \phi(a)(r) . \qquad (4")$$

The matrix for $\phi(a)$ relative to $\{r,s\}$ is the following:

$$\phi(a) \longleftrightarrow \begin{bmatrix} f_r(a) & -f_s(a) \\ f_s(a) & f_r(a) \end{bmatrix} .$$

Clearly, relations (c6') and (c6") yield that:

(c7) $$\phi(a+b) = \phi(a)\phi(b),$$

for any real numbers a and b. Thus, ϕ is a homomorphism, carrying R to $\mathcal{O}^+$. It follows that $\phi(0) = I$, which is consistent with the fact that $f(0) = r$. Moreover, for each real number a, $\phi(-a) = \phi(a)^{-1}$, so that:

$$\begin{bmatrix} f_r(-a) & -f_s(-a) \\ f_s(-a) & f_r(-a) \end{bmatrix} = \begin{bmatrix} f_r(a) & f_s(a) \\ -f_s(a) & f_r(a) \end{bmatrix},$$

which is consistent with relations (c5') and (c5").

Finally, let us note that relation (c3) may be reformulated as follows:

(c8) $$\lim_{c\to 0}(1/c).(\phi(c)(r) - r) = s.$$

Our next objective is to demonstrate that s lies in the range of f. The significance of this fact will soon be evident.

Let R^+ be the set of all positive real numbers. Let us suppose for the moment that the values of f_r on R^+ are all positive. Since $f_s' = f_r$, it would follow that f_s is strictly increasing on R^+. Since $f_s(0) = 0$, we could infer that the values of f_s on R^+ are all positive. From relation (c6'), we obtain:

$$f_r(2a) = (f_r(a) - f_s(a))(f_r(a) + f_s(a)),$$

where a is any real number. Hence, for each positive real number a, we would conclude that $f_r(a) - f_s(a)$ is positive, so that:

$$f_s(a) < f_s'(a).$$

Now, by elementary argument (based upon the mean value theorem), one could infer that f_s is unbounded, a contradiction. Hence, our supposition that all the values of f_r on

R^+ are positive is untenable. Bearing in mind that $f_r(0) = 1$, we may apply the intermediate value theorem, to conclude that there exists some positive real number c such that $f_r(c) = 0$. Let γ be the infimum of the set of all such numbers:

$$\gamma = \inf\{c \in R^+ : f_r(c) = 0\}.$$

Since $f_r(0) = 1$ and since f_r is continuous at 0, we must have: $0 < \gamma$. Since f_r is continuous at γ, we obtain: $f_r(\gamma) = 0$. Clearly, the values of f_r on the interval $(0,\gamma)$ are positive. Since $f_s' = f_r$, f_s must be strictly increasing on $(0,\gamma)$. Since $f_s(0) = 0$, the values of f_s on the interval $(0,\gamma)$ are positive. Since $f_r' = -f_s$, f_r must be strictly decreasing on $(0,\gamma)$. Since $f_r(\gamma) = 0$ and since $|f(\gamma)| = 1$, we obtain: $f_s(\gamma) = 1$. Therefore, $f(\gamma) = s$.

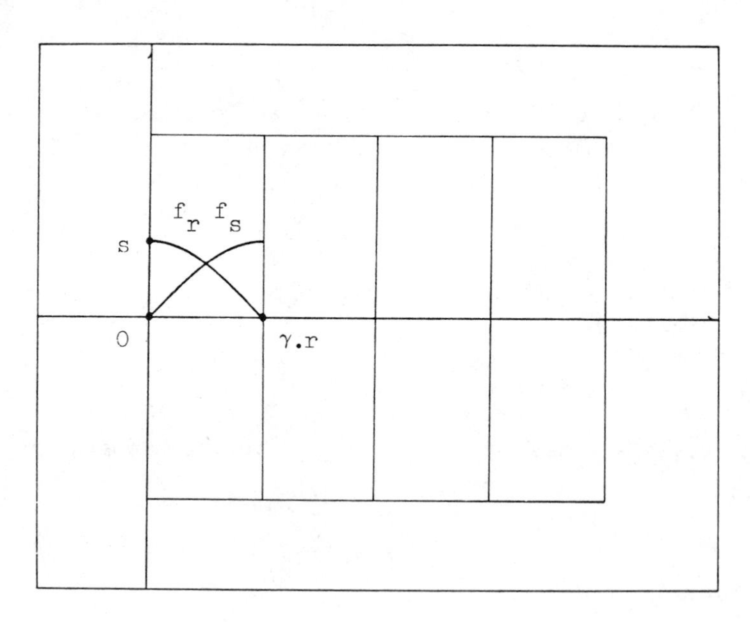

Figure 5

Let a be any real number in the interval $[0,\gamma]$. Let n be any positive integer. For each integer m such that $0 \leq m \leq n$, let u_m stand for $\phi(ma/n)(r)$. Clearly:

$$u_m = \phi(a/n)(u_{m-1}), \qquad 0 < m \leq n; \tag{5}$$

hence:

$$\begin{aligned} |u_{m+1} - u_m| &= |u_m - u_{m-1}|, \qquad 0 < m < n, \\ &= |u_1 - u_0| \\ &= |\phi(a/n)(r) - r| . \end{aligned} \tag{6}$$

Since f_r is strictly decreasing and f_s is strictly increasing on $(0,\gamma)$, we are justified in depicting the members u_m $(0 \leq m \leq n)$ of Δ in the following manner.

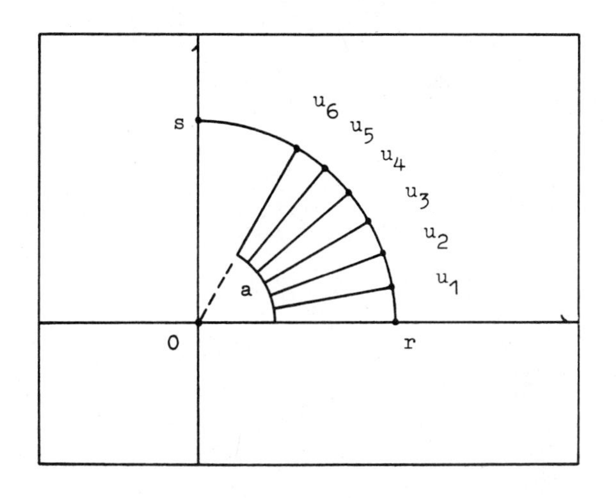

Figure 6

Applying relation (c8), we find that:

$$\begin{aligned} &\lim_{n\to\infty} \sum_{m=1}^{n} |u_m - u_{m-1}| \\ &\qquad = \lim_{n\to\infty} n|\phi(a/n)(r) - r| \end{aligned} \tag{7}$$

$$= a(\lim_{n\to\infty} |(n/a).(\phi(a/n)(r) - r)|)$$
$$= a|s|$$
$$= a .$$

Hence, we may make the provisional judgement that a is the measure of the "arc" joining r to $\phi(a)(r)$, and so it is the measure of the "angle of rotation" associated with $\phi(a)$. In particular, γ measures the arc joining r to s. With respect to conventional notation, we shall denote 2γ by the familiar symbol π: $\gamma = \pi/2$.

Since $\phi(\pi/2) = s$ and since ϕ is a homomorphism, we obtain:

$$\begin{aligned}
\phi(\pi/2) &\longleftrightarrow \begin{bmatrix} 0 & -1 \\ 1 & 0 \end{bmatrix}, \\
\phi(\pi) &\longleftrightarrow \begin{bmatrix} -1 & 0 \\ 0 & -1 \end{bmatrix}, \\
\phi(3\pi/2) &\longleftrightarrow \begin{bmatrix} 0 & 1 \\ -1 & 0 \end{bmatrix}, \\
\phi(2\pi) &\longleftrightarrow \begin{bmatrix} 1 & 0 \\ 0 & 1 \end{bmatrix},
\end{aligned} \qquad (8)$$

relative to $\{r,s\}$, and hence:

$$\begin{aligned}
f(\pi/2) &= s , \\
f(\pi) &= -r , \\
f(3\pi/2) &= -s , \\
f(2\pi) &= r .
\end{aligned} \qquad (9)$$

Now relations (c6') and (c6") yield that:

(c9') $\qquad f_r(a + (\pi/2)) = -f_s(a)$,

(c9") $\qquad f_s(a + (\pi/2)) = f_r(a)$,

(c10') $f_r(a+2\pi) = f_r(a)$,

(c10") $f_s(a+2\pi) = f_s(a)$,

where a is any real number. These relations make it possible to determine the qualitative behavior of f_r and f_s on R. They imply that the graphs of f_r and f_s on the interval $[0, \pi/2]$ produce the graphs of f_r and f_s on R, by translation and by replication, and that f_r and f_s are periodic, with period 2π.

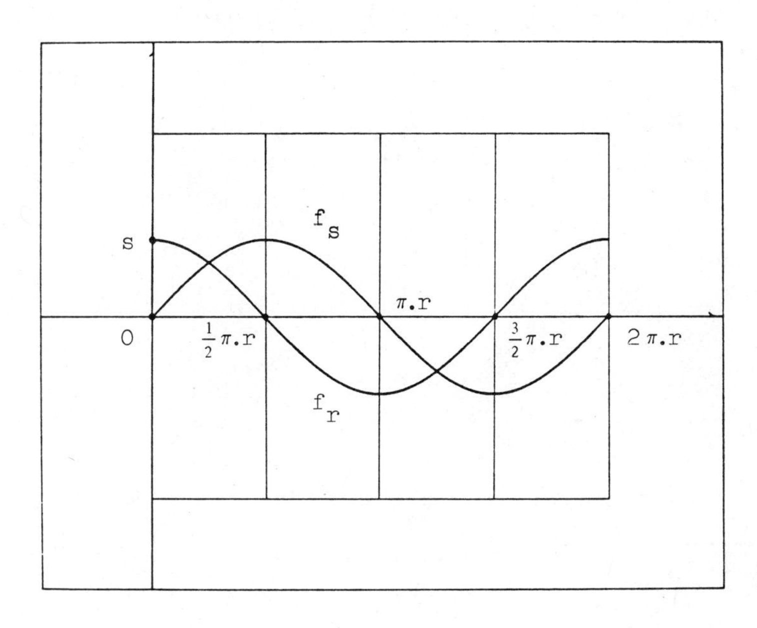

Figure 7

Implicit in the foregoing discussion is the implication that the range of f is Δ. Informally, one may say that, as the real number a runs through the interval $[0, 2\pi]$, the image $f(a)$ traces out the unit circle Δ, in counter-clockwise motion.

The following basic proposition is now evident.

<u>Proposition</u> <u>7</u>. The homomorphism ϕ carrying R to $\mathcal{O}^+$ is surjective, and its kernel is $2\pi Z$.

Now let W be any rotation on R^2, and let b be any real number. We shall say that b is a <u>measure</u> of W provided that $\phi(b) = W$. We shall refer to the particular real number a for which $0 \leq a < 2\pi$ and $\phi(a) = W$, as the <u>principal</u> <u>measure</u> of W. The argument contained in relations (5), (6), and (7) shows that the principal measure of W is a measure of the "amount of turning" involved in applying W to the members of R^2.

The central objective of this article is now within reach.

<u>Proposition</u> <u>8</u>. Every finite subgroup of $\mathcal{O}^+$ is cyclic. Moreover, for each positive integer n, there is precisely one (finite) subgroup of $\mathcal{O}^+$ having order n.

Proof. Let $\mathcal{P}$ be a finite subgroup of $\mathcal{O}^+$, and let n be the order of $\mathcal{P}$. Let us assume that $1 < n$. We plan to prove that $\mathcal{P}$ is cyclic. Let

$$W_0,\ W_1,\ \ldots\ ,\ W_{n-1}$$

be an enumeration of the members of $\mathcal{P}$, where $W_0 = I$. Let

$$a_0,\ a_1,\ \ldots\ ,\ a_{n-1}$$

be the corresponding principal measures of these rotations. [Of course, $a_0 = 0$.] We may as well assume that a_1 is the smallest among a_1, a_2, ... , and a_{n-1}. Let m be any integer for which $1 \leq m < n$. We may present a_m in the form: $a_m = ka_1 + b$, where k is a positive integer and where b is a real number drawn from $[0, a_1)$. Clearly:

$$\begin{aligned} W_m &= \phi(a_m) \\ &= \phi(a_1)^k \phi(b) \end{aligned}$$

$$= W_1{}^k\phi(b).$$

Hence, $\phi(b)$ is a member of $\mathscr{P}$, and its principal measure is b. Since $b < a_1$, b must be 0, so $a_m = ka_1$ and $W_m = W_1{}^k$. Hence, every member of $\mathscr{P}$ is a power of W_1, which means that $\mathscr{P}$ is cyclic. Since W_1 must have order n, it is clear that a_1 cannot be less than $2\pi/n$. Since each of the $n-1$ distinct members $a_1, a_2, \ldots, $ and a_{n-1} of $[0, 2\pi)$ is a positive integral multiple of a_1, it follows that a_1 cannot be greater than $2\pi/n$. Hence, $a_1 = 2\pi/n$. Now it is plain that any two finite subgroups of $\mathscr{O}^+$ having the same order, must be the same.

Finally, for each positive integer n, the subgroup of $\mathscr{O}^+$ generated by $\phi(2\pi/n)$ is finite and has order n. ///

4° Hereafter, we shall denote the components of the circular mapping f by cos and sin. Hence, for each real number a:

$$f(a) = \begin{bmatrix} \cos a \\ \sin a \end{bmatrix}, \tag{1}$$

$$\phi(a) \longleftrightarrow \begin{bmatrix} \cos a & -\sin a \\ \sin a & \cos a \end{bmatrix}. \tag{2}$$

Of course, all the foregoing relations involving f should now be rewritten, to accommodate the conventional notation.

5° For later reference, let us examine certain relations governing rotations and reflections on R^2, and let us determine the conjugacy classes of members of $\mathscr{O}$ and the conjugacy classes of finite subgroups of $\mathscr{O}$.

Let W and W' be rotations, and let X and X' be reflections. Let V be an arbitrary member of $\mathscr{O}$.

From Proposition 7, it is plain that $\mathscr{O}^+$ is abelian. Hence:

$$WW'W^{-1} = W'. \tag{1}$$

Clearly, XW' is a reflection, so that, by Proposition 3, $X^2 = I$ and $(XW')^2 = I$. Hence:

$$XW'X^{-1} = W'^{-1}. \tag{2}$$

Applying Proposition 5, together with the fact that VXV^{-1} is a reflection, we obtain:

$$M_{VXV^{-1}} = V(M_X). \tag{3}$$

Now let W'' be either of the two rotations for which $W''(M_X) = M_{X'}$. One may define such rotations by introducing normal members y of M_X and y' of $M_{X'}$ and by declaring W'' to be the rotation which carries y to y', or else y to $-y'$.

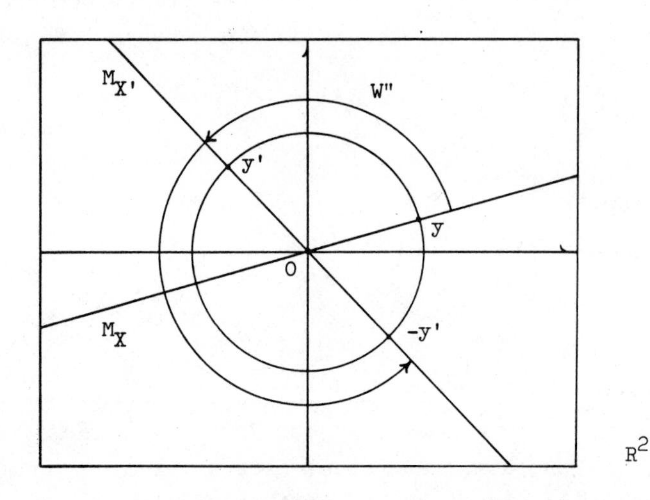

Figure 8

By relation (3), we have:

$$W''XW''^{-1} = X'. \tag{4}$$

Combining relations (2) and (4), we obtain:

$$X'X = W''^2. \tag{5}$$

Hence, the two rotations which transform M_X to $M_{X'}$ can be described as "square roots" of the rotation $X'X$. In general, for any rotation W, there are precisely two rotations $\overline{W}$ such that $\overline{W}^2 = W$. In fact, $\overline{W}$ must be either $\phi(a/2)$ or $\phi((a/2)+\pi)$, where a is the principal measure of W. One calls $\phi(a/2)$ the <u>principal square root</u> of W, and denotes it by $W^{1/2}$. Of course, the other square root is $-W^{1/2}$.

Assuming for the moment that $X' = WX$, we may derive from relation (5) the following useful fact:

$$M_{WX} = W^{1/2}(M_X). \tag{6}$$

Now let us describe the conjugacy classes of members of $\mathcal{O}$. From relations (1) and (2), it is evident that, for each rotation W, the conjugacy class in $\mathcal{O}$ containing W is $\{W, W^{-1}\}$. Moreover, by relation (4), $\mathcal{O}^-$ is itself a conjugacy class in $\mathcal{O}$.

Let us turn to a description of the conjugacy classes of finite subgroups of $\mathcal{O}$.

For each positive integer n, let W_n stand for the rotation $\phi(2\pi/n)$. Let $\mathcal{Z}_n$ denote the subgroup of $\mathcal{O}$ generated by W_n. Clearly, $\mathcal{Z}_n$ consists of the following rotations:

$$\mathcal{Z}_n: \qquad I,\ W_n^{\,1},\ W_n^{\,2},\ \ldots\ldots,\ W_n^{\,n-1}.$$

Of course, $\mathcal{Z}_n$ has order n. For each reflection X, let $\mathcal{D}_n(X)$ denote the subgroup of $\mathcal{O}$ generated by W_n and X. Using relation (2), one can easily verify that the members of $\mathcal{D}_n(X)$ are the following:

$$\mathcal{D}_n(X): \quad \begin{array}{l} I,\ W_n^{\,1},\ W_n^{\,2},\ \ldots\ldots,\ W_n^{\,n-1}, \\ X,\ W_n^{\,1}X,\ W_n^{\,2}X,\ \ldots\ldots,\ W_n^{\,n-1}X. \end{array}$$

Clearly, $\mathscr{D}_n(X)$ has order $2n$. In virtue of relation (6), we are justified in depicting the reflections in $\mathscr{D}_n(X)$ as follows.

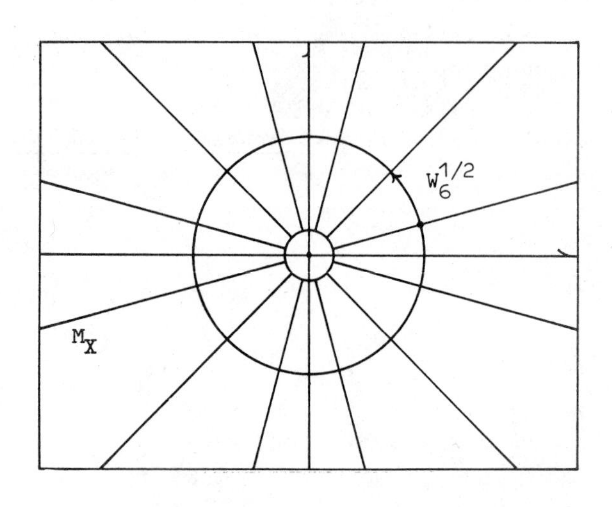

Figure 9

Let $\mathscr{P}$ be any finite subgroup of $\mathscr{O}$. Let $\mathscr{P}^+$ stand for $\mathscr{P} \cap \mathscr{O}^+$. By Proposition 8, $\mathscr{P}^+$ must be $\mathscr{Z}_n$, where n is the order of $\mathscr{P}^+$. It may happen that $\mathscr{P} = \mathscr{P}^+$; in that case, $\mathscr{P} = \mathscr{Z}_n$. Let us assume that $\mathscr{P} \neq \mathscr{P}^+$. Accordingly, we may introduce a reflection X in $\mathscr{P}$. For each reflection X' in $\mathscr{P}$, $X'X^{-1}$ is contained in $\mathscr{P}^+$, so $X' = WX$, where W is a member of $\mathscr{P}^+$. It follows that:

$$\mathscr{P} = \mathscr{Z}_n \cup \{WX : W \varepsilon \mathscr{Z}_n\},$$

which is to say that $\mathscr{P} = \mathscr{D}_n(X)$.

Now let n' and n'' be any positive integers, and let X' and X'' be any reflections. Since $\mathscr{Z}_{n'}$ and $\mathscr{Z}_{n''}$ have orders n' and n'', respectively, it is clear that $\mathscr{Z}_{n'}$ and $\mathscr{Z}_{n''}$ are not conjugate in $\mathscr{O}$ unless $n' = n''$. The same may be said of $\mathscr{D}_{n'}(X')$ and $\mathscr{D}_{n''}(X'')$. Moreover, $\mathscr{Z}_{n'}$ and $\mathscr{D}_{n''}(X'')$ cannot be conjugate in $\mathscr{O}$, because the latter contains a reflection while the former does not. Finally, by relation (4), X' and X'' are conjugate in $\mathcal{O}$: $X'' = WX'W^{-1}$, where W is a suitable rotation. It follows that, when $n' = n''$, then $\mathcal{D}_{n''}(X'') = W\mathcal{D}_{n'}(X')W^{-1}$, so that $\mathcal{D}_{n'}(X')$ and $\mathcal{D}_{n''}(X'')$ are conjugate in $\mathcal{O}$.

Let us summarize the foregoing discussion.

Proposition 9. For each finite subgroup $\mathcal{P}$ of $\mathcal{O}$, either there is a positive integer n such that $\mathcal{P} = \mathcal{Z}_n$ or there are a positive integer n and a reflection X such that $\mathcal{P} = \mathcal{D}_n(X)$. Moreover:

(a) for each positive integer n, the conjugacy class of subgroups of $\mathcal{O}$ containing $\mathcal{Z}_n$ is $\{\mathcal{Z}_n\}$;

(b) for each positive integer n, $\{\mathcal{D}_n(X) : X \in \mathcal{O}^-\}$ is itself a conjugacy class of subgroups of $\mathcal{O}$.

1.3 THE EUCLIDEAN PLANE

1° By a euclidean plane, we shall mean a set E together with a (nonempty) family $\mathcal{K}$ of bijective mappings carrying E to R^2, such that:

(e1) for any mappings K' and K'' in $\mathcal{K}$, $K''K'^{-1}$ is a cartesian transformation on R^2;

(e2) for any mapping K in $\mathcal{K}$ and for any cartesian transformation C on R^2, CK is a member of $\mathcal{K}$.

We shall refer to the mappings in $\mathcal{K}$ as cartesian coordinate mappings for E. For any mappings K' and K'' in $\mathcal{K}$, we shall call $K''K'^{-1}$ the cartesian coordinate transformation linking K' and K''.

Informally, one should view a euclidean plane E as a flat blank surface of infinite extent, such as would serve an idealized draftsman. The draftsman would be equipped not only with straight-edge and compass but also with ruler and protractor. One should interpret a cartesian coordinate map-

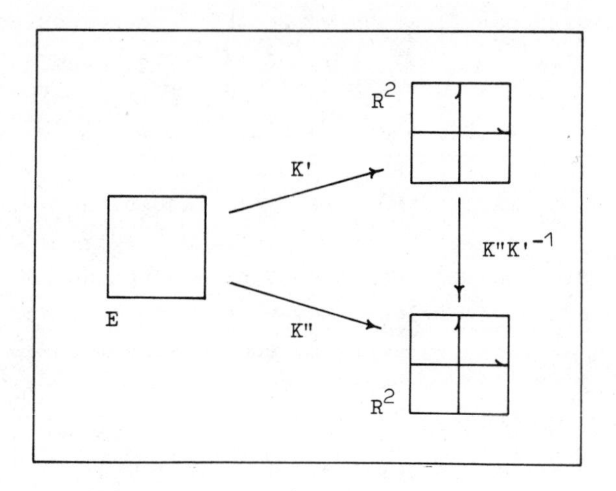

Figure 10

ping as the result of constructing a cartesian coordinate system in E (that is, an ordered pair of directed lines, perpendicular to one another), with respect to which the points of E can be identified as ordered pairs of real numbers; and one should view cartesian (coordinate) transformations as mappings which relate the coordinatizations of E by two cartesian coordinate systems.

The foregoing definition of a euclidean plane simply isolates in analytic terms the basic properties of the naive process of coordinatization.

2° Now let E be a euclidean plane. Using the cartesian coordinate mappings for E, we shall describe the metric structure, and the basic groups of transformations on E.

Let α and β be any points in E. Let K' and K" be any cartesian coordinate mappings for E, and let C be the cartesian coordinate transformation linking K' and K": $C = K''K'^{-1}$. We have:

$$|K''(\alpha) - K''(\beta)| = |C(K'(\alpha)) - C(K'(\beta))| \tag{1}$$
$$= |K'(\alpha) - K'(\beta)|.$$

Hence, the real number $|K(\alpha) - K(\beta)|$ is the same for all cartesian coordinate mappings K. We shall refer to it as the <u>euclidean distance</u> between α and β, and denote it by $\delta(\alpha,\beta)$. Of course, the mapping

$$\delta((\alpha,\beta) \longmapsto \delta(\alpha,\beta))$$

carrying $E \times E$ to R is a metric, the <u>euclidean metric</u> on E.

Now let θ be any mapping carrying E to itself. For each cartesian coordinate mapping K, we shall denote by θ^K the mapping $K\theta K^{-1}$ carrying R^2 to itself, and refer to it as the <u>cartesian form</u> for θ relative to K. For any cartesian coordinate mappings K' and K'', we have:

$$\begin{aligned} \theta^{K''} &= K''\theta K''^{-1} \\ &= (CK')\theta(CK')^{-1} \\ &= C\,\theta^{K'}C^{-1}, \end{aligned} \tag{2}$$

where $C = K''K'^{-1}$. In diagrammatic form, relation (2) appears as follows.

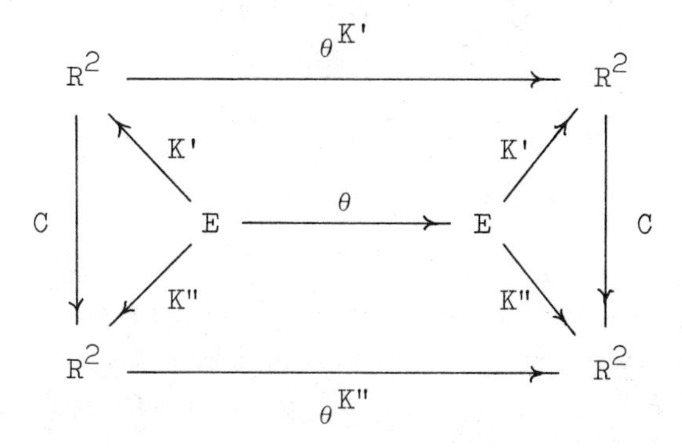

Clearly, $\theta^{K'}$ is an affine transformation on R^2 iff $\theta^{K''}$ is such. We are led to define an <u>affine transformation</u> on E to be any mapping θ carrying E to itself such that, for some (and hence for any) cartesian coordinate mapping

K, θ^K is an affine transformation on R^2. We shall denote the family of all affine transformations on E by **A**. It is of course a group of transformations, the affine group on E.

In analogous manner, one defines the group **T** of all translations, and the group **E** of all euclidean transformations on E. In particular, one defines a euclidean transformation on E to be any mapping θ carrying E to itself such that, for some (and hence for any) cartesian coordinate mapping K, θ^K is a cartesian transformation on R^2.

One should note that each affine transformation on E is a homeomorphism, and that each euclidean transformation on E is an isometry.

3° Let θ be an affine transformation on E. Let K' and K" be cartesian coordinate mappings, and let C stand for $K''K'^{-1}$. Let us display the translational and linear parts of $\theta^{K'}$, $\theta^{K''}$, and C:

$$\theta^{K'} = [t', L'], \qquad \theta^{K''} = [t'', L''], \qquad C = [u, U].$$

Since $\theta^{K''} = C\,\theta^{K'}C^{-1}$, we may invoke relation (5) in 1.2° to obtain the following fundamental relations:

$$t'' = u + U(t') - (UL'U^{-1})(u), \tag{1}$$

$$L'' = UL'U^{-1}. \tag{2}$$

4° Now let **F** be any subgroup of the affine group **A** on E. For each cartesian coordinate mapping K, we shall denote by $\mathbf{F}^K$ the cartesian form for **F** relative to K:

$$\mathbf{F}^K = \{\theta^K : \theta \in \mathbf{F}\}.$$

It is a subgroup of the affine group $\mathscr{A}$ on R^2. With reference to 1.4°, we shall denote by T^K the translational part, and by $\mathscr{G}^K$ the linear part of $\mathbf{F}^K$, obtaining the sequence:

$$T^K \xrightarrow{\ q\ } \mathbf{F}^K \xrightarrow{\ p\ } \mathscr{G}^K.$$

The foregoing sequence will be called the <u>cartesian presentation</u> of $\mathbf{F}$ relative to K.

Let K' and K" be cartesian coordinate mappings for E. The relation between the cartesian presentations of $\mathbf{F}$ relative to K' and K" can be calculated easily. Thus, let C stand for $K''K'^{-1}$, and let V be the linear part of C. Let J_C stand for the inner automorphism

$$(A \longmapsto CAC^{-1})$$

of $\mathscr{A}$, and let J_V stand for the inner automorphism

$$(L \longmapsto VLV^{-1})$$

of $\mathscr{L}$. One finds that:

$$\begin{aligned} T^{K''} &= V(T^{K'}), \\ \mathbf{F}^{K''} &= J_C(\mathbf{F}^{K'}), \\ \mathscr{G}^{K''} &= J_V(\mathscr{G}^{K'}). \end{aligned} \tag{1}$$

These facts can be organized in the following diagram.

$$\begin{array}{ccccc} T^{K'} & \xrightarrow{\ q\ } & \mathbf{F}^{K'} & \xrightarrow{\ p\ } & \mathscr{G}^{K'} \\ V\big\downarrow & & \big\downarrow J_C & & \big\downarrow J_V \\ T^{K''} & \xrightarrow{\ q\ } & \mathbf{F}^{K''} & \xrightarrow{\ p\ } & \mathscr{G}^{K''} \end{array}$$

5° Let us now introduce the basic typology of euclidean transformations. We shall show that every euclidean transformation (excepting the identity) can be identified as a translation, rotation, reflection, or glide-reflection, and that these cases are mutually exclusive. Of course, the identity itself may be regarded either as a translation or as a rotation.

Let θ be any euclidean transformation on E, distinct from the identity. Let K be a cartesian coordinate mapping, and let the cartesian form for θ relative to K be presented as follows:

$$\theta^K = [t,V].$$

We shall entertain four conditions:

(a) $V = I$;

(b) V is a rotation, distinct from I ;

(c) V is a reflection, and $V(t) = -t$;

(d) V is a reflection, and $V(t) \neq -t$.

Each of these conditions is persistent over cartesian coordinate mappings, in the sense that it is valid relative to all cartesian coordinate mappings, if it is valid relative to one such mapping.

Indeed, let K' and K" be any cartesian coordinate mappings, and let C stand for $K''K'^{-1}$: $C = [u,U]$. Let us apply relations (1) and (2) in 3°. Thus, if V' is a rotation then V" is a rotation. In fact, since U must be either a rotation or a reflection, V" would equal either V' or V'^{-1}. Hence, both conditions (a) and (b) are persistent.

Now let us assume that V' is a reflection, and that $V'(t') = -t'$. Then V" is also a reflection, and:

$$\begin{aligned} V''(t'') &= (UV'U^{-1})(u + U(t') - (UV'U^{-1})(u)) \\ &= (UV'U^{-1})(u) + (UV')(t') - u \\ &= (UV'U^{-1})(u) - U(t') - u \\ &= -t''. \end{aligned}$$

Hence, both conditions (c) and (d) are persistent.

When condition (a) is valid, then θ is a translation on E, as defined in 2°.

Let us assume that condition (b) is satisfied. From Proposition 2, it is plain that $I - V$ is injective. It must therefore be invertible. Accordingly, we may introduce a member u of R^2 such that $(I - V)(u) = -t$. Let K' be the cartesian coordinate mapping $[u,I]K$. We find that $\theta^{K'} = [0,V]$. Hence, with respect to the "well chosen" cartesian coordinate mapping K', the cartesian form for θ is a rotation on R^2.

When condition (b) is valid, one says that θ is a rotation on E.

Now let us assume that V is a reflection. Let u be $(-1/2).t$, and let K' be $[u,I]K$. Let $\theta^{K'}$ be presented as follows: $\theta^{K'} = [t', V']$. We have:

$$V' = V ;$$

$$t' = u + t - V(u)$$

$$= (1/2).(t + V(t)) ;$$

$$V'(t') = t' .$$

When condition (c) is satisfied, then $t' = 0$, so that $\theta^{K'} = [0,V]$. That is, relative to K', the cartesian form for θ is a reflection on R^2. When condition (d) is satisfied, then $t' \neq 0$ and $V'(t') = t'$, so that t' is a non-zero member of the axis of reflection for V'. Thus, relative to K', the cartesian form for θ is a "glide-reflection" on R^2. [See Figure 11.]

When condition (c) is valid, one refers to θ as a reflection on E. For condition (d), one says that θ is a glide-reflection on E.

6° We shall conclude this section with comments on the questions of existence and uniqueness of euclidean planes.

Let us consider first the question of uniqueness. For any two euclidean planes E' and E", one can introduce a

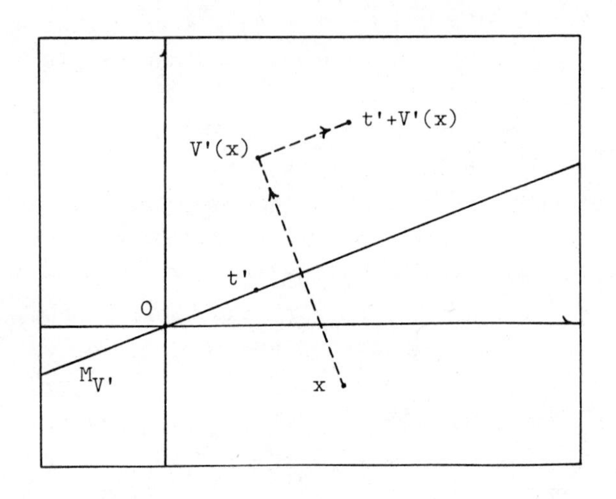

Figure 11

bijective mapping H carrying E' to E" such that, for each mapping K carrying E" to R^2, K is a cartesian coordinate mapping for E" iff KH is a cartesian coordinate mapping for E'. In fact, one may take H to be $K''^{-1}K'$, where K' and K" are any particular cartesian coordinate mappings for E' and E", respectively. The mapping H is an "isomorphism," in the sense that it sets in bijective correspondence the relevant structures on E' and E", namely, the cartesian coordinate mappings, the euclidean metrics, and the basic groups of transformations. As a result, whatever one can prove for E', one can also prove for E", and conversely. Hence, any two euclidean planes are essentially the same.

The question of existence of euclidean planes can be resolved easily. Thus, one may introduce a set E, and a bijective mapping K_o carrying E to R^2. One may then define $\mathscr{K}$ to consist of all mappings of the form: CK_o, where C runs through the group $\mathscr{C}$ of all cartesian transformations on R^2. Clearly, the conditions (e1) and (e2) for a euclidean plane would be satisfied.

For example, one could take E to be R^2 itself,

and K_0 to be the identity mapping on R^2. Then $\mathscr{K}$ would be $\mathscr{C}$.

With the foregoing example in mind, one might contend that the distinction between a euclidean plane and the cartesian plane is at best merely notational and at worst entirely pedantic. In response, let us note that, for our subsequent analysis of plane ornaments, we shall introduce:

(a) euclidean transformations on E, to measure the symmetry structure of a given plane ornament;

(b) cartesian coordinate mappings, to translate the analysis of symmetry structure from geometric into arithmetic terms;

(c) cartesian coordinate transformations, designed to reduce the arithmetic analysis to the simplest possible form.

Hence, for conceptual clarity, we require a presentation of euclidean geometry which distinguishes the three basic classes of mappings: **E**, $\mathscr{K}$, and $\mathscr{C}$. We shall persist in using terminology and notation which preserve the distinctions.

7° Hereafter, we shall settle upon one particular euclidean plane E.

1.4 PROBLEMS

1. Let ϕ be a mapping carrying R to $\mathscr{O}^{+}$, satisfying conditions (c7) and (c8) in 2.3°. Let f be the mapping $\omega^{-1}\phi$ carrying R to Δ. Prove that f is circular.

2. Let W be a rotation on R^2, with principal measure a. Note that the trace of W is 2cosa. Show that the trace of W is an integer iff a is one of the following real numbers:

$$0, \quad \pi/3, \quad \pi/2, \quad 2\pi/3, \quad \pi, \quad 4\pi/3, \quad 3\pi/2, \quad 5\pi/3 .$$

Conclude that the trace of W is an integer iff the order of W is finite and equals 1, 2, 3, 4, or 6.

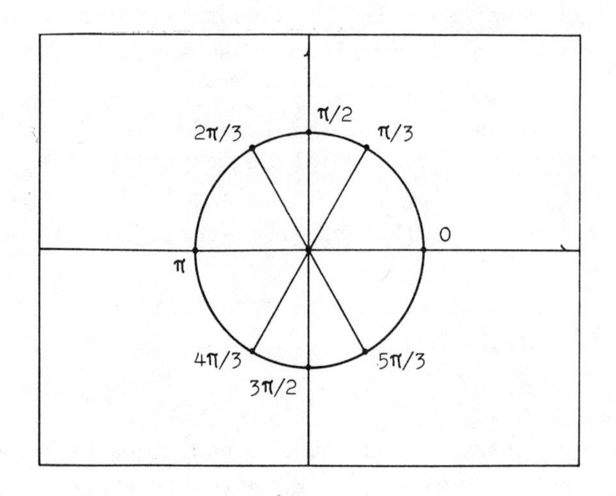

Figure 12

3. Let $\mathscr{U}$ be a subgroup of $\mathscr{C}$, and let $\mathscr{U}^{\triangle}$ stand for the normalizer of $\mathscr{U}$ in $\mathscr{C}$. Prove that if $\mathscr{U}$ is abelian and if $\mathscr{T} \subseteq \mathscr{U}^{\triangle}$ then $\mathscr{U} \subseteq \mathscr{T}$. Conclude that $\mathscr{T}$ is the only normal subgroup of $\mathscr{C}$ which equals its own centralizer in $\mathscr{C}$.

4. Prove that:

(a) for any two finite subgroups $\mathscr{P}'$ and $\mathscr{P}''$ of $\mathscr{O}$, if $\mathscr{P}'$ and $\mathscr{P}''$ are conjugate in $\mathscr{L}$ then they are conjugate in $\mathscr{O}$;
(b) for any finite subgroup $\mathscr{P}'$ of $\mathscr{L}$, there exists a (finite) subgroup $\mathscr{P}''$ of $\mathscr{O}$ such that $\mathscr{P}'$ and $\mathscr{P}''$ are conjugate in $\mathscr{L}$.

As a result, each conjugacy class of finite subgroups

of $\mathscr{L}$ includes precisely one conjugacy class of finite subgroups of $\mathscr{O}$.
One can prove (a) by reviewing the proof of Proposition 9. For the proof of (b), one may proceed as follows.
Let k be the order of $\mathscr{P}'$. For any members x and y of R^2, let $x*y$ stand for the sum:

$$x*y = (1/k)\sum_{L\in\mathscr{P}'} L(x)\bullet L(y) .$$

Prove that the mapping

$$((x,y) \longmapsto x*y)$$

carrying $R^2 \times R^2$ to R is an inner product on R^2. Prove that $\mathscr{P}'$ preserves the new inner product, in the sense that, for any member L of $\mathscr{P}'$ and for any members x and y of R^2, $L(x)*L(y) = x*y$. Let $\{u,v\}$ be a basis for R^2 which is orthonormal with respect to the new inner product. Let M be the member of $\mathscr{L}$ defined by the conditions: $M(u) = r$, $M(v) = s$. Let $\mathscr{P}''$ stand for the subgroup $M\mathscr{P}'M^{-1}$ of $\mathscr{L}$. Prove that $\mathscr{P}''$ is included in $\mathscr{O}$.

5. By a <u>similarity transformation</u> on R^2, we mean any affine transformation of the form:

$$[t, a.V],$$

where t is any member of R^2, where V is any member of $\mathscr{O}$, and where a is any positive real number. Note that the family of all such transformations is a group, the <u>similarity group</u> on R^2. Let it be denoted by $\mathscr{S}$. With reference to 3.2°, formulate the definitions of <u>similarity transformation</u>, and of the <u>similarity group</u> **S** on E. Prove that:

(a) the normalizer of **S** in **A** is **S**;
(b) the normalizer of **E** in **A** is **S**.

[One should first identify those which have order four, among all linear mappings on R^2 of the form: a.V, where V is any member of $\mathcal{O}$ and where a is any positive real number.]

6. Let θ be any rotation on E, distinct from the identity. Prove that there exists exactly one point α in E which is fixed under θ: $\theta(\alpha) = \alpha$. One refers to α as the <u>center of rotation</u> for θ.

7. By a <u>line</u> in R^2, one means any subset of R^2 of the form:

$$M_{u,v} = \{(1-a).u + a.v : a \in R\},$$

where u and v are any two distinct members of R^2. Show that, for each line M in R^2 and for any affine transformation A on R^2, A(M) is a line in R^2. Following the procedure in 3.2°, define a <u>line</u> in E to be any subset $\mathring{M}$ of E such that, for some (and hence for any) cartesian coordinate mapping K, $K(\mathring{M})$ is a line in R^2.

8. Let θ be a reflection on E, and let M_θ be the set of fixed points in E under θ:

$$M_\theta = \{\alpha \in E : \theta(\alpha) = \alpha\}.$$

Prove that M_θ is a line in E. One calls M_θ the <u>axis of reflection</u> for θ. Prove that the mapping $(\theta \longmapsto M_\theta)$ carries the family of all reflections on E bijectively to the family of all lines in E. Let θ' and θ'' be any two reflections on E: $\theta' \neq \theta''$. Show that:

(t) if $M_{\theta'} \cap M_{\theta''}$ is empty then $\theta''\theta'$ is a translation on E;
(r) if $M_{\theta'} \cap M_{\theta''}$ is not empty then $\theta''\theta'$ is a rotation on E.

Now let **E*** denote the family of all euclidean transformations of the form:

$$\theta''\theta'\theta,$$

where each of θ, θ', and θ'' is either a reflection on E or else the identity transformation on E. Prove that: **E*** = **E**.

9. Let θ be a mapping carrying E to itself. Prove that θ is a similarity transformation iff:

(s) there exists a positive real number a such that, for any two points α and β in E, $\delta(\theta(\alpha),\theta(\beta)) = a\delta(\alpha,\beta)$.

Prove that θ is a euclidean transformation iff:

(e) for any two points α and β in E, $\delta(\theta(\alpha),\theta(\beta)) = \delta(\alpha,\beta)$.

One can readily derive the former result from the latter. For the latter, one should introduce the cartesian form θ^K for θ (relative to some cartesian coordinate mapping K), and make use of the following identity:

$$u \bullet v = (1/2)(|u|^2 + |v|^2 - |u - v|^2),$$

where u and v are any members of R^2.

10. Let h be a mapping carrying R to itself, satisfying the conditions:

    (1) $h(1) \neq 0$;
    (2) for any real numbers a and b, $h(a+b) = h(a) + h(b)$;
    (3) for each real number c, $h(c^2) = h(c)^2$.

    Prove that h must be the identity mapping on R. This result is a technical preliminary to the following problem on affine transformations.

11. Let θ be a bijective mapping carrying E to itself. Prove that θ is an affine transformation iff it preserves lines, in the sense that, for each line $\mathring{M}$ in E, $\theta(\mathring{M})$ is a line in E.
    In cartesian terms, what one must prove is that, for any mapping F carrying R^2 to itself, if

    (1) F is bijective
    and (2) F preserves lines

    then F is an affine transformation on R^2. Thus, let F satisfy conditions (1) and (2). One can readily design an affine transformation A on R^2 such that $A(0) = F(0)$, $A(r) = F(r)$, and $A(s) = F(s)$. Let B stand for the mapping $A^{-1}F$. Clearly, B satisfies conditions (1) and (2); moreover, $B(0) = 0$, $B(r) = r$, and $B(s) = s$.
    One can now prove that $B = I$, from which it will follow that $F = A$. Thus, let M_r and M_s stand for the lines $M_{0,r}$ and $M_{0,s}$ in R^2. They are of course the coordinate axes in R^2. Show that $B(M_r) = M_r$ and $B(M_s) = M_s$. Let h be the mapping carrying R to itself, defined by the condition that, for each real number a, $B(a.r) = h(a).r$. Clearly, h is bijective; moreover, $h(0) = 0$ and $h(1) = 1$. Prove that, for each real number a, $B(a.s) = h(a).s$. By imitating

the classical geometric constructions for addition and multiplication, prove that, for any real numbers a, b, and c, $h(a+b) = h(a) + h(b)$ and $h(c^2) = h(c)^2$.
By the preceding problem, h must be the identity mapping on R.
Now one can easily prove that $B = I$.

12. The object of this problem is to introduce the ideas of linear and angular measure on E. In the next problem, one will find characterizations of similarity and of euclidean transformations on E, in terms of such measures.
By a <u>line segment</u> in R^2, one means any subset of R^2 of the form:

$$J_{u,v} = \{(1-a).u + a.v : \quad a \in R, \quad 0 \leq a \leq 1\};$$

by a <u>ray</u> in R^2, one means any subset of R^2 of the form:

$$N_{u,v} = \{(1-a).u + a.v : \quad a \in R, \quad 0 \leq a\},$$

where u and v are any two distinct members of R^2.
Let A be any affine transformation on R^2. Show that, for any line segment J in R^2, $A(J)$ is a line segment in R^2, and that, for any ray N in R^2, $A(N)$ is a ray in R^2. Using these facts, formulate the definitions of <u>line segment</u> and of <u>ray</u> in E. Show that it is meaningful to speak of the <u>end-points</u> of a given line segment, and of the <u>vertex</u> of a given ray in E.
For any line segment $\mathring{J}$ in E, one defines the <u>measure</u> of $\mathring{J}$ to be the (positive) real number $\delta(\alpha,\beta)$, where α and β are the end-points of $\mathring{J}$.
By an <u>angle</u> in E, one means any pair $\{\mathring{N}', \mathring{N}''\}$ of rays in E having common vertex.
Given an angle $\{\mathring{N}', \mathring{N}''\}$ in E, show that there exist

rotations θ_1 and θ_2 on E such that $\theta_1(\mathring{N}') = \mathring{N}''$ and $\theta_2(\mathring{N}'') = \mathring{N}'$. Note that these rotations are unique and that $\theta_2 = \theta_1^{-1}$. Note also that the common center of rotation for θ_1 and θ_2 coincides with the common vertex for $\mathring{N}'$ and $\mathring{N}''$.
Now let K' and K" be any cartesian coordinate mappings, and let W_1' and W_1'' be the linear parts of $\theta_1^{K'}$ and $\theta_1^{K''}$, respectively, and W_2' and W_2'', the linear parts of $\theta_2^{K'}$ and $\theta_2^{K''}$, respectively. Prove that: $\{W_1', W_2'\} = \{W_1'', W_2''\}$.
Finally, let K be any cartesian coordinate mapping, and let W_1 and W_2 be the linear parts of θ_1^{K} and θ_2^{K}, respectively. Let a stand for the smaller of the principal measures of W_1 and W_2. By the foregoing result, the number a is the same for all K. One refers to a as the <u>measure</u> of the angle $\{\mathring{N}',\mathring{N}''\}$. When $a = 0$, then $\mathring{N}' = \mathring{N}''$, and $\{\mathring{N}',\mathring{N}''\}$ is a <u>null</u> angle. When $a = \pi$, then $\mathring{N}' \cup \mathring{N}''$ is a line in E, and $\{\mathring{N}',\mathring{N}''\}$ is a <u>straight</u> angle.

13. Let θ be an affine transformation on E. Prove that θ is a similarity transformation iff it preserves angular measure: for each angle $\{\mathring{N}',\mathring{N}''\}$ in E, the measures of $\{\mathring{N}',\mathring{N}''\}$ and $\{\theta(\mathring{N}'),\theta(\mathring{N}'')\}$ are equal. Prove that θ is a euclidean transformation iff it preserves linear measure: for each line segment $\mathring{J}$ in E, the measures of $\mathring{J}$ and $\theta(\mathring{J})$ are equal. The latter result follows immediately from problem 9.

14. For each similarity transformation Σ on E, let J_Σ stand for the automorphism $(\theta \longmapsto \Sigma\theta\Sigma^{-1})$ of **E**. [See problem 5.] Prove that the homomorphism $(\Sigma \longmapsto J_\Sigma)$ carrying **S** to the automorphism group of **E**, is bijective. Hence, one may identify **S** with the automorphism group of **E**.
In substance, what one must prove is that, for each automorphism Ω of **E**, there exists a similarity transformation Σ on E such that $\Omega = J_\Sigma$. The fol-

lowing procedure will establish that result.
Let $\mathbf{X}$ be the set of all reflections on E, and $\mathbf{W}$, the set of all rotations having order two. Show that $\Omega(\mathbf{X}) = \mathbf{X}$ and that $\Omega(\mathbf{W}) = \mathbf{W}$.
For each member W of $\mathbf{W}$, let c_W stand for the center of rotation for W. Note that the mapping $(W \longmapsto c_W)$ carrying $\mathbf{W}$ to E is bijective. Let Σ be the mapping carrying E to itself, defined by the relation:

$$\Sigma(c_W) = c_{\Omega(W)},$$

where W is any member of $\mathbf{W}$. Clearly, Σ is bijective.
Show that, for any members W of $\mathbf{W}$ and X of $\mathbf{X}$, $X(c_W) = c_W$ iff $WX = XW$. Using that result, prove that $\Sigma(M_X) = M_{\Omega(X)}$. Conclude that Σ preserves lines. By problem 11, Σ must be an affine transformation on E.
Now let X be any member of $\mathbf{X}$. Let W' and W" be any members of $\mathbf{W}$. Prove that $X(c_{W'}) = c_{W''}$ iff there exists a member Y of $\mathbf{X}$ such that $W'Y = YW'$, $W''Y = YW''$, $XY = YX$, and $W''XY = XYW'$. In this context, prove that Y is unique. Conclude that:

$$X(c_{W'}) = c_{W''} \quad \text{iff} \quad \Omega(X)(c_{\Omega(W')}) = c_{\Omega(W'')}.$$

When untangled, the foregoing equivalence asserts that $\Omega(X) = \Sigma X \Sigma^{-1}$.
Now apply problems 8 and 5, to conclude that $J_\Sigma = \Omega$ and that Σ is a similarity transformation on E.

15. Let the definition of a euclidean plane be modified, by replacing the cartesian group $\mathscr{C}$ by the similarity group $\mathscr{S}$. Thus, let us require that any two cartesian coordinate mappings be linked by a similarity transformation, and let us admit any similarity transformation

as a coordinate transformation.
Show that the foregoing discussion of euclidean geometry would remain essentially the same. In particular, show that, while the euclidean metric would be lost, one could introduce the euclidean (metrizable) topology in its place; that one could still define the basic groups of transformations; and that one could still characterize euclidean transformations as those mappings which "preserve equal distances."

16. By a quadratic polynomial on R^2, one means any function p defined on R^2, of the form:

$$p(x) = c(x,x) + b(x) + a, \qquad x \in R^2,$$

where c is any nontrivial symmetric bilinear functional on R^2, where b is any linear functional on R^2, and where a is any real number. The components c, b, and a are uniquely determined by p; they are the quadratic, the linear, and the constant parts of p, respectively.
Prove that, for every function p defined on R^2 and for any cartesian transformation C on R^2, p is a quadratic polynomial iff pC^{-1} is a quadratic polynomial.
Now let ψ be any function defined on E. For each cartesian coordinate mapping K, let ψ^K stand for the cartesian form for ψ relative to K, that is, for ψK^{-1}. Obviously, for any cartesian coordinate mappings K' and K'', $\psi^{K''} = \psi^{K'}C^{-1}$, where C stands for $K''K'^{-1}$. Hence, $\psi^{K'}$ is a quadratic polynomial on R^2 iff $\psi^{K''}$ is such. Following the now familiar pattern, we define a quadratic polynomial on E to be any function ψ defined on E such that, for some (and hence for any) cartesian coordinate mapping K, ψ^K is a quadratic polynomial on R^2.
Let p be a quadratic polynomial on R^2, with quadratic part c. Let L be the unique symmetric linear mapping

carrying R^2 to itself, which generates c, in the sense that, for each member x of R^2, $c(x,x) = x\bullet L(x)$. One refers to $\det L$ as the discriminant of p.

Prove that, for every quadratic polynomial ψ on E and for any cartesian coordinate mappings K' and K'', the discriminants of $\psi^{K'}$ and $\psi^{K''}$ are equal. One refers to that common value as the discriminant of ψ. We shall denote it by d_ψ.

Now let ψ be any quadratic polynomial on E. Let N_ψ be the null set of ψ:

$$N_\psi = \{a \in E: \psi(a) = 0\}.$$

(+) Assume that d_ψ is positive. It may happen that N_ψ is empty or that it consists of just one point. Setting those cases aside, show that there is some cartesian coordinate mapping K such that $K(N_\psi)$ has the form:

$$K(N_\psi) = \{x \in R^2: (x_r/a)^2 + (x_s/b)^2 = 1\},$$

where a and b are positive real numbers. This is the case of the ellipse. [See Figure 13.]

(0) Assume that d_ψ is 0. Excluding the cases in which N_ψ is empty or consists of one or two points, show that there is some cartesian coordinate mapping K such that $K(N_\psi)$ has the form:

$$K(N_\psi) = \{x \in R^2: (ax_r)^2 - x_s = 0\},$$

where a is a positive real number. This is the case of the parabola. [See Figure 14.]

(-) Finally, assume that d_ψ is negative. Show that there must be some cartesian coordinate mapping K

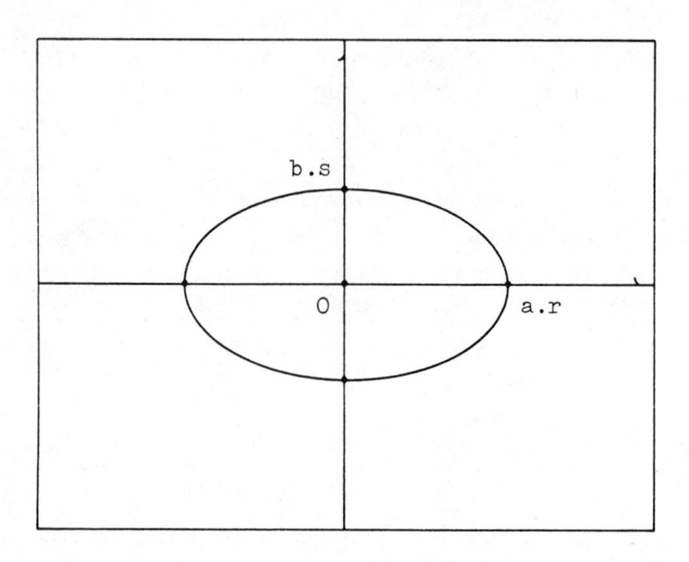

Figure 13

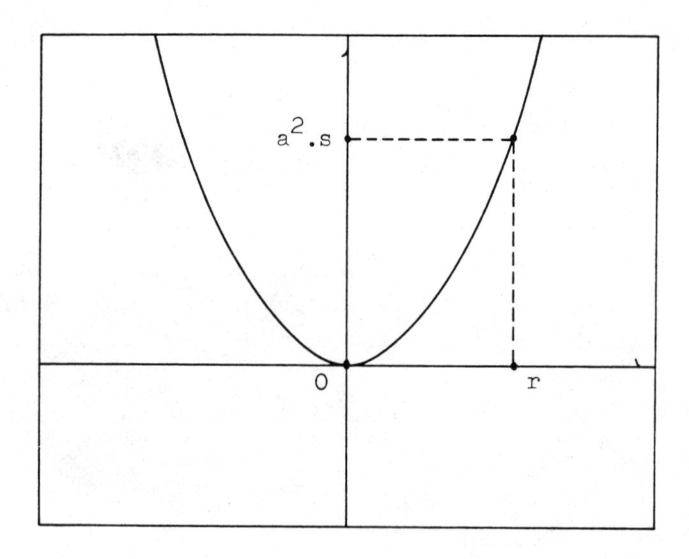

Figure 14

such that $K(N_\psi)$ has one of the following two forms:

$$K(N_\psi) = \{x \in R^2: (x_r/a)^2 - (x_s/b)^2 = 1\},$$

$$K(N_\psi) = \{x \in R^2: (x_r/a)^2 - (x_s/b)^2 = 0\},$$

where a and b are positive real numbers. These are the cases of the hyperbola, the latter being degenerate. [See Figure 15.]

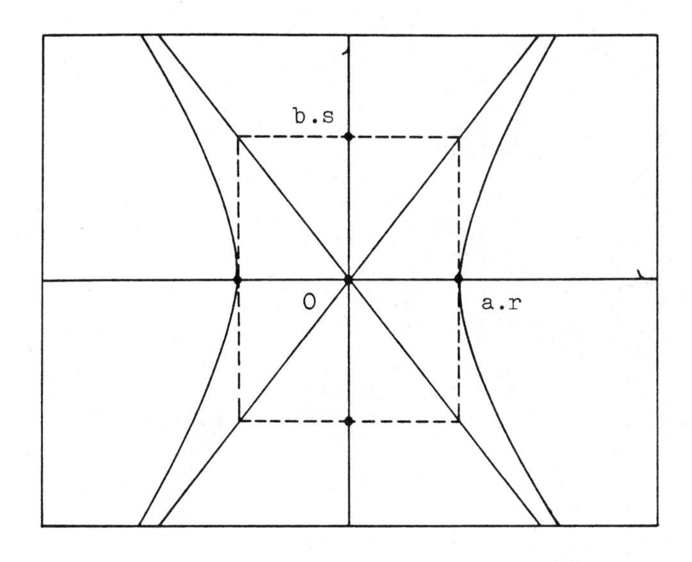

Figure 15

Chapter 2

PLANE ORNAMENTS

2.0 INTRODUCTION

In this chapter, we shall present the theory of ornamental subgroups of the euclidean group, and the classification (by the relation of isomorphism) of such subgroups into 17 types. These groups arise as the symmetry groups of certain mosaics in the euclidean plane, called plane ornaments, and also as the symmetry groups of plane crystals. We shall use both these sources to motivate and to illustrate the theory.

Our enumeration of the 17 ornamental classes proceeds by identifying, in successive steps, the geometric and the arithmetic classes, then by solving explicit group extension problems. This procedure rests upon the fundamental theorem of L. Bieberbach, that two ornamental groups are isomorphic iff they are conjugate in the affine group.

In the last section of this chapter, we shall present the theorem of H. Zassenhaus, characterizing in abstract terms those groups which can be realized within isomorphism as ornamental subgroups of the euclidean group.

2.1 MOSAICS

1° By a <u>tile</u> in the euclidean plane, we mean any compact subset of E having non-empty interior. As examples of tiles, let us cite regular n-gons, stars, crosses, swastikas, and the like.

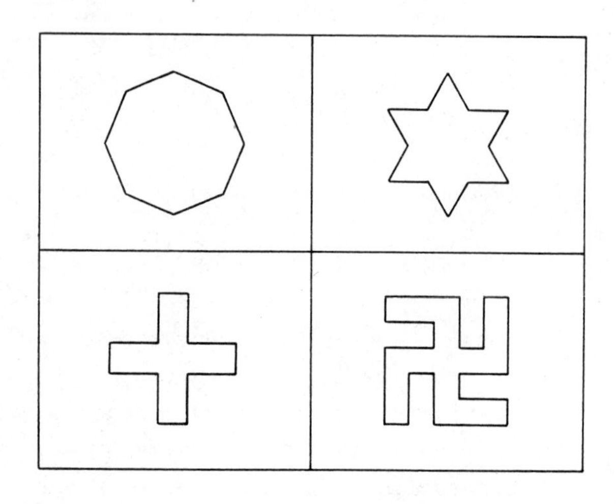

Figure 16

On behalf of the artisan, one may wish to restrict attention to tiles τ such that $int(\tau)$ is connected and such that $\tau = cl(int(\tau))$. The former condition demands that tiles consist of one piece, the latter, that they have no "filaments." However, for our purposes, these additional conditions would prove to be redundant.

2° By a <u>mosaic</u> in the euclidean plane, we mean any family Ω of tiles in E such that:

(a) for each point α in E, there exists some tile τ in Ω which contains α;

(b) for any two tiles τ' and τ'' in Ω, if $\tau' \neq \tau''$ then $int(\tau') \cap int(\tau'')$ is empty.

Hence, a mosaic in E is essentially a partition of E by tiles, with the proviso that some of the tiles involved may overlap at their boundaries.

To be realistic, one may wish to require that Ω be locally finite, which would mean that, for every bounded subset B of E, the subset of Ω consisting of the tiles τ for which $\tau \cap B$ is not empty, is finite. However, this condition would also prove to be redundant.

Mosaics are legion. For illustration, let us consider the following examples. The first two (Figures 17, 18) are the common quadratic and hexagonal mosaics, which play the role of reference frames in ornamental design. The next two (Figures 19, 20) are simple brick pavements.

The next four mosaics (Figures 21, 22, 23, 24) have been chosen from collections of Egyptian, Chinese, and Islamic ornamental art. The first appears on the ceiling of a tomb at Thebes (XIX Dynasty), the second is derived from a window lattice in a private home in Chengtu, Szechwan (1875 A.D.), and the third and fourth stand for wall ornaments at the Alhambra in Granada, Spain (c, 1300 A.D.). The collections from which these mosaics have been drawn are the ones assembled by P. Fořtová-Šámalová, D.S. Dye, O. Jones, and J. Bourgoin, listed in the bibliography.

The last two examples (Figures 25, 26) represent certain special classes of pattern: the "mosaics of Archimedes," composed of regular polygons, and the "mosaics of Kepler," involving regular star-polygons. One can find careful descriptions of these classes in the treatise on tilings and patterns by B. Grünbaum and G.C. Shephard.

We plan to study a particular class of mosaics, called plane ornaments. They are distinguished among mosaics in general by the property of two-dimensional periodicity. For a first impression of this property, one should contrast the ten examples of mosaics just mentioned with the three depicted in Figures 27, 28, and 29.

In order to state the condition of two-dimensional periodicity precisely, and to lay a foundation for the classification of plane ornaments, we must define the symmetry group associated with any given mosaic.

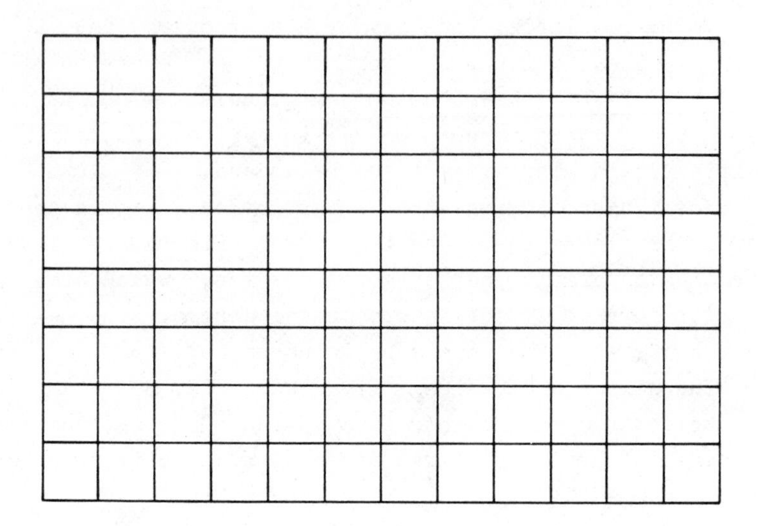

Figure 17

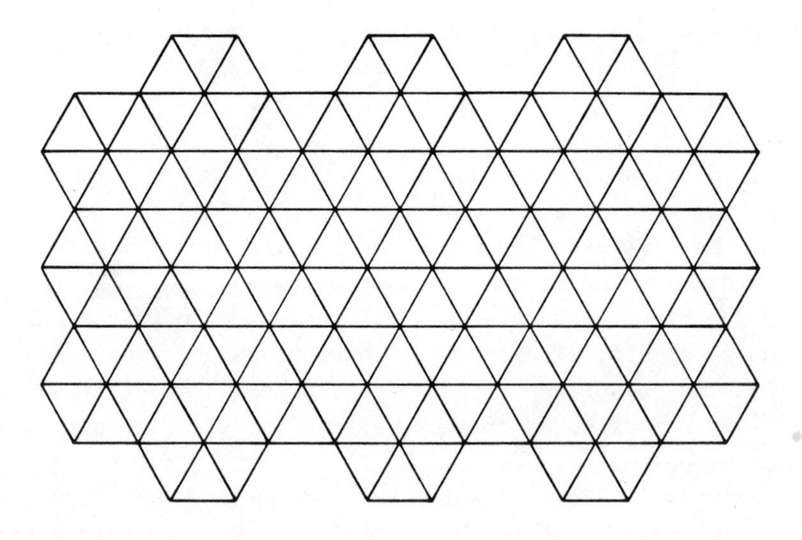

Figure 18

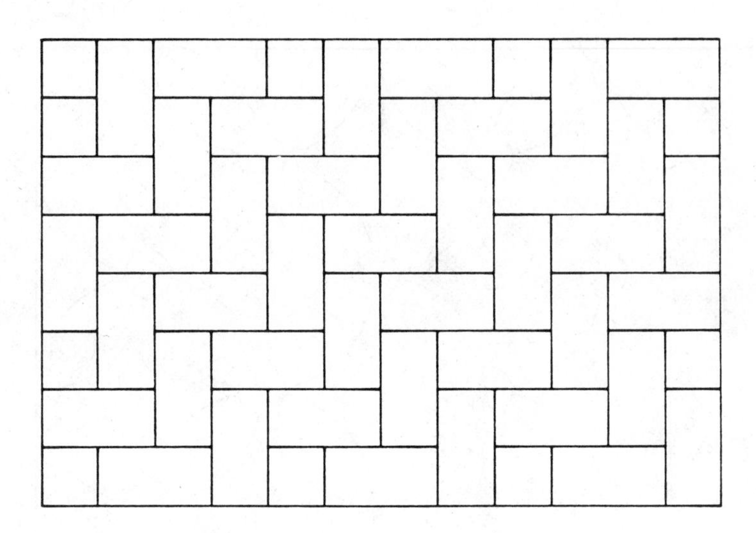

Figure 19

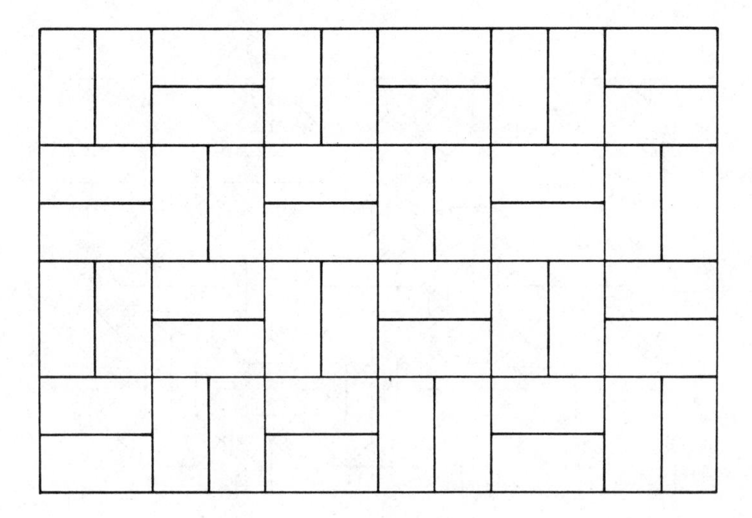

Figure 20

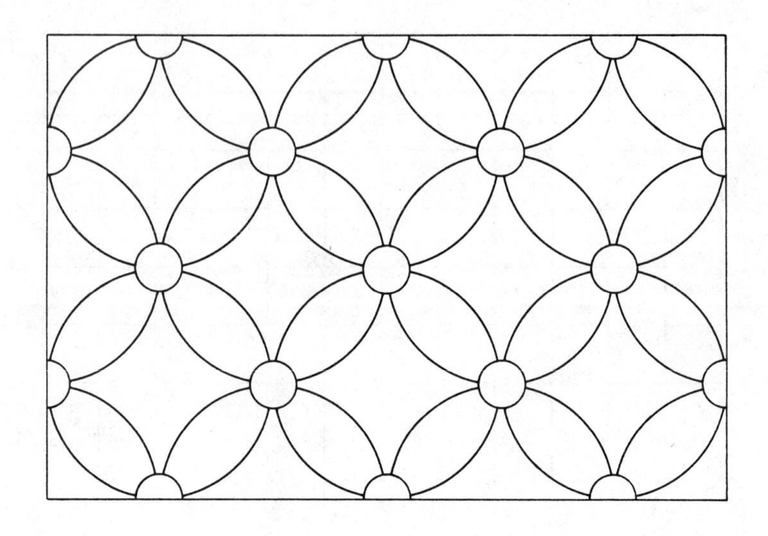

Figure 21

Figure 22

Figure 23

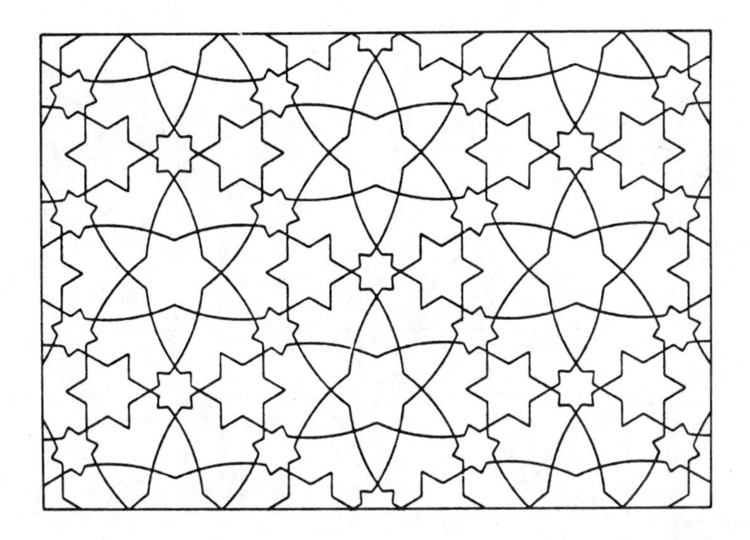

Figure 24

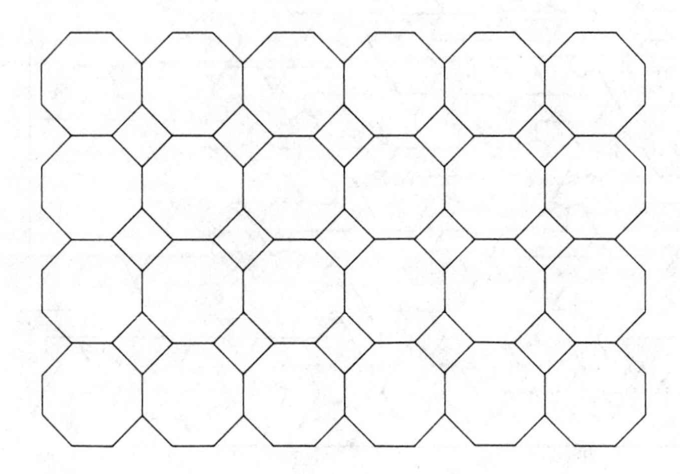

Figure 25

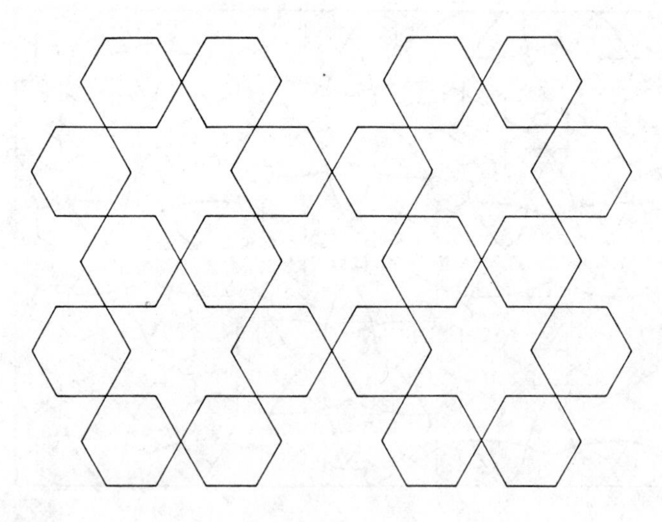

Figure 26

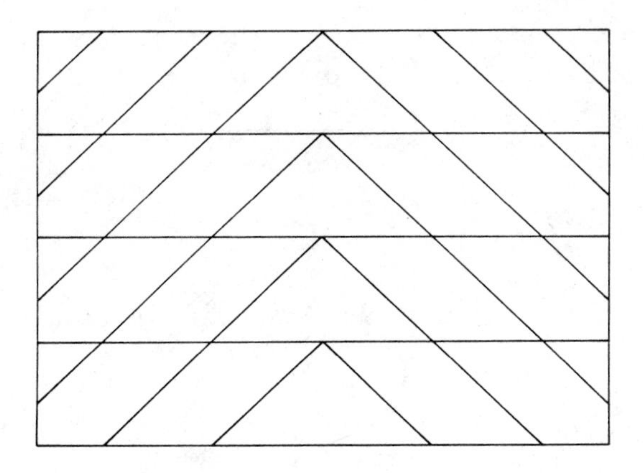

Figure 27

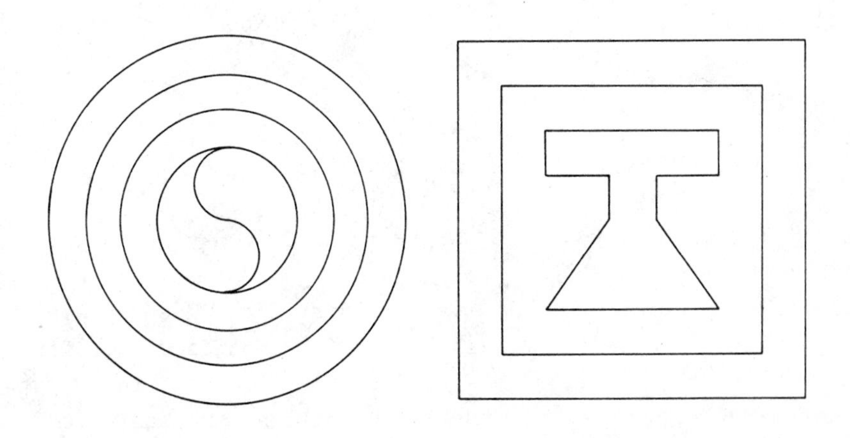

Figure 28 Figure 29

3° Let θ be any euclidean transformation on E. For each tile τ in E, $\theta(\tau)$ is also a tile in E. We shall refer to it as the <u>transform</u> of τ under θ, and denote it by $\theta.\tau$.

For each mosaic Ω in E, the family of all transforms under θ of the various tiles in Ω is itself a mosaic in E. We shall refer to it as the <u>transform</u> of Ω under θ, and denote it by $\theta.\Omega$:

$$\theta.\Omega = \{\theta.\tau : \tau \varepsilon \Omega\}.$$

For any euclidean transformations θ' and θ'' on E, we have:

$$(\theta'\theta'').\Omega = \theta'.(\theta''.\Omega).$$

Moreover, the transform of Ω under the identity transformation ι on E is Ω itself. Hence, the foregoing definition yields an action of the euclidean group **E** on the family of all mosaics in E.

For a given mosaic Ω and a given euclidean transformation θ, it may happen that Ω is invariant under θ:

$$\theta.\Omega = \Omega.$$

Of course, this condition entails that:

for each tile τ in Ω, $\theta.\tau$ is also a tile in Ω,

though in general τ and $\theta.\tau$ are distinct. One can easily verify that the latter condition in turn implies that, for each tile τ'' in Ω, there exists some tile τ' in Ω such that $\tau'' = \theta.\tau'$. Hence, the two conditions are equivalent.

Now let Ω be any mosaic in E. By a <u>symmetry</u> of Ω, we shall mean any euclidean transformation θ on E under which Ω is invariant.

Clearly, the family of all symmetries of Ω coincides with the stabilizer of Ω under the action of the euclidean group **E**. Hence, it is a subgroup of **E**. We shall refer to it as the <u>symmetry group</u> of Ω, and denote it by $\mathbf{E}_{\Omega}$:

$$\mathbf{E}_{\Omega} = \{ \theta \varepsilon \mathbf{E} : \theta.\Omega = \Omega \}.$$

4° The foregoing discussion may be generalized, to apply to broader groups of transformations on E. Thus, one may define the transform of a given mosaic Ω in E under an affine transformation, and even under an arbitrary homeomorphism on E. Hence, one may refer to "affine" symmetries and even to "topological" symmetries of Ω. On occasion, we shall find it useful to do so.

5° For illustration, let us describe the symmetry groups of the mosaics depicted in Figures 19, 23, and 24.

In Figure 23, there are tiles shaped like the letter **s**, but none like the mirror image of **s**. Hence, the mosaic admits no reflections and no glide-reflections among its symmetries. But one can identify certain translations as symmetries, namely, the ones which transform the center of a star to the center of a star. Moreover, all rotations which have order six and which leave fixed the center of a star, are symmetries. In the following copy of Figure 23, we have indicated two such translational symmetries by bold arrows, and one such rotational symmetry by a bold arc-arrow. [See Figure 30.] These symmetries are actually sufficient to generate the symmetry group of the mosaic. In particular, appropriate products of them (and of their inverses) yield the evident rotational symmetries having order three and leaving fixed a point at which three of the **s**-shaped tiles meet, and the rotational symmetries having order two and leaving fixed the geometric center of an **s**-shaped tile.

For Figure 24, the same symmetry structure appears, but augmented by reflectional symmetries. In the following copy of Figure 24, we have indicated (as before) one rotational and two translational symmetries, and we have identified by a bold line the axis of a particular reflectional symmetry. [See Figure 31.] The indicated symmetries generate the symmetry group of the mosaic.

Treating Figure 19 in similar manner, one can determine certain translational and rotational symmetries (the latter having order two). There are no reflectional symmetries, but certain glide-reflectional symmetries appear. Using a copy of Figure 19, we have indicated a set of symmetries sufficient to generate the symmetry group. [See Figure 32.] The glide-reflection is represented by a bold crossed-arrow, superimposed upon a bold line.

For a given mosaic Ω, the group $\mathbf{E}_\Omega$ provides a measure of the inherent symmetry of Ω, with respect not to the particular tiles which comprise Ω but to the relative situation of those tiles in the plane. For the case in which Ω is a plane ornament, we shall find that $\mathbf{E}_\Omega$ reveals the essential structure of Ω, and provides a fruitful classification of such mosaics.

Figure 30

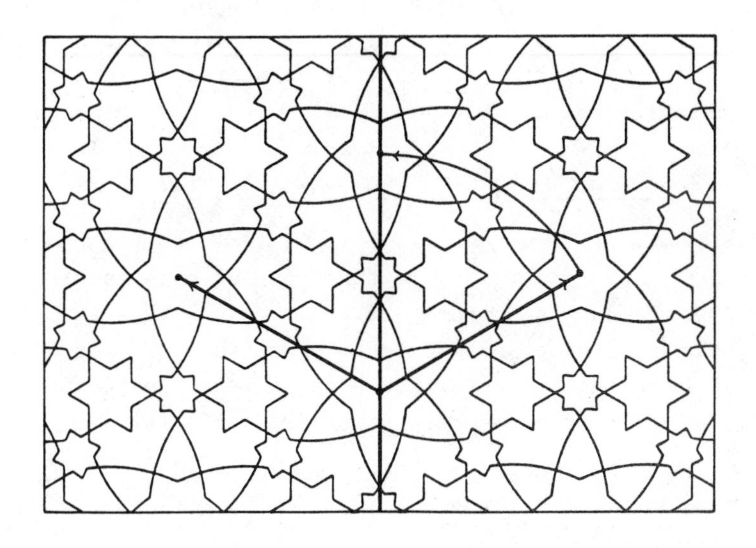

Figure 31

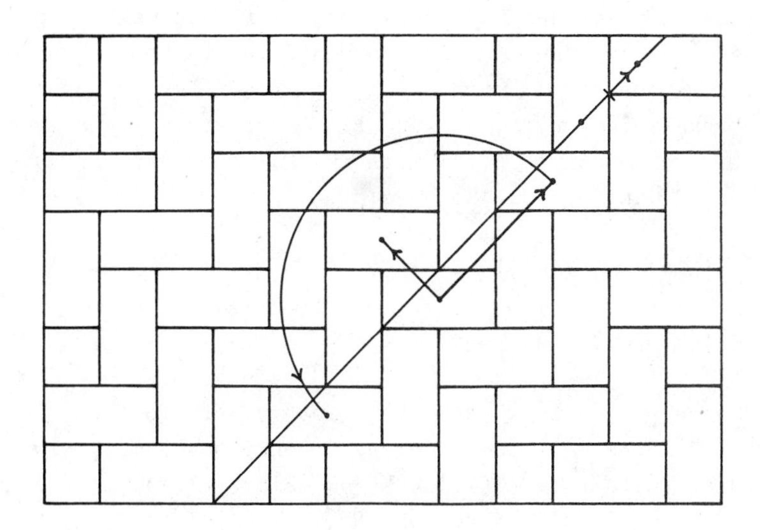

Figure 32

6° Now let Ω be any mosaic in E. For each cartesian coordinate mapping K, we obtain the cartesian form $\mathbf{E}_{\Omega}^{K}$ for $\mathbf{E}_{\Omega}$ relative to K, and the cartesian presentation of $\mathbf{E}_{\Omega}$ relative to K:

$$T_{\Omega}^{K} \xrightarrow{\quad q \quad} \mathbf{E}_{\Omega}^{K} \xrightarrow{\quad p \quad} \mathcal{G}_{\Omega}^{K} .$$

Of course, T_{Ω}^{K} and $\mathcal{G}_{\Omega}^{K}$ are the translational and linear parts of the cartesian form $\mathbf{E}_{\Omega}^{K}$.

Our present objective is to prove that T_{Ω}^{K} is <u>discrete</u>, in the sense that there exists some positive real number a such that, for each member t of T_{Ω}^{K}, if $|t| < a$ then $t = 0$.

<u>Proposition 10</u>. The subgroup T_{Ω}^{K} of R^2 is discrete.

Proof. Let τ be any tile in Ω. Since $\mathrm{int}(\tau)$ is not empty, we may introduce a member α of $\mathrm{int}(\tau)$ and a positive real number a such that, for each member β of E, if $\delta(\alpha,\beta) < a$ then β is contained in $\mathrm{int}(\tau)$. Now let t be any member of T_{Ω}^{K} for which $|t| < a$. Let θ stand for the translation $K^{-1}[t,I]K$ on E. Clearly, θ is a symmetry of Ω, so the tile $\theta(\tau)$ lies in Ω. Moreover, $\theta(\alpha)$ is a member of $\mathrm{int}(\theta(\tau))$. We have:

$$|K(\alpha) - K(\theta(\alpha))| \;=\; |t|,$$

so $\quad \delta(\alpha,\theta(\alpha)) < a,$

and hence $\theta(\alpha)$ is a member of $\mathrm{int}(\tau)$. It follows that $\mathrm{int}(\tau) \cap \mathrm{int}(\theta(\tau))$ is not empty, hence that $\theta(\tau) = \tau$. Now we infer that $[t,I](K(\tau)) = K(\tau)$. Since $K(\tau)$ is a compact (and hence bounded) subset of R^2, we conclude that t must be 0. ///

While elementary, the foregoing proposition is crucial to understanding the structure of the symmetry groups of mosaics. Moreover, it is the only one of the results of this chapter which requires attention to specific topological properties of tiles. Accordingly, one may wish to note that the proposition would remain valid under definitions of tile and mosaic substantially more general than the ones which we have adopted. For example, one might define a "tile" to be any closed subset of E having nonempty interior, then define a "discrete" tile to be any tile which does not admit "arbitrarily small" translational symmetries. Obviously, every bounded tile would be discrete, since it has no translational symmetries at all, other than the identity. One might then define a "mosaic" as we have done, but with the additional condition that it contain at least one discrete tile. With minor modifications, the foregoing argument would still prove the proposition.

We adopted the more restrictive definitions of tile and mosaic because, ad initium, they are easier to understand. In retrospect, however, one might wish to substitute the more comprehensive definitions, in order to increase the scope of the theory.

As an example of a mosaic which would fall under the new definitions, but not under the old, one should consider the Chinese brick-work depicted in Figure 33. We have drawn it, in modified form, from D.S. Dye's collection.

2.2 LATTICES

1° In this section, we shall characterize the discrete subgroups of R^2. The particular case of lattices in R^2 will correspond to plane ornaments.

Clearly, the trivial subgroup $\{0\}$ of R^2 is discrete.

For each nonzero member u of R^2, the subgroup

$$\{m.u : m \in Z\}$$

of R^2 is discrete. We shall refer to it as a <u>chain</u>.

For any linearly independent members u and v of R^2,

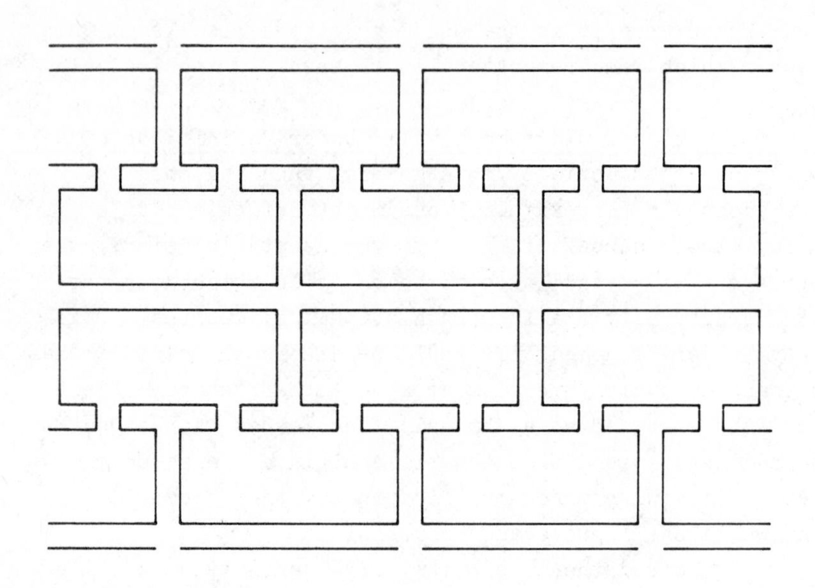

Figure 33

the subgroup

$$\{m.u + n.v : \quad m \in Z, \quad n \in Z\}$$

of R^2 is discrete. One refers to it as a <u>lattice</u> in R^2, and one refers to $\{u,v\}$ as a <u>basis</u> for the lattice. [See Figure 34.]

<u>Proposition 11</u>. Every nontrivial discrete subgroup of R^2 is either a chain or a lattice.

Proof. Let T be any nontrivial discrete subgroup of R^2. We shall prove first that, for each bounded subset B of R^2, $B \cap T$ is finite.

Thus, let a be a positive real number such that, for each member t of T, if $|t| < a$ then $t = 0$. It follows that, for any members t' and t'' of T, if $|t' - t''| < a$ then $t' = t''$. Hence, every subset of R^2 having diameter less than a will contain at most one member of T. Since B

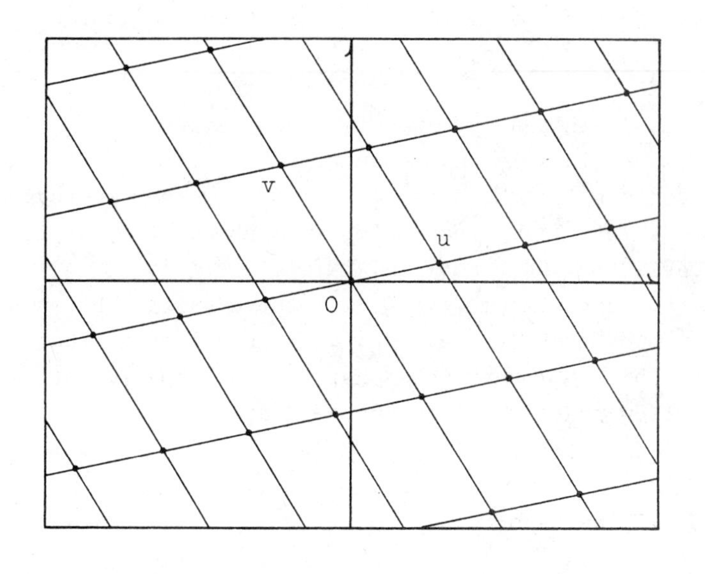

Figure 34

can be covered by a finite number of subsets of R^2 having diameter less than a, $B \cap T$ must be finite.

Let w be any member of $T \setminus \{0\}$. By the preceding result, the subset

$$\{t \in T : \quad 0 < |t| \leq |w|\}$$

of T is finite. Consequently, $T \setminus \{0\}$ must contain members having minimum norm. Let u be such a member. Let T_u stand for the chain in R^2 determined by u: $T_u = Z.u$. It may be that $T = T_u$, in which case T is a chain. Let us assume that $T \neq T_u$. We shall prove that T is a lattice in R^2.

We contend that, for each real number a, if $a.u$ is contained in T then a must be an integer; that is, $T \cap R.u = T_u$. Indeed, letting $[a]$ and (a) stand for the integral and fractional parts of a, we have:

$$(a).u = a.u - [a].u .$$

Hence, if $a.u$ is contained in T then $(a).u$ is contained in T. However, $|(a).u| = (a)|u| < |u|$. By definition of u, it follows that $(a) = 0$, hence that a is an integer.

Now it is clear that, for each member x of $T \setminus T_u$, u and x are linearly independent.

Just as for the case of $T \setminus \{0\}$, we may introduce a member v of $T \setminus T_u$ having minimum norm. Of course, $\{u,v\}$ is a basis for R^2, and $|u| \leq |v|$.

For each member t of T, there exist real numbers a and b such that $t = a.u + b.v$. We shall prove that $(a) = 0$ and $(b) = 0$, that is, that a and b are integers. It will follow that T is a lattice, and that $\{u,v\}$ is a basis for T.

Let y stand for $(a).u + (b).v$. Clearly, y is a member of T. We have:

$$(u + v) - y = (1-(a)).u + (1-(b)).v$$

and $1-(b) \neq 0$.

Hence, $(u + v) - y$ is a member of $T \setminus T_u$. Now it follows that:

$$\begin{aligned} |v| &\leq |(u + v) - y| \\ &\leq (1-(a))|u| + (1-(b))|v| \\ &\leq (2 - ((a) + (b)))|v|, \end{aligned}$$

and hence that:

$$(a) + (b) \leq 1.$$

By the definitions of u and v, it is plain that if one of (a) and (b) is positive then both are positive. Let us suppose that both (a) and (b) are positive. It would follow that y is contained in $T \setminus T_u$. By Cauchy's inequality, we have:

$$u \bullet v < |u||v|.$$

Hence, we would obtain:

$$\begin{aligned} |y|^2 &= |(a).u + (b).v|^2 \\ &= (a)^2|u|^2 + 2(a)(b)\, u \bullet v + (b)^2|v|^2 \\ &< (a)^2|v|^2 + 2(a)(b)|v|^2 + (b)^2|v|^2 \\ &= ((a) + (b))^2|v|^2 \\ &\le |v|^2; \end{aligned}$$

therefore: $|y| < |v|$, which would contradict the definition of v.

We infer that both (a) and (b) equal 0. ///

The foregoing argument contains an explicit procedure for generating a basis for a given lattice T. One first selects a member u of $T \setminus \{0\}$ having minimum norm. One then selects a member v of $T \setminus Z.u$ having minimum norm. The resulting pair $\{u,v\}$ is a basis for T.

2° Now let us consider the possible structures of lattices in R^2, subject to particular symmetry conditions. We shall in fact develop a hierarchy of lattice structures, designed to be of use in our subsequent description of geometric and arithmetic classes.

Let T be a lattice in R^2, and let $\mathcal{G}$ be a subgroup of $\mathcal{O}$. Let us assume that T is invariant under $\mathcal{G}$. [See 1.1.4°.] We plan to prove that $\mathcal{G}$ must be finite, and that every rotation in $\mathcal{G}$ must have order 1, 2, 3, 4, or 6.

Let $\{u,v\}$ be a basis for T. Let a be the larger of $|u|$ and $|v|$, and let B be the subset of T consisting of all members t for which $|t| \le a$. By the first step in the proof of Proposition 11, B must be finite. Clearly, for each member V of $\mathcal{G}$, V(u) and V(v) lie in B. Hence, we may form the mapping

$$(V \longmapsto (V(u),\, V(v))$$

carrying $\mathcal{G}$ to $B \times B$. Obviously, this mapping is injective.

Therefore, $\mathcal{G}$ must be finite.

Now let W be any rotation in $\mathcal{G}$. Since W(u) and W(v) lie in T, the matrix for W relative to the basis {u,v} must have integral entries. It follows that the trace of W is an integer. Therefore, W must have order 1, 2, 3, 4, or 6. [See 1.4, problem 2.]

In conjunction with Proposition 9, these results yield the following proposition.

Proposition 12. For each lattice T in R^2 and for each subgroup $\mathcal{G}$ of $\mathcal{O}$, if T is invariant under $\mathcal{G}$ then $\mathcal{G}$ is finite. If $\mathcal{G}$ contains only rotations then it must be one of the following groups:

$$\mathcal{Z}_1, \quad \mathcal{Z}_2, \quad \mathcal{Z}_3, \quad \mathcal{Z}_4, \quad \mathcal{Z}_6.$$

If $\mathcal{G}$ contains a reflection then it must be one of the following groups:

$$\mathcal{D}_1(X), \quad \mathcal{D}_2(X), \quad \mathcal{D}_3(X), \quad \mathcal{D}_4(X), \quad \mathcal{D}_6(X),$$

where X is any reflection in $\mathcal{G}$.

The heart of the foregoing proposition is the assertion that rotational symmetries of a given lattice must have order 1, 2, 3, 4, or 6. This condition is usually called the crystallographic restriction.

3° Let T be any lattice in R^2. Let $\mathcal{O}_T$ be the family of all orthogonal linear mappings V under which T is invariant: V(T) = T. Clearly, $\mathcal{O}_T$ is a subgroup of $\mathcal{O}$. We shall refer to it as the symmetry group of T.

By Proposition 12, $\mathcal{O}_T$ is finite. In fact, it must be one of the ten (types of) groups listed in that proposition. We plan to determine which of the ten (types of) groups can serve as the symmetry groups of lattices, and to show how such groups dictate the structure of the underlying lattices.

Let us first note that $\mathcal{O}_T$ must contain -I. Hence, $\mathcal{O}_T$ cannot be $\mathcal{Z}_1$, $\mathcal{Z}_3$, $\mathcal{D}_1(X)$, or $\mathcal{D}_3(X)$.

Let W stand for the rotation having principal measure

$\pi/2$. Of course, W has order 4. Let us assume that W is contained in $\mathcal{O}_T$, which is to say that $\mathcal{Z}_4$ is included in $\mathcal{O}_T$. Let u be a member of $T \setminus \{0\}$ having minimum norm, and let v stand for W(u). Then v is a member of $T \setminus Z.u$ having minimum norm, so $\{u,v\}$ is a basis for T. Let X be the reflection for which $X(u) = u$. Since $W^3 = W^{-1}$ and $W^2 = -I$, we have:

$$\begin{aligned} X(W(u)) &= W^{-1}(X(u)) \\ &= -W(u). \end{aligned}$$

Hence, $X(v) = -v$. It follows that X is contained in $\mathcal{O}_T$. Now there is no alternative but that $\mathcal{O}_T = \mathcal{D}_4(X)$. In this case, we shall say that the lattice T is <u>quadratic</u>. [See Figure 35.] With respect to the basis $\{u,v\}$ for T, the matrices for W and X are the following:

$$W \longleftrightarrow \begin{bmatrix} 0 & -1 \\ 1 & 0 \end{bmatrix}, \qquad X \longleftrightarrow \begin{bmatrix} 1 & 0 \\ 0 & -1 \end{bmatrix}.$$

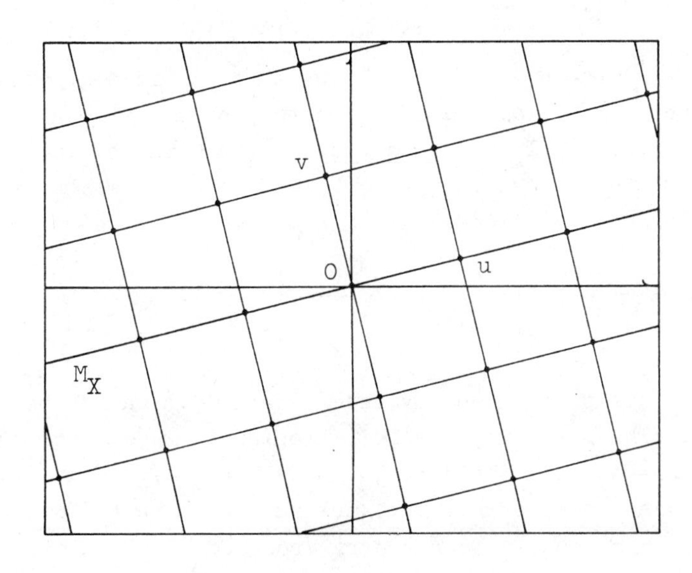

Figure 35

Now let W stand for the rotation having principal measure $\pi/3$, so that the order of W is 6. Let us assume that W^2 is contained in $\mathcal{O}_T$. Accordingly, $\mathcal{Z}_3$ is included in $\mathcal{O}_T$. Let u be a member of $T\setminus\{0\}$ having minimum norm, and let v stand for $W^2(u)$. Then v is a member of $T\setminus Z.u$ having minimum norm, so $\{u,v\}$ is a basis for T. Since $W^3 = -I$, we have:

$$\begin{aligned} 0 &= I + W^3 \\ &= (I + W)(I - W + W^2). \end{aligned}$$

Since $I + W$ is invertible, we obtain:

$$W = I + W^2.$$

From this relation, one can easily prove that $W(T) = T$, so that W is a member of $\mathcal{O}_T$. Let X be the reflection for which $X(u) = u$. Since $W^4 = (W^2)^{-1}$, we have:

$$\begin{aligned} X(W^2(u)) &= (W^2)^{-1}(X(u)) \\ &= -W(u). \end{aligned}$$

Hence, $X(v) = -(u+v)$. It follows that X is contained in $\mathcal{O}_T$. Therefore, $\mathcal{O}_T = \mathcal{D}_6(X)$. In this case, we shall say that the lattice T is <u>hexagonal</u>. [See Figure 36.] With respect to the basis $\{u,v\}$ for T, the matrices for W and X appear as follows:

$$W \longleftrightarrow \begin{bmatrix} 1 & -1 \\ 1 & 0 \end{bmatrix}, \qquad X \longleftrightarrow \begin{bmatrix} 1 & -1 \\ 0 & -1 \end{bmatrix}.$$

At this point, we may infer that $\mathcal{O}_T$ cannot be $\mathcal{Z}_4$ or $\mathcal{Z}_6$. The surviving possibilities are $\mathcal{D}_4(X)$ and $\mathcal{D}_6(X)$, which we have just considered, as well as $\mathcal{Z}_2$ and $\mathcal{D}_2(X)$.

It is a simple matter to design T so that $\mathcal{O}_T$ equals $\mathcal{Z}_2$. In this case, the symmetry group of T is minimal. We shall say that the lattice T is <u>arbitrary</u>.

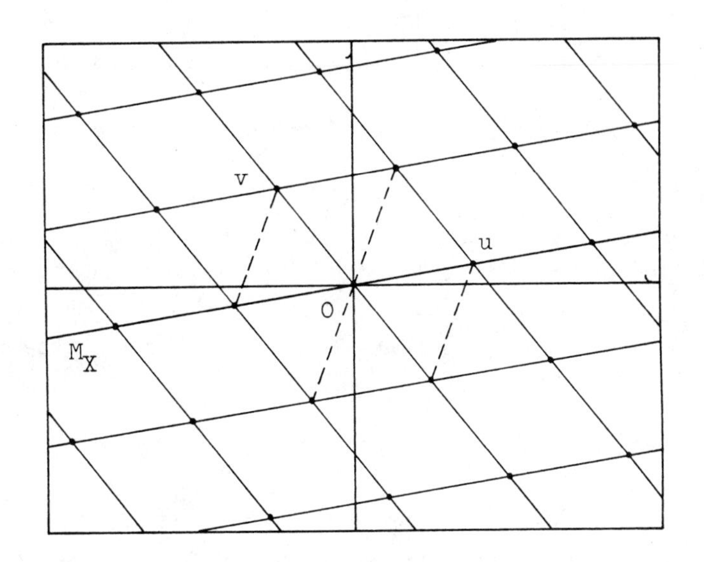

Figure 36

Finally, let us assume that $\mathcal{O}_T$ contains a reflection. Let X stand for such a reflection. Of course, -X must be a member of $\mathcal{O}_T$ as well; let it be denoted by Y. For each member t of T, t + X(t) lies in M_X and t + Y(t) lies in M_Y. Selecting w from the (obviously nonempty) subset $T \setminus (M_X \cup M_Y)$ of R^2, we obtain a nonzero member w + X(w) of M_X and a nonzero member w + Y(w) of M_Y. Hence, we may introduce a member u of $M_X \cap (T \setminus \{0\})$ having minimum norm, and a member v of $M_Y \cap (T \setminus \{0\})$ having minimum norm. Again let t be any member of T. Again, t + X(t) lies in M_X and t + Y(t) lies in M_Y. By imitating the proof of Proposition 11, we obtain integers m and n such that:

$$t + X(t) = m.u,$$

$$t + Y(t) = n.v.$$

Hence:

$$t = (m/2).u + (n/2).v,$$

$$X(t) = (m/2).u - (n/2).v.$$

Either both m and n are odd, or else both m and n are even. Otherwise, either (1/2).u or (1/2).v would lie in T, contrary to the definitions of u and v.

Let us assume that, for each member t of T, both m and n are even. It follows that {u,v} is a basis for T. In this case, we shall say that the lattice T is <u>orthogonal</u>. [See Figure 37.] With respect to {u,v}, the matrix for X is the following:

$$X \longleftrightarrow \begin{bmatrix} 1 & 0 \\ 0 & -1 \end{bmatrix}.$$

Now let us assume that there exists some member t* of T for which both m* and n* are odd. Let u* and v* stand for (1/2).(u - v) and (1/2).(u + v), respectively. One can easily verify that u* and v* lie in T, and that,

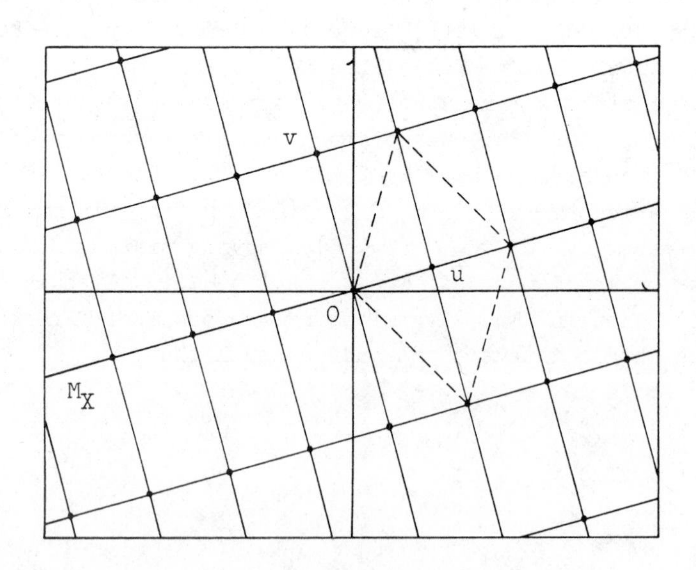

Figure 37

for each member t of T:

$$t = ((m-n)/2).u^* + ((m+n)/2).v^* .$$

It follows that $\{u^*,v^*\}$ is a basis for T. In this case, we shall say that the lattice T is rhombic. [See Figure 38.] The matrix for X relative to $\{u^*,v^*\}$ appears as follows:

$$X \longleftrightarrow \begin{bmatrix} 0 & 1 \\ 1 & 0 \end{bmatrix}.$$

In either of the foregoing two cases, one will obtain $\mathcal{O}_T = \mathcal{D}_2(X)$, provided that $|u| \neq |v|$.

We conclude that the possible symmetry groups of lattices are $\mathcal{I}_2$, $\mathcal{D}_2(X)$, $\mathcal{D}_4(X)$, and $\mathcal{D}_6(X)$. They yield the following hierarchy of bravais lattice structures:

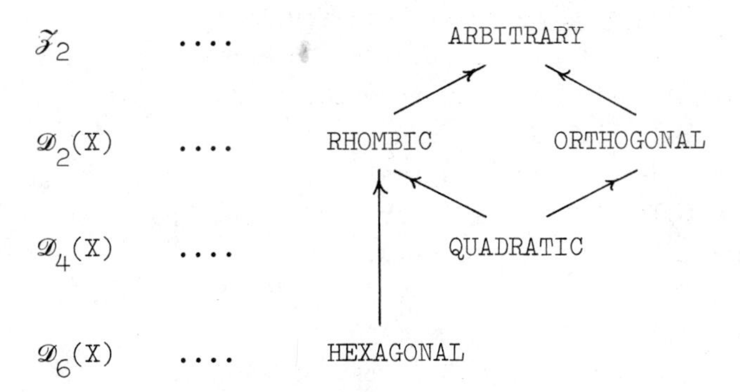

The arrows indicate relations of implication.

For later reference, let us summarize the preceding results.

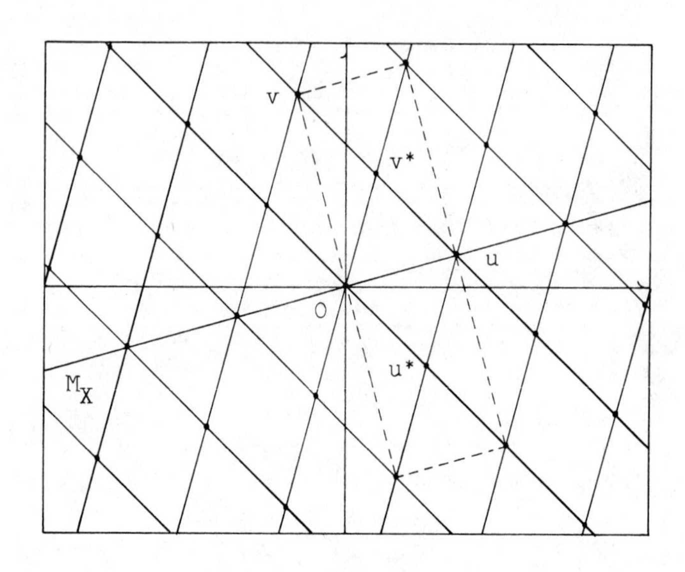

Figure 38

<u>Proposition</u> <u>13</u>. For each lattice T in R^2:

(a) if T is invariant under a reflection then T is either orthogonal or rhombic;

(b) if T is invariant under a rotation having order three then T is hexagonal;

(c) if T is invariant under a rotation having order four then T is quadratic.

2.3 PLANE ORNAMENTS

1° Let Ω be a mosaic in E. For any cartesian coordinate mappings K' and K" for E, the corresponding cartesian presentations of $\mathbf{E}_\Omega$ are related as follows:

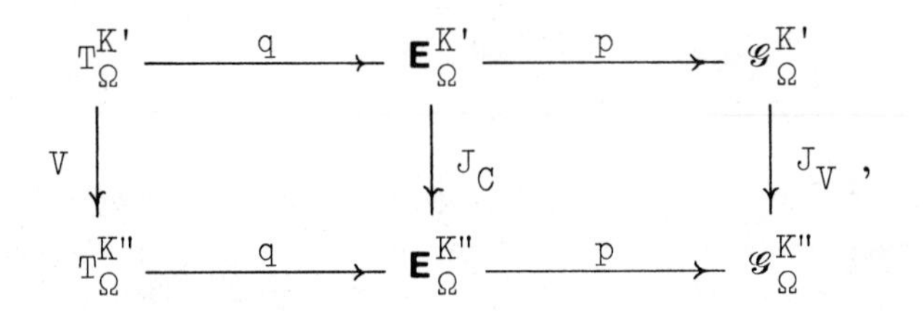

where C stands for $K''K'^{-1}$ and V, for the linear part of C, and where J_C and J_V are the appropriate (inner) automorphisms. [See 1.3.4°.] Since V is a linear isomorphism, it is plain that $T_\Omega^{K'}$ is a lattice in R^2 iff $T_\Omega^{K''}$ is a lattice in R^2. Hence, such conditions can serve to provide an explicit formulation of the idea of two-dimensional periodicity in the structure of Ω.

We now define a <u>plane ornament</u> in the euclidean plane to be any mosaic Ω in E such that, for some (and hence for any) cartesian coordinate mapping K, T_Ω^K is a lattice in R^2.

From the foregoing diagram, it is plain that the various lattices T_Ω^K underlying Ω are mutually equivalent with respect to $\mathcal{O}$. That is, for any cartesian coordinate mappings K' and K'', there exists a member V of $\mathcal{O}$ (namely, the linear part of $K''K'^{-1}$) such that $T_\Omega^{K''} = V(T_\Omega^{K'})$. Hence, we may unambiguously identify the lattice structure of Ω, in the sense of the preceding section: we may say that the lattic structure of Ω is the arbitrary, the orthogonal, the rhombic, the quadratic, or the hexagonal.

Figures 17 - 26 present examples of plane ornaments. In particular, the lattice structure for Figure 19 is the orthogonal, for Figures 23 and 24, the hexagonal. [See Figures 32, 30, and 31.]

In contrast, Figure 27 provides an example of a mosaic Ω for which the various translational parts T_Ω^K are chains. This situation is characteristic of "border ornaments." [See

8, problem 20.] For Figures 28 and 29, the various translational parts are trivial.

2° In parallel with 1°, let us now formulate the definition of ornamental subgroup of the euclidean group.

Let **F** be any subgroup of **E**. For any cartesian coordinate mappings K' and K", the cartesian presentations of **F** are related as follows:

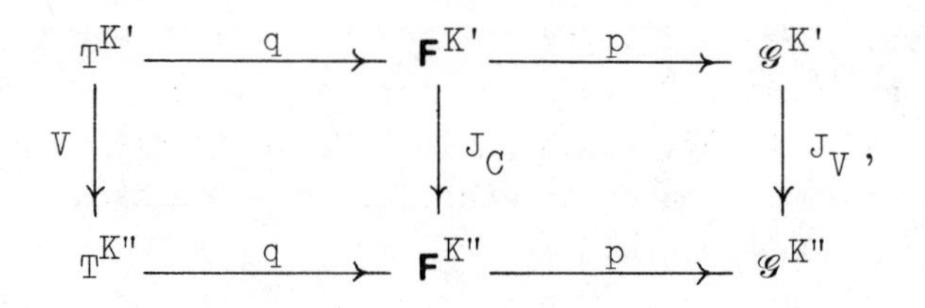

where C stands for $K''K'^{-1}$ and V, for the linear part of C. In this general context, $T^{K'}$ and $T^{K''}$ need not be discrete. However, motivated by the definition of plane ornament we shall focus attention upon the cases in which $T^{K'}$ and $T^{K''}$ are not only discrete but are in fact lattices in R^2.

We shall say that **F** is an <u>ornamental</u> subgroup of **E** provided that, for some (and hence for any) cartesian coordinate mapping K, T^K is a lattice in R^2.

Obviously, for each mosaic Ω in E, Ω is a plane ornament iff $\mathbf{E}_\Omega$ is an ornamental subgroup of **E**. We assert that the complementary result is also true, that is, that every ornamental subgroup of **E** is the symmetry group of some plane ornament. However, this assertion is not obvious. We shall prove it later, in the course of analyzing and classifying all ornamental subgroups of **E**.

3° The basic objective of this chapter is to classify all ornamental subgroups of **E** by the relation of isomorphism. Thus, we shall regard two ornamental subgroups $\mathbf{F}_1$ and $\mathbf{F}_2$ of **E** as <u>equivalent</u> iff they are isomorphic. The equivalence classes which follow from this relation will be called <u>ornamental classes</u>.

We shall regard two plane ornaments Ω' and Ω'' in E as <u>equivalent</u> iff the corresponding symmetry groups $\mathbf{E}_{\Omega'}$ and $\mathbf{E}_{\Omega''}$ are isomorphic. Hence, the classification of ornamental subgroups of $\mathbf{E}$ will provide just as well a classification of plane ornaments in E.

Of course, the relation of isomorphism is hardly canonical. There are many reasonable alternatives to it, from which other classifications would follow. Among the problems at the end of this chapter, one will find a few examples of such alternatives. However, the relation of isomorphism is in itself sufficiently discriminating to provide an informative classification. Moreover, every viable equivalence relation would surely imply the relation of isomorphism, so that every viable classification would in principle be derivable by refinement of that to be obtained in this chapter. For these reasons, the relation of isomorphism will play the central role in our exposition. One can find a general discussion of this matter, bearing upon arbitrary tilings and patterns, in the treatise by B. Grünbaum and G.C. Shephard.

4° Now let us develop the principal theorem of our subject, due to L. Bieberbach. This remarkable result will serve as the base for our subsequent calculation of all ornamental classes.

<u>THEOREM</u> <u>1</u>. For any ornamental subgroups $\mathbf{F}_1$ and $\mathbf{F}_2$ of $\mathbf{E}$, $\mathbf{F}_1$ and $\mathbf{F}_2$ are isomorphic iff they are conjugate in the affine group $\mathbf{A}$.

For our proof of this theorem, we require two preliminary propositions.

<u>Proposition</u> <u>14</u>. For any finite subgroup $\mathscr{P}$ of $\mathscr{A}$, there exists some member w of R^2 such that, for each member A of $\mathscr{P}$, $A(w) = w$.

Proof. Let k be the order of $\mathscr{P}$, and let w stand for

$$\Sigma_{B \in \mathscr{P}}(1/k).B(0) .$$

In effect, w is the average of the translational parts of the various members of $\mathscr{P}$. Let A be any member of $\mathscr{P}$, and let t and L stand for the translational and linear parts of A. We have:

$$\begin{aligned} A(w) &= t + L(w) \\ &= \Sigma_{B \in \mathscr{P}}(1/k).t + \Sigma_{B \in \mathscr{P}}(1/k).L(B(0)) \\ &= \Sigma_{B \in \mathscr{P}}(1/k).(t + L(B(0))) \\ &= \Sigma_{B \in \mathscr{P}}(1/k).AB(0) \\ &= w. \quad /// \end{aligned}$$

Let $\mathscr{F}$ be any subgroup of $\mathscr{C}$, and let T and $\mathscr{G}$ stand for the translational and linear parts of $\mathscr{F}$. Let $\mathscr{U}$ stand for for $\mathscr{T} \cap \mathscr{F}$, so that $q(T) = \mathscr{U}$.

<u>Proposition</u> <u>15</u>. If T is a lattice in R^2 then the subgroup $\mathscr{U}$ of $\mathscr{F}$ satisfies, and is uniquely characterized by the following conditions:

(a) $\mathscr{U}$ is a normal subgroup of $\mathscr{F}$;

(b) the centralizer of $\mathscr{U}$ in $\mathscr{F}$ is $\mathscr{U}$ itself.

Proof. Clearly, $\mathscr{U}$ is a normal subgroup of $\mathscr{F}$. Moreover, $\mathscr{U}$ is abelian, so it is included in its centralizer in $\mathscr{F}$. Let C be any member of the centralizer of $\mathscr{U}$ in $\mathscr{F}$. Let u and V be the translational and linear parts of C, so that $C = [u,V]$. For each member t of T, we have:

$$[u,V][t,I] = [t,I][u,V],$$

and hence:

$$V(t) = t.$$

Since T is a lattice, it follows that $V = I$. Hence, C is contained in $\mathscr{U}$. We infer that the centralizer of $\mathscr{U}$ in

$\mathcal{F}$ is $\mathcal{U}$.

Now let $\mathcal{V}$ be any normal subgroup of $\mathcal{F}$, which equals its centralizer in $\mathcal{F}$. We shall prove that $\mathcal{V} = \mathcal{U}$. Let C be any member of $\mathcal{V}$, and let u and V be the translational and linear parts of C. Let t be any member of T. We find that:

$$[t,I][u,V][t,I]^{-1}[u,V]$$
$$= [u,V][t,I][u,V][t,I]^{-1},$$

which entails that:

$$t + u - V(t) + V(u)$$
$$= u + V(t) + V(u) - V^2(t);$$

hence:

$$V^2(t) - 2.V(t) + I(t) = 0.$$

Since T is a lattice, we may infer that:

$$V^2 - 2.V + I = 0 ; \tag{1}$$

that is:

$$(V - I)^2 = 0 . \tag{2}$$

Of course, V must be either a rotation or a reflection. If V were a reflection then V^2 would be I; by relation (1), V itself would be I, a contradiction. Hence, V must be a rotation. If V were not I then $V - I$ would be invertible, which is denied by relation (2). Hence, V must be I. It follows that C is contained in $\mathcal{U}$. Hence, $\mathcal{V}$ is a subset of $\mathcal{U}$.

Since $\mathcal{U}$ is abelian, it now follows that $\mathcal{U}$ is a subset of the centralizer of $\mathcal{V}$ in $\mathcal{F}$. We conclude that $\mathcal{V} = \mathcal{U}$. ///

Let us return to THEOREM 1.

Proof. Let K' and K" be any cartesian coordinate mappings, and let $\mathcal{F}_1$ and $\mathcal{F}_2$ stand for $\mathbf{F}_1^{K'}$ and $\mathbf{F}_2^{K''}$ respectively. We shall prove THEOREM 1, by proving the following equivalent form of it:

$\mathcal{F}_1$ and $\mathcal{F}_2$ are isomorphic iff they are conjugate in the affine group $\mathcal{A}$.

Obviously, if $\mathcal{F}_1$ and $\mathcal{F}_2$ are conjugate in $\mathcal{A}$ then they are isomorphic. Let us assume that $\mathcal{F}_1$ and $\mathcal{F}_2$ are isomorphic, and let us introduce an isomorphism Φ carrying $\mathcal{F}_1$ to $\mathcal{F}_2$. We shall prove that there exists an affine transformation A such that Φ is (implemented by) J_A, in the sense that, for each member F of $\mathcal{F}_1$, $\Phi(F) = AFA^{-1}$. Of course, it would follow that $\mathcal{F}_1$ and $\mathcal{F}_2$ are conjugate in $\mathcal{A}$.

Let T_1, T_2, $\mathcal{G}_1$, and $\mathcal{G}_2$ stand for the translational and linear parts of $\mathcal{F}_1$ and $\mathcal{F}_2$, respectively. We obtain the following diagram:

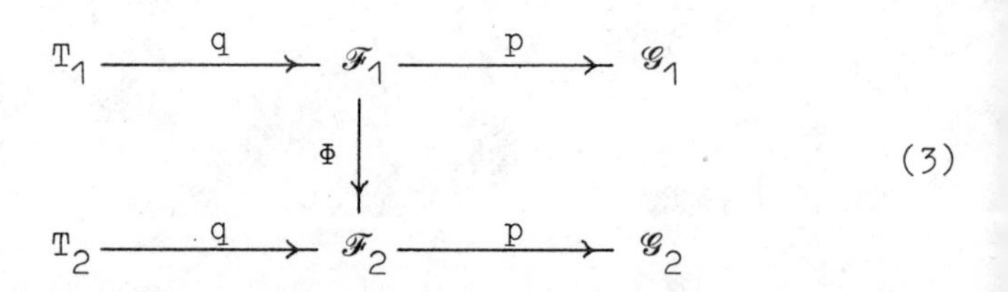

(3)

Let $\mathcal{U}_1$ and $\mathcal{U}_2$ stand for $q(T_1)$ and $q(T_2)$, respectively. By Proposition 15, $\mathcal{U}_1$ is a normal subgroup of $\mathcal{F}_1$, which equals its centralizer in $\mathcal{F}_1$. Since Φ is an isomorphism, the same must be true of $\Phi(\mathcal{U}_1)$ as a subgroup of $\mathcal{F}_2$. Again by Proposition 15, $\Phi(\mathcal{U}_1)$ must equal $\mathcal{U}_2$. Hence, there exists an isomorphism L_o carrying T_1 to T_2 such that $\Phi q = qL_o$. Since T_1 and T_2 are lattices, there must exist an invertible linear mapping L on R^2 such that, for each member t of T_1, $L_o(t) = L(t)$.

Now let M_1 be any member of $\mathscr{G}_1$. Let u_1 be a member of R^2 such that $[u_1,M_1]$ is contained in $\mathscr{F}_1$, and let $[u_2,M_2]$ stand for $\Phi([u_1,M_1])$. For each member t of T_1, we have:

$$[u_1,M_1][t,I][u_1,M_1]^{-1} = [M_1(t),I];$$

applying Φ to both sides of the foregoing equality, we obtain:

$$[u_2,M_2][L(t),I][u_2,M_2]^{-1} = [L(M_1(t)),I],$$

so that:

$$[M_2(L(t)),I] = [L(M_1(t)),I].$$

Hence:

$$(M_2L)(t) = (LM_1)(t).$$

Since T_1 is a lattice, we infer that:

$$M_2L = LM_1.$$

At this point, we have shown that there exists a member L of $\mathscr{L}$ such that:

$$L(T_1) = T_2 \quad \text{and} \quad L\mathscr{G}_1L^{-1} = \mathscr{G}_2. \tag{4}$$

The diagrammatic relation (3) may now be augmented, as follows:

$$\begin{array}{ccccc} T_1 & \xrightarrow{q} & \mathscr{F}_1 & \xrightarrow{p} & \mathscr{G}_1 \\ L\big\downarrow & & \Phi\big\downarrow & & \big\downarrow J_L, \\ T_2 & \xrightarrow{q} & \mathscr{F}_2 & \xrightarrow{p} & \mathscr{G}_2 \end{array} \tag{5}$$

where J_L is the inner automorphism of $\mathscr{L}$ determined by L.

Let us also interpret J_L as the inner automorphism of $\mathscr{A}$ determined by [0,L]. Let $\mathscr{F}^*$ stand for $J_L(\mathscr{F}_1)$. Obviously, the translational and linear parts of $\mathscr{F}^*$ are T_2 and $\mathscr{G}_2$. Let Φ^* stand for the isomorphism ΦJ_L^{-1} carrying $\mathscr{F}^*$ to $\mathscr{F}_2$. Now one can readily justify the following diagram:

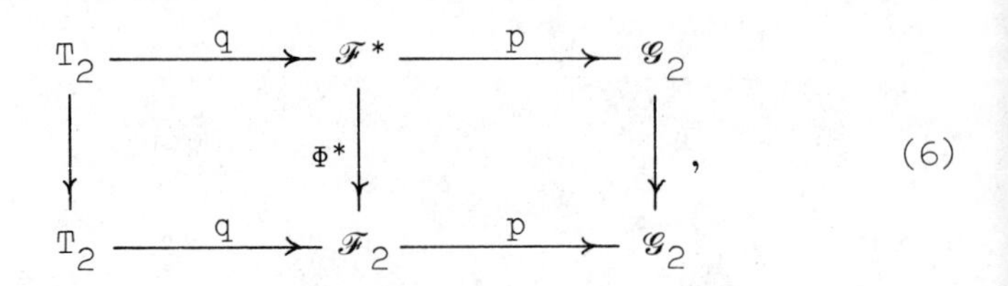

(6)

where the two unmarked arrows indicate identity mappings.

We intend to prove that there exists a member t of R^2 such that Φ^* is (implemented by) $J_{[t,I]}$. Having done so, we may then take A to be the affine transformation [t,L], with the effect that Φ is (implemented by) J_A, as desired. The proof of THEOREM 1 will then be complete.

For each member M of $\mathscr{G}_2$, let us introduce a member u_M of R^2 for which $[u_M,M]$ is contained in $\mathscr{F}^*$. In particular, let us take u_I to be 0. The diagrammatic relation (6) yields that, for each member M of $\mathscr{G}_2$, the linear part of $\Phi^*([u_M,M])$ equals M; let v_M stand for the translational part, so that:

$$\Phi^*([u_M,M]) = [v_M,M].$$

Clearly, $v_I = 0$.

Let us consider the subset $\mathscr{P}$ of the affine group $\mathscr{A}$, consisting of all transformations of the form:

$$[v_M - u_M, M],$$

where M runs through $\mathscr{G}_2$. Of course, $\mathscr{P}$ is a finite set. We contend that $\mathscr{P}$ is a subgroup of $\mathscr{A}$.

Taking M to be I, we find that $[0,I]$ is a member of $\mathscr{P}$.

For each member M of $\mathscr{G}_2$, $[u_{M^{-1}}, M^{-1}][u_M, M]$ is a translation in $\mathscr{F}^*$. Hence:

$$\Phi^*([u_{M^{-1}}, M^{-1}][u_M, M]) = [u_{M^{-1}}, M^{-1}][u_M, M].$$

It follows that:

$$[u_{M^{-1}}, M^{-1}][u_M, M] = [v_{M^{-1}}, M^{-1}][v_M, M],$$

so that:

$$-M^{-1}(v_M - u_M) = v_{M^{-1}} - u_{M^{-1}} .$$

Hence:

$$[v_M - u_M, M]^{-1} = [v_{M^{-1}} - u_{M^{-1}}, M^{-1}].$$

Accordingly, $\mathscr{P}$ is closed under inversion.

For any members M and N of $\mathscr{G}_2$, the product

$$[u_{MN}, MN][u_N, N]^{-1}[u_M, M]^{-1}$$

is a translation in $\mathscr{F}^*$. Hence:

$$\Phi^*([u_{MN}, MN][u_N, N]^{-1}[u_M, M]^{-1})$$
$$= [u_{MN}, MN][u_N, N]^{-1}[u_M, M]^{-1}.$$

It follows that:

$$[u_{MN}, MN][u_N, N]^{-1}[u_M, M]^{-1}$$
$$= [v_{MN}, MN][v_N, N]^{-1}[v_M, M]^{-1},$$

so that:

$$v_{MN} - u_{MN} = M(v_N - u_N) + (v_M - u_M).$$

Hence:

$$[v_M - u_M, M][v_N - u_N, N] = [v_{MN} - u_{MN}, MN].$$

Accordingly, $\mathscr{P}$ is closed under multiplication.

We conclude that $\mathscr{P}$ is a finite subgroup of $\mathscr{A}$.

Let us now apply Proposition 14, to obtain a member t of R^2 such that, for each member M of $\mathscr{G}_2$:

$$[v_M - u_M, M](t) = t,$$

that is:

$$v_M - u_M + M(t) = t\ .$$

Let F be any member of $\mathscr{F}^*$. Let u and M be the translational and linear parts of F; let u' stand for $u - u_M$, so that:

$$F = [u',I][u_M, M].$$

Clearly, $[u',I]$ is a translation in $\mathscr{F}^*$, so u' is a member of T_2. We have:

$$\begin{aligned} \Phi^*(F) &= \Phi^*([u',I][u_M, M]) \\ &= \Phi^*([u',I])\Phi^*([u_M, M]) \\ &= [u',I][v_M, M] \\ &= [u',I][t + u_M - M(t), M] \\ &= [t,I][u',I][t,I]^{-1}[t,I][u_M, M][t,I]^{-1} \\ &= [t,I][u',I][u_M, M][t,I]^{-1} \\ &= [t,I]\,F\,[t,I]^{-1}. \end{aligned}$$

Hence, Φ^* is (implemented by) $J_{[t,I]}$. ///

The heart of THEOREM 1 is the assertion that if two ornamental subgroups of $\mathbf{E}$ are isomorphic then they are conjugate in $\mathbf{A}$. The foregoing argument actually proves a stronger assertion, namely, that every isomorphism carrying one ornamental subgroup of $\mathbf{E}$ to another is (implemented by) an inner automorphism of $\mathbf{A}$.

5° With THEOREM 1 in mind, let us now introduce the concept of arithmetic class. This concept makes it possible to organize the calculation of all ornamental classes in the form of specific group extension problems.

Let $\mathbf{F}$ be an ornamental subgroup of $\mathbf{E}$, and let K be any cartesian coordinate mapping. Of course, T^K is a lattice in R^2 and $\mathscr{G}^K$ is a subgroup of $\mathscr{O}$. Moreover, $\mathbf{F}^K$ is compatible with $(T^K, \mathscr{G}^K)$. [See 1.1.4°.] By Proposition 1, T^K is invariant under $\mathscr{G}^K$.

Let us use the term <u>arithmetic group</u> to refer to any ordered pair $(T, \mathscr{G})$, where T is a lattice in R^2 and where $\mathscr{G}$ is a subgroup of $\mathscr{O}$ under which T is invariant. Obviously, $(T^K, \mathscr{G}^K)$ is an arithmetic group. We shall call it the arithmetic group "determined" by $\mathbf{F}$ relative to K.

Let $\mathbf{F}_1$ and $\mathbf{F}_2$ be ornamental subgroups of $\mathbf{E}$, and let us assume that $\mathbf{F}_1$ and $\mathbf{F}_2$ are isomorphic, so that they belong to the same ornamental class. Let K' and K" be any cartesian coordinate mappings. By the proof of THEOREM 1 there must exist a member L of $\mathscr{L}$ such that:

$$T_2^{K''} = L(T_1^{K'})$$

and $$\mathscr{G}_2^{K''} = L\,\mathscr{G}_1^{K'}L^{-1}.$$

[See relation (4) in 4°.] This fact suggests that we regard two arithmetic groups $(T', \mathscr{G}')$ and $(T'', \mathscr{G}'')$ as <u>equivalent</u> iff there exists some member L of $\mathscr{L}$ such that:

$$T'' = L(T'), \tag{1}$$

$$\mathcal{G}'' = L\,\mathcal{G}'\,L^{-1}. \tag{2}$$

The resulting equivalence classes will be called arithmetic classes.

In particular, the arithmetic groups $(T_1^{K'}, \mathcal{G}_1^{K'})$ and $(T_2^{K''}, \mathcal{G}_2^{K''})$ determined by $\mathbf{F}_1$ relative to K' and by $\mathbf{F}_2$ relative to K", are equivalent.

Now it is plain that every ornamental class determines a unique arithmetic class, namely, that which contains the arithmetic groups $(T^K, \mathcal{G}^K)$ determined by the various members $\mathbf{F}$ of the given ornamental class, relative to arbitrary cartesian coordinate mappings K. Hence, we may expect to solve the problem of calculating all ornamental classes, by the following procedure:

(a) describe all arithmetic classes;

(b) for each arithmetic class, describe all ornamental classes which determine it.

In section 4, we shall carry out step (a), in section 5, step (b). As we shall see, step (b) involves the group extension problems mentioned earlier.

2.4 ARITHMETIC CLASSES

1° The problem of describing all arithmetic classes can itself be resolved, by means of the concept of geometric class.

Thus, let $(T, \mathcal{G})$ be an arithmetic group. By Proposition 12, $\mathcal{G}$ must be finite; indeed, it must be one of the following ten (types of) groups:

$$\mathcal{Z}_1, \quad \mathcal{Z}_2, \quad \mathcal{Z}_3, \quad \mathcal{Z}_4, \quad \mathcal{Z}_6,$$

$$\mathcal{D}_1(X), \quad \mathcal{D}_2(X), \quad \mathcal{D}_3(X), \quad \mathcal{D}_4(X), \quad \mathcal{D}_6(X),$$

where X is any reflection in $\mathscr{G}$. These various groups are called geometric groups.

Let $(T', \mathscr{G}')$ and $(T'', \mathscr{G}'')$ be arithmetic groups, contained in the same arithmetic class. In particular, $\mathscr{G}'$ and $\mathscr{G}''$ must be conjugate in $\mathscr{L}$. [See relation (2) in 3.5°.] It follows that $\mathscr{G}'$ and $\mathscr{G}''$ are actually conjugate in $\mathscr{O}$. [See 1.4, problem 4.] Hence, each arithmetic class "determines" precisely one of the following ten conjugacy classes of (finite) subgroups of $\mathscr{O}$:

$$\mathbb{Z}_k = \{\mathscr{Z}_k\},$$

$$\mathbb{D}_k = \{\mathscr{D}_k(X) : X \in \mathscr{O}^-\},$$

where k equals 1, 2, 3, 4, or 6. These conjugacy classes are called geometric classes.

2° Now let us enumerate the arithmetic classes. To achieve the enumeration, we shall consider in turn each of the geometric classes; for each such class, we shall describe all the arithmetic classes which determine it.

Let us consider first the geometric class $\mathbb{Z}_1$. For any lattice T in R^2, T is invariant under $\mathscr{Z}_1$, so $(T, \mathscr{Z}_1)$ is an arithmetic group. Moreover, for any lattices T' and T'' in R^2, there exists a member L of $\mathscr{L}$ such that $T'' = L(T')$. Clearly, $\mathscr{Z}_1 = L\mathscr{Z}_1 L^{-1}$, as well. Hence, $(T', \mathscr{Z}_1)$ and $(T'', \mathscr{Z}_1)$ are equivalent. We conclude that there is just one arithmetic class which determines $\mathbb{Z}_1$. Let it be denoted by $\mathbb{Z}_1^a$.

The geometric class $\mathbb{Z}_2$ may be treated in similar manner. Again, there is just one arithmetic class which determines $\mathbb{Z}_2$. Let it be denoted by $\mathbb{Z}_2^a$.

Let us consider the geometric class $\mathbb{Z}_3$. Let W be the rotation having principal measure $\pi/3$, so that W^2 generates $\mathscr{Z}_3$. Let u be any nonzero member of R^2, and let v stand for $W^2(u)$. Clearly, $\{u,v\}$ is a basis for R^2. Let T be the lattice determined by $\{u,v\}$. By the discussion of lattice and symmetry groups in 2.3°, it is plain that T is invarian under $\mathcal{Z}_3$, hence that $(T, \mathcal{Z}_3)$ is an arithmetic group. Now let T' and T'' be any lattices in R^2 which are invariant under $\mathcal{Z}_3$. Again by the discussion in 2.3^o, one may introduce bases $\{u',v'\}$ and $\{u'',v''\}$ for T' and T'', respectively, such that the matrices for W^2 relative to those bases both equal the following:

$$W^2 \longleftrightarrow \begin{bmatrix} 0 & -1 \\ 1 & -1 \end{bmatrix}.$$

Let L be the member of $\mathscr{L}$ for which $L(u') = u''$ and $L(u'') = v''$. From the foregoing matrix condition, we infer that $LW^2L^{-1} = W^2$. Consequently, $L(T') = T''$ and $L\mathcal{Z}_3L^{-1} = \mathcal{Z}_3$. Hence, $(T', \mathcal{Z}_3)$ and $(T'', \mathcal{Z}_3)$ are equivalent. Once again, we conclude that there is just one arithmetic class determining the geometric class $\mathbb{Z}_3$. We shall denote it by $\mathbb{Z}_3^h$.

By analogous arguments, based upon the discussion in 2.3^o, one may treat the geometric classes $\mathbb{Z}_4$ and $\mathbb{Z}_6$. In each case, one obtains one arithmetic class. Let them be denoted by $\mathbb{Z}_4^q$ and $\mathbb{Z}_6^h$, respectively.

The superscripts employed in the notation for arithmetic classes are intended to indicate the type of the underlying lattice.

Let us turn to the geometric class $\mathbb{D}_4$. Let W be the rotation with principal measure $\pi/4$, so that W generates $\mathcal{Z}_4$. Let u be any nonzero member of R^2, and let v be $W(u)$. Then $\{u,v\}$ is a basis for R^2, and we may introduce the lattice T determined by $\{u,v\}$. Let X be the reflection for which $X(u) = u$. By 2.3^o, T is invariant under W and X, hence under $\mathscr{D}_4(X)$. As a result, we obtain the arithmetic group $(T, \mathscr{D}_4(X))$. Now let $(T', \mathscr{D}_4(Y'))$ and $(T'', \mathscr{D}_4(Y''))$ be arithmetic groups, with second components drawn from $\mathbb{D}_4$. By 2.3^o, there are bases $\{u',v'\}$ and $\{u'',v''\}$ for T' and T'', respectively, such that the matrices for W relative to those bases both equal the following:

$$W \longleftrightarrow \begin{bmatrix} 0 & -1 \\ 1 & 0 \end{bmatrix}.$$

Let X' and X'' be the reflections for which $X'(u') = u'$ and $X''(u'') = u''$. By 2.3^o, the matrices for X' and X'' relative to $\{u',v'\}$ and $\{u'',v''\}$, respectively, both equal the following:

$$X',\ X'' \longleftrightarrow \begin{bmatrix} 1 & 0 \\ 0 & -1 \end{bmatrix}.$$

Moreover, $\mathscr{D}_4(Y')$ must equal $\mathscr{D}_4(X')$ and $\mathscr{D}_4(Y'')$ must equal $\mathscr{D}_4(X'')$. Let L be the member of $\mathscr{L}$ for which $L(u') = u''$ and $L(v') = v''$. By the matrix conditions just cited, $LWL^{-1} = W$ and $LX'L^{-1} = X''$. Hence, we have both $L(T') = T''$ and $L\,\mathscr{D}_4(X')\,L^{-1} = \mathscr{D}_4(X'')$, so that $(T', \mathscr{D}_4(Y'))$ and $(T'', \mathscr{D}_4(Y''))$ are equivalent. We conclude that $\mathbb{D}_4$ yields just one arithmetic class. Let it be denoted by $\mathbb{D}_4^q$.

In similar manner, one may show that the geometric class $\mathbb{D}_6$ yields just one arithmetic class. It will be denoted by $\mathbb{D}_6^h$.

Now let us turn to the somewhat more subtle cases of $\mathbb{D}_1$ and $\mathbb{D}_2$.

Let X be a reflection. One can easily design bases $\{u^o,v^o\}$ and $\{u^r,v^r\}$ for R^2 such that $X(u^o) = u^o$, $X(v^o) = -v^o$, $X(u^r) = v^r$, and $X(v^r) = u^r$. Let T^o and T^r be the lattices in R^2 determined by $\{u^o,v^o\}$ and $\{u^r,v^r\}$, respectively. Of course, T^o is orthogonal and T^r is rhombic. Both lattices are invariant under X, with the result that $(T^o, \mathscr{D}_1(X))$, $(T^o, \mathscr{D}_2(X))$, $(T^r, \mathscr{D}_1(X))$, and $(T^r, \mathscr{D}_2(X))$ are arithmetic groups.

Let Y be any reflection, and let T be any lattice in R^2 invariant under Y. By 2.3^o, one of the following assertions is true:

(o) there is a basis $\{u', v'\}$ for T such that $Y(u') = u'$ and $Y(v') = -v'$;

(r) there is a basis $\{u'', v''\}$ for T such that $Y(u'') = v''$ and $Y(v'') = u''$.

In case (o), one could introduce the member L of $\mathscr{L}$ such that $L(u') = u^o$ and $L(v') = v^o$, obtaining $L(T) = T^o$ and $LYL^{-1} = X$; it would follow that $L\mathscr{D}_1(Y)L^{-1} = \mathscr{D}_1(X)$ and $L\mathscr{D}_2(X)L^{-1} = \mathscr{D}_2(X)$, hence that $(T, \mathscr{D}_1(Y))$ and $(T^o, \mathscr{D}_1(X))$ are equivalent and that $(T, \mathscr{D}_2(Y))$ and $(T^o, \mathscr{D}_2(X))$ are equivalent. In case (r), one would obtain (by similar argument) that $(T, \mathscr{D}_1(Y))$ and $(T^r, \mathscr{D}_1(X))$ are equivalent and that $(T, \mathscr{D}_2(Y))$ and $(T^r, \mathscr{D}_2(X))$ are equivalent.

We contend now that $(T^o, \mathscr{D}_1(X))$ and $(T^r, \mathscr{D}_1(X))$ are not equivalent. Let us suppose to the contrary that there is some member L of $\mathscr{L}$ such that $L(T^o) = T^r$ and $L\mathscr{D}_1(X)L^{-1} = \mathscr{D}_1(X)$. Of course, the latter equality entails that:

$$LXL^{-1} = X. \tag{1}$$

Let a, b, c, and d be the real numbers for which:

$$L(u^o) = a.u^r + b.v^r ;$$

$$L(v^o) = c.u^r + d.v^r .$$

By problem 8 in 8, we know that a, b, c, and d are integers and that $ad - bc = \pm 1$. By relation (1):

$$\begin{bmatrix} a & c \\ b & d \end{bmatrix}\begin{bmatrix} 1 & 0 \\ 0 & -1 \end{bmatrix} = \begin{bmatrix} 0 & 1 \\ 1 & 0 \end{bmatrix}\begin{bmatrix} a & c \\ b & d \end{bmatrix},$$

which, by simple calculation, yields that $a = b$ and $c = -d$. But then $ad - bc = 2ab$, which cannot equal ± 1. This contradiction shows that our original contention, namely, that $(T^o, \mathscr{D}_1(X))$ and $(T^r, \mathscr{D}_1(X))$ are not equivalent, is true.

By similar argument, one may show that $(T^o, \mathscr{D}_2(X))$

and $(T^r, \mathcal{D}_2(X))$ are not equivalent. In the course of that argument, one would replace relation (1) by the following:

$$LXL^{-1} = X \quad \text{or} \quad LXL^{-1} = -X . \tag{2}$$

Each of these alternatives yields the same contradictory conclusion, that $ad - bc$ both equals and does not equal ± 1.

We infer that each of the geometric classes $\mathbb{D}_1$ and $\mathbb{D}_2$ is determined by precisely two arithmetic classes. For $\mathbb{D}_1$, we denote the arithmetic classes by $\mathbb{D}_1^o$ and $\mathbb{D}_1^r$, the former being the class containing $(T^o, \mathcal{D}_1(X))$, and the latter, $(T^r, \mathcal{D}_1(X))$; for $\mathbb{D}_2$, by $\mathbb{D}_2^o$ and $\mathbb{D}_2^r$, which contain $(T^o, \mathcal{D}_2(X))$ and $(T^r, \mathcal{D}_2(X))$, respectively.

Finally, let us consider the geometric class $\mathbb{D}_3$. Let W be the rotation with principal measure $\pi/3$, so that W^2 generates $\mathcal{Z}_3$. Let u^* be a nonzero member of R^2, and let v^* be $W^2(u^*)$. Clearly, $\{u^*,v^*\}$ is a basis for R^2. Let T^* be the lattice in R^2 determined by $\{u^*,v^*\}$. Let X' be the reflection for which $X'(u^*) = u^*$, and let X'' stand for the reflection $-X'$. Then:

$$\mathcal{D}_3(X') = \mathcal{Z}_3 \cup \{X', W^2X', W^4X'\} ,$$

$$\mathcal{D}_3(X'') = \mathcal{Z}_3 \cup \{X'', W^2X'', W^4X''\} .$$

[See $1.2.5^o$.] [See Figure 39.] Clearly, T^* is invariant under both $\mathcal{D}_3(X')$ and $\mathcal{D}_3(X'')$, so that $(T^*, \mathcal{D}_3(X'))$ and $(T^*, \mathcal{D}_3(X''))$ are arithmetic groups. With respect to the basis $\{u^*,v^*\}$, the matrices for W^2, X', and X'' are the following:

$$W^2 \longleftrightarrow \begin{bmatrix} 0 & -1 \\ 1 & -1 \end{bmatrix}, \qquad X' \longleftrightarrow \begin{bmatrix} 1 & -1 \\ 0 & -1 \end{bmatrix}; \tag{3)'$$

$$W^2 \longleftrightarrow \begin{bmatrix} 0 & -1 \\ 1 & -1 \end{bmatrix} \qquad X'' \longleftrightarrow \begin{bmatrix} -1 & 1 \\ 0 & 1 \end{bmatrix}. \tag{3)''$$

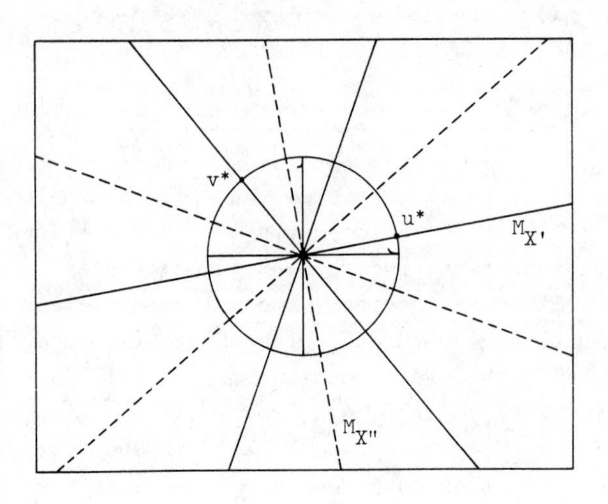

Figure 39

Let Y be any reflection, and let T be any lattice in R^2 invariant under $\mathscr{D}_3(Y)$. In particular, T is invariant under W^2. Hence, by 2.3^o, the symmetry group of T is $\mathscr{D}_6(X)$, where X is a suitably defined reflection. In fact, there is a basis $\{u,v\}$ for T with respect to which the matrices for W and X are as follows:

$$W \longleftrightarrow \begin{bmatrix} 1 & -1 \\ 1 & 0 \end{bmatrix}, \qquad X \longleftrightarrow \begin{bmatrix} 1 & -1 \\ 0 & -1 \end{bmatrix}. \tag{4}$$

Let Y' stand for X, and Y'' for $-X$. Now $\mathscr{D}_3(Y)$ must equal either $\mathscr{D}_3(Y')$ or $\mathscr{D}_3(Y'')$. Relation (4) yields that the matrices for W^2 and Y' relative to $\{u,v\}$ are the same as those for W^2 and X' relative to $\{u^*,v^*\}$, as presented in relation (3)', and that the matrices for W^2 and Y'' relative to $\{u,v\}$ are the same as those for W^2 and X'' relative to $\{u^*,v^*\}$, as presented in relation (3)". By the now familiar pattern of argument, it follows that $(T, \mathscr{D}_3(Y))$ must be equivalent either to $(T^*, \mathscr{D}_3(X'))$ or to $(T^*, \mathscr{D}_3(X''))$.

Let $\mathbb{D}_3'$ and $\mathbb{D}_3''$ stand for the arithmetic classes containing $(T^*, \mathscr{D}_3(X'))$ and $(T^*, \mathscr{D}_3(X''))$, respectively. We contend that they are distinct. Let us suppose to the contrary that there exists some member L of $\mathscr{L}$ such that $L(T^*) = T^*$ and $L\mathscr{D}_3(X')L^{-1} = \mathscr{D}_3(X'')$. Clearly, either $LW^2L^{-1} = W^2$ or $LW^2L^{-1} = W^4$. Replacing L by X"L, if necessary, we may guarantee that:

$$LW^2L^{-1} = W^2. \tag{5}$$

Moreover, $LX'L^{-1}$ would have to be X", W^2X'', or W^4X''. Replacing L by $W^{-1}L$ or by $W^{-2}L$, as necessary, we may guarantee that:

$$LX'L^{-1} = X''. \tag{6}$$

By problem 8 in 8, we may introduce integers a, b, c, and d such that:

$$L(u^*) = a.u^* + b.v^*,$$

$$L(v^*) = c.u^* + d.v^*,$$

and such that $ad - bc = \pm 1$. Relations (5) and (6) yield:

$$\begin{bmatrix} a & c \\ b & d \end{bmatrix}\begin{bmatrix} 0 & -1 \\ 1 & -1 \end{bmatrix} = \begin{bmatrix} 0 & -1 \\ 1 & -1 \end{bmatrix}\begin{bmatrix} a & c \\ b & d \end{bmatrix},$$

$$\begin{bmatrix} a & c \\ b & d \end{bmatrix}\begin{bmatrix} 1 & -1 \\ 0 & -1 \end{bmatrix} = \begin{bmatrix} -1 & 1 \\ 0 & 1 \end{bmatrix}\begin{bmatrix} a & c \\ b & d \end{bmatrix}.$$

These relations entail that:

$$b = -c, \quad d = a + c, \quad b = 2a,$$

hence that:

$$ad - bc = 3a^2,$$

ARITHMETIC CLASS	LATTICE TYPE	GENERATORS W	X
$\mathbb{Z}_1^a$	arbitrary	$\begin{pmatrix} 1 & 0 \\ 0 & 1 \end{pmatrix}$	-
$\mathbb{Z}_2^a$	arbitrary	$\begin{pmatrix} -1 & 0 \\ 0 & -1 \end{pmatrix}$	-
$\mathbb{Z}_3^h$	hexagonal	$\begin{pmatrix} 0 & -1 \\ 1 & -1 \end{pmatrix}$	-
$\mathbb{Z}_4^q$	quadratic	$\begin{pmatrix} 0 & -1 \\ 1 & 0 \end{pmatrix}$	-
$\mathbb{Z}_6^h$	hexagonal	$\begin{pmatrix} 1 & -1 \\ 1 & 0 \end{pmatrix}$	-
$\mathbb{D}_1^r$	rhombic	$\begin{pmatrix} 1 & 0 \\ 0 & 1 \end{pmatrix}$	$\begin{pmatrix} 0 & 1 \\ 1 & 0 \end{pmatrix}$
$\mathbb{D}_1^o$	orthogonal	$\begin{pmatrix} 1 & 0 \\ 0 & 1 \end{pmatrix}$	$\begin{pmatrix} 1 & 0 \\ 0 & -1 \end{pmatrix}$
$\mathbb{D}_2^r$	rhombic	$\begin{pmatrix} -1 & 0 \\ 0 & -1 \end{pmatrix}$	$\begin{pmatrix} 0 & 1 \\ 1 & 0 \end{pmatrix}$
$\mathbb{D}_2^o$	orthogonal	$\begin{pmatrix} -1 & 0 \\ 0 & -1 \end{pmatrix}$	$\begin{pmatrix} 1 & 0 \\ 0 & -1 \end{pmatrix}$
$\mathbb{D}_3'$	hexagonal	$\begin{pmatrix} 0 & -1 \\ 1 & -1 \end{pmatrix}$	$\begin{pmatrix} 1 & -1 \\ 0 & -1 \end{pmatrix}$
$\mathbb{D}_3''$	hexagonal	$\begin{pmatrix} 0 & -1 \\ 1 & -1 \end{pmatrix}$	$\begin{pmatrix} -1 & 1 \\ 0 & 1 \end{pmatrix}$
$\mathbb{D}_4^q$	quadratic	$\begin{pmatrix} 0 & -1 \\ 1 & 0 \end{pmatrix}$	$\begin{pmatrix} 1 & 0 \\ 0 & -1 \end{pmatrix}$
$\mathbb{D}_6^h$	hexagonal	$\begin{pmatrix} 1 & -1 \\ 1 & 0 \end{pmatrix}$	$\begin{pmatrix} 1 & -1 \\ 0 & -1 \end{pmatrix}$

Table 1

which cannot equal ± 1. The contradiction just obtained shows that the arithmetic classes $\mathbb{D}_3'$ and $\mathbb{D}_3''$ are distinct.

3° In summary, there are 13 arithmetic classes. We present them in the table on the opposite page, Table 1, with sufficient data to construct a representative arithmetic group for each of them. To determine such a representative, we have displayed generators for the underlying geometric group; the generators appear in matrix form, calculated relative to a suitable basis for the underlying lattice. [See Figures 34 - 38.]

2.5 ORNAMENTAL CLASSES

1° From a given arithmetic class, let us select a representative arithmetic group $(T, \mathscr{G})$. We expect to describe all ornamental classes which determine the given arithmetic class, by calculating all subgroups of $\mathscr{C}$ which are compatible with $(T, \mathscr{G})$ and by classifying all such subgroups under the relation of conjugacy in $\mathscr{A}$. The following proposition shows that the results will be the same, whatever the representative arithmetic group initially chosen. Hence, we will be justified in using the particular arithmetic groups presented in Table 1.

<u>Proposition</u> <u>16</u>. For each arithmetic group $(T, \mathscr{G})$, for any ornamental subgroup $\mathbf{F}$ of $\mathbf{E}$, and for any cartesian coordinate mapping K, if $(T, \mathscr{G})$ and $(T^K, \mathscr{G}^K)$ are equivalent then there exists a subgroup $\mathscr{F}$ of $\mathscr{C}$ such that $\mathscr{F}$ is compatible with $(T, \mathscr{G})$ and such that $\mathscr{F}$ and $\mathbf{F}^K$ are conjugate in $\mathscr{A}$.

Proof. By hypothesis, there exists some member L of $\mathscr{L}$ such that $L(T^K) = T$ and $L\mathscr{G}^K L^{-1} = \mathscr{G}$. Clearly, $J_L(\mathbf{F}^K)$ must be a subgroup of $\mathscr{C}$ and must be compatible with $(T, \mathscr{G})$. To prove the proposition, one need only take $\mathscr{F}$ to be $J_L(\mathbf{F}^K)$. ///

2° The next proposition will help to simplify the calculations necessary to obtain all ornamental classes.

Let $(T, \mathscr{G})$ be an arithmetic group. We shall say that a subgroup $\mathscr{F}$ of $\mathscr{C}$ is <u>consonant</u> with $(T, \mathscr{G})$ iff $\mathscr{F}$ is compatible with $(T, \mathscr{G})$ and, for each rotation W in $\mathscr{G}$, $[0,W]$ is contained in $\mathscr{F}$.

Obviously, $T \rtimes \mathscr{G}$ is consonant with $(T, \mathscr{G})$.

We can show the geometric significance of the condition of consonance, by viewing $\mathscr{F}$ as a cartesian form of the symmetry group of some plane ornament. For example, let Ω be the plane ornament depicted in Figure 19. Let that figure be augmented, by introducing two points α and β. [See Figure 40.] The latter point is placed "at random," while the former occupies a preferred position. Let K be any cartesian coordinate mapping. Let us take $(T, \mathscr{G})$ to be $(T^K, \mathscr{G}^K)$ and $\mathscr{F}$ to be $\mathbf{E}_{\Omega}^{K}$. One can readily verify that if $K(\beta) = 0$ then $\mathscr{F}$ is not consonant with $(T, \mathscr{G})$, while if $K(\alpha) = 0$ then $\mathscr{F}$ is consonant. Informally speaking, one may expect the condition of consonance to be satisfied when the origin of the cartesian coordinate system defining K has been set at a point in E of maximum rotational symmetry relative to Ω.

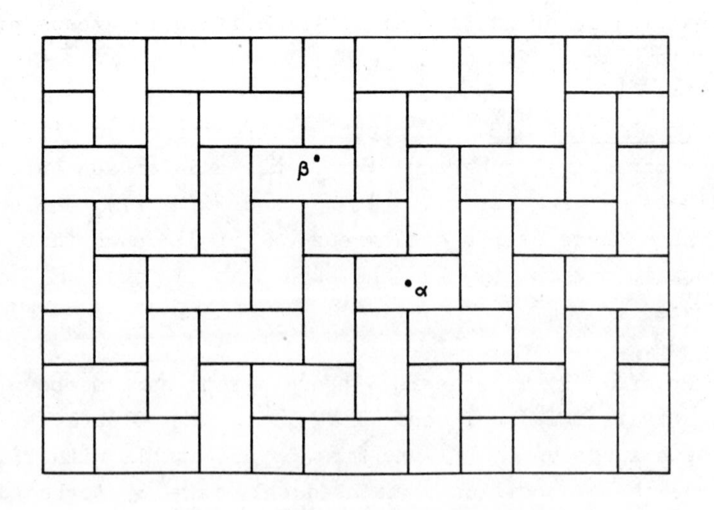

Figure 40

The preceding discussion suggests that, with respect to a given arithmetic group, one should be able to render a compatible subgroup consonant, by applying a suitable coordinate transformation.

<u>Proposition</u> <u>17</u>. For each arithmetic group $(T, \mathscr{G})$ and for any subgroup $\mathscr{F}'$ of $\mathscr{C}$ compatible with $(T, \mathscr{G})$, there exists a subgroup $\mathscr{F}''$ of $\mathscr{C}$ consonant with $(T, \mathscr{G})$ such that $\mathscr{F}'$ and $\mathscr{F}''$ are conjugate in $\mathscr{A}$. In fact, the conjugacy of $\mathscr{F}'$ and $\mathscr{F}''$ in $\mathscr{A}$ can be implemented by a translation.

Proof. Let $\mathscr{G}^+$ stand for $\mathscr{G} \cap \mathscr{O}^+$, and let k be the order of $\mathscr{G}^+$. Let W be the rotation having principal measure $2\pi/k$, so that W generates $\mathscr{G}^+$. We may as well assume that $1 < k$, so that $W \neq I$, because the proposition is obviously true when $k = 1$.

Since $\mathscr{F}'$ is compatible with $(T, \mathscr{G})$, we may select a member u of R^2 such that $[u,W]$ is contained in $\mathscr{F}'$. Of course, $I - W$ is invertible, so there must exist a member t of R^2 such that:

$$(I - W)(t) = -u.$$

It follows that:

$$[t,I][u,W][t,I]^{-1} = [0,W].$$

Hence, for each integer j $(1 \leq j \leq k)$:

$$[t,I][u,W]^j[t,I]^{-1} = [0,W^j]. \tag{1}$$

Now let $\mathscr{F}''$ stand for $J_{[t,I]}(\mathscr{F}')$. Clearly, $\mathscr{F}''$ is compatible with $(T, \mathscr{G})$. By relation (1), it is in fact consonant with $(T, \mathscr{G})$. ///

3° Supported by the foregoing two propositions, we may now present a concrete procedure for computing all ornamental classes:

(b1) for each of the arithmetic groups described in Table 1, calculate all subgroups of $\mathscr{C}$ which are consonant with it;

(b2) for the family of all such subgroups of $\mathscr{C}$, exhibit one representative from each equivalence class following the relation of conjugacy in $\mathscr{A}$.

The resulting list of subgroups of $\mathscr{C}$ will comprise, in cartesian form, one representative ornamental subgroup of **E** from each of the ornamental classes.

4° For each arithmetic class, the computational procedure just described must yield at least one ornamental class. In particular, when following step (b1), one will always find the appropriate semi-direct product $T \rtimes \mathscr{G}$. One refers, in turn, to the corresponding ornamental class as the <u>symmorphic case</u>, among all ornamental classes which determine the given arithmetic class.

5° Before beginning the calculations, we must present one more proposition, designed to reduce step (b1) of the computational procedure to a simple routine.

Let $(T, \mathscr{G})$ be an arithmetic group, and let us assume that $\mathscr{G}$ contains a reflection. With reference to 1.2.5°, we may introduce a reflection X and a rotation W having finite order k, which together generate $\mathscr{G}$. The members of $\mathscr{G}$ are as follows:

$$\begin{matrix} I, & W, & W^2, & \ldots, & W^{k-1}, \\ X, & WX, & W^2X, & \ldots, & W^{k-1}X. \end{matrix}$$

<u>Proposition</u> <u>18</u>. For each subgroup $\mathscr{F}$ of $\mathscr{C}$ consonant with $(T, \mathscr{G})$ and for any member x of R^2, if $[x,X]$ is contained in $\mathscr{F}$ then:

$$(\rho) \qquad x - W(x) \in T,$$
$$(\xi) \qquad x + X(x) \in T.$$

For each member x of R^2, if x satisfies conditions (ρ) and (ξ) then there exists exactly one subgroup $\mathscr{F}$ of $\mathscr{C}$ which is consonant with $(T, \mathscr{G})$ and which contains $[x,X]$. Moreover, for each reflection Y in $\mathscr{G}$, $[x,Y]$ is contained in $\mathscr{F}$.

Proof. Let $\mathscr{F}$ be any subgroup of $\mathscr{C}$ consonant with $(T, \mathscr{G})$, and let x be any member of R^2 for which $[x,X]$ is a member of $\mathscr{F}$. Clearly:

$$[x,X]^2 = [x + X(x), I],$$
$$[x,X][0,W^{-1}][x,X]^{-1}[0,W^{-1}] = [x - W(x), I].$$

From these relations, it is plain that conditions (ρ) and (ξ) are satisfied.

Now let x be any member of R^2 for which conditions (ρ) and (ξ) are satisfied. Let $\mathscr{F}$ be the subset of $\mathscr{C}$ consisting of all members having one of the following two forms:

$$\begin{aligned}&[s,I][0,W^i],\\&[t,I][0,W^j][x,X],\end{aligned} \tag{1}$$

where s and t are any members of T and where i and j are any integers $(0 \leq i,j < k)$. We contend that $\mathscr{F}$ is in fact a subgroup of $\mathscr{C}$. It would immediately follow that $\mathscr{F}$ is consonant with $(T, \mathscr{G})$ and that it contains $[x,X]$. Thus, for any integers i and j $(0 \leq i,j < k)$, we have:

$$x - W^{i+1}(x) = (x - W^i(x)) + W^i(x - W(x)),$$
$$x + (W^{j+1}X)(x) = (x - W^{j+1}(x)) + W^{j+1}(x + X(x)).$$

Arguing recursively from conditions (ρ) and (ξ), we obtain:

$(\rho)^*$ $\quad x - W^i(x) \ \varepsilon \ T,$

$(\xi)^*$ $\quad x + (W^jX)(x) \ \varepsilon \ T.$

By straightforward application of these conditions, one can now verify that $\mathscr{F}$ is a subgroup of $\mathscr{C}$.

Every subgroup of $\mathscr{C}$, consonant with $(T, \mathscr{G})$ and containing $[x,X]$, must include $\mathscr{F}$. It is then obvious that such a subgroup must equal $\mathscr{F}$. Hence, $\mathscr{F}$ is unique.

By item (1) and by condition $(\rho)^*$, it is clear that, for each reflection Y in $\mathscr{G}$, $[x,Y]$ is contained in $\mathscr{F}$. ///

Given a member x of R^2 satisfying conditions (ρ) and (ξ), we shall denote by $\mathscr{F}_x$ the unique subgroup of $\mathscr{C}$ consonant with $(T, \mathscr{G})$ and containing $[x,X]$.

Clearly, $\mathscr{F}_0 = T \rtimes \mathscr{G}$.

For any members x and y of R^2, if x satisfies conditions (ρ) and (ξ) and if $y - x$ is a member of T then y also satisfies conditions (ρ) and (ξ). Moreover, if both x and y satisfy conditions (ρ) and (ξ) then $\mathscr{F}_x = \mathscr{F}_y$ iff $y - x$ is a member of T. Hence, for the purpose of calculating all subgroups of $\mathscr{C}$ consonant with $(T, \mathscr{G})$, one need examine only one member x of each coset of T in R^2. Relative to a given basis $\{u,v\}$ for T, one would examine all members of R^2 having the form:

$$x = a.u + b.v,$$

where a and b are any real numbers for which $0 \le a < 1$ and $0 \le b < 1$.

6° Now let us carry out the calculations prescribed in 3°.

For the first five arithmetic classes: $\mathbb{Z}_1^a$, $\mathbb{Z}_2^a$, $\mathbb{Z}_3^h$, $\mathbb{Z}_4^q$, and $\mathbb{Z}_6^h$, the calculations are quite simple. Thus, let $(T, \mathscr{G})$ be a representative arithmetic group for any one of them. [See Table 1.] Since $\mathscr{G}$ contains only rotations, there is just one subgroup of $\mathscr{C}$ consonant with $(T, \mathscr{G})$, namely, $T \rtimes \mathscr{G}$. As a result, each of the foregoing arithmetic classes yields just one ornamental class, the corresponding symmorphic case. We shall denote these ornamental classes by: $\mathbb{Z}_1^{a}\llcorner$, $\mathbb{Z}_2^{a}\llcorner$, $\mathbb{Z}_3^{h}\llcorner$, $\mathbb{Z}_4^{q}\llcorner$, and $\mathbb{Z}_6^{h}\llcorner$.

In Figures 41 - 45, we present plane ornaments illustrating the foregoing ornamental classes. For each of them, the symmetry group is a representative of the corresponding ornamental class. In each case, we have indicated generators for the symmetry group, in the usual manner. [See 1.5°.]

Hereafter, we shall make use of Proposition 18.

Let us consider the arithmetic class $\mathbb{D}_1^r$. Let $(T, \mathscr{G})$ be an arithmetic group in $\mathbb{D}_1^r$, as described in Table 1. We may introduce a basis $\{u,v\}$ for T and generators W and X for $\mathscr{G}$ such that, relative to $\{u,v\}$, the matrices for W and X are as follows:

$$W \longleftrightarrow \begin{bmatrix} 1 & 0 \\ 0 & 1 \end{bmatrix}, \qquad X \longleftrightarrow \begin{bmatrix} 0 & 1 \\ 1 & 0 \end{bmatrix}.$$

Let x be any member of R^2 of the form:

$$x = a.u + b.v,$$

where $0 \leq a < 1$ and $0 \leq b < 1$. Let us assume that x satisfies conditions (ρ) and (ξ) in Proposition 18. Since $W = I$, condition (ρ) yields no information. However:

$$x + X(x) = (a + b).(u + v),$$

so condition (ξ) means that $a + b$ is a member of Z. It follows that $a + b$ must be either 0 or 1. In the former case, both a and b would be 0, yielding $x = 0$. Let us examine the latter case, in which $a + b = 1$. [See Figure 46.] We contend that $\mathscr{F}_0$ and $\mathscr{F}_x$ are conjugate in $\mathscr{A}$, in fact, by a translation. Let t stand for $b.v$. We have:

$$\begin{aligned} t - X(t) &= b.v - b.u \\ &= -u + x . \end{aligned}$$

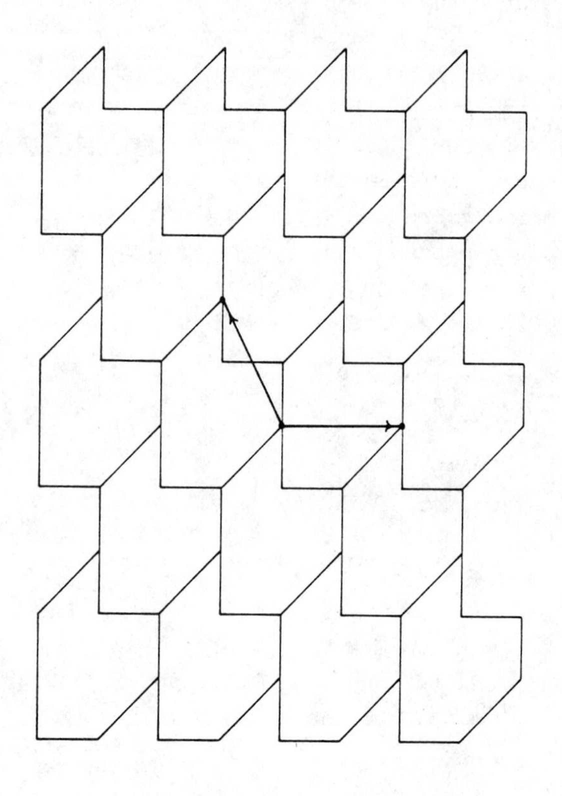

Figure 41

$\mathbb{Z}_1^a$┗•

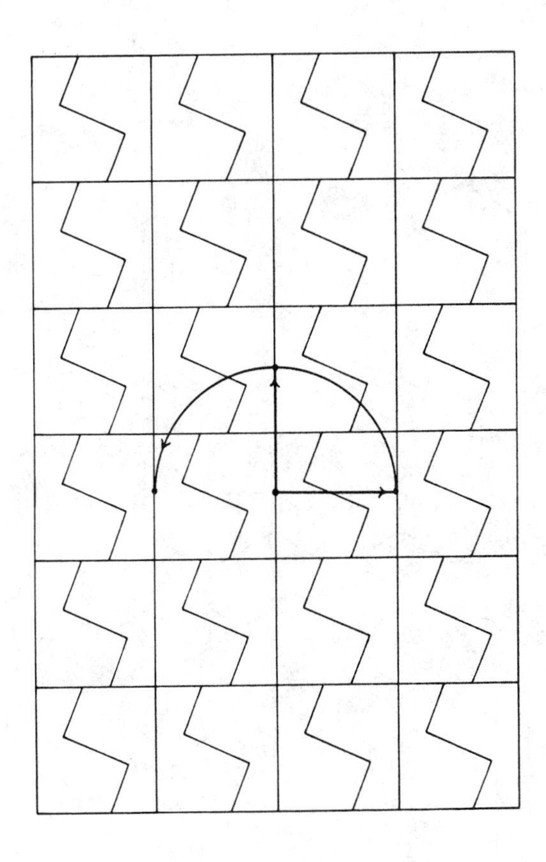

Figure 42

$\mathbb{Z}_2^a \llcorner$

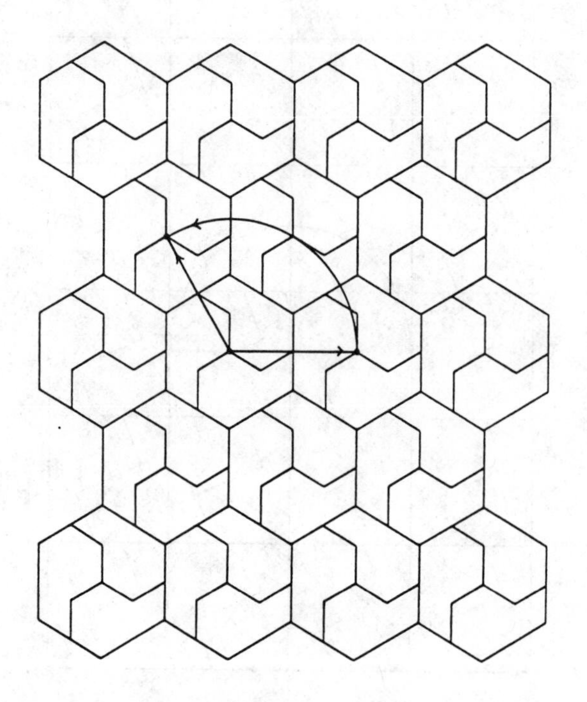

Figure 43

$\mathbb{Z}_3^h$

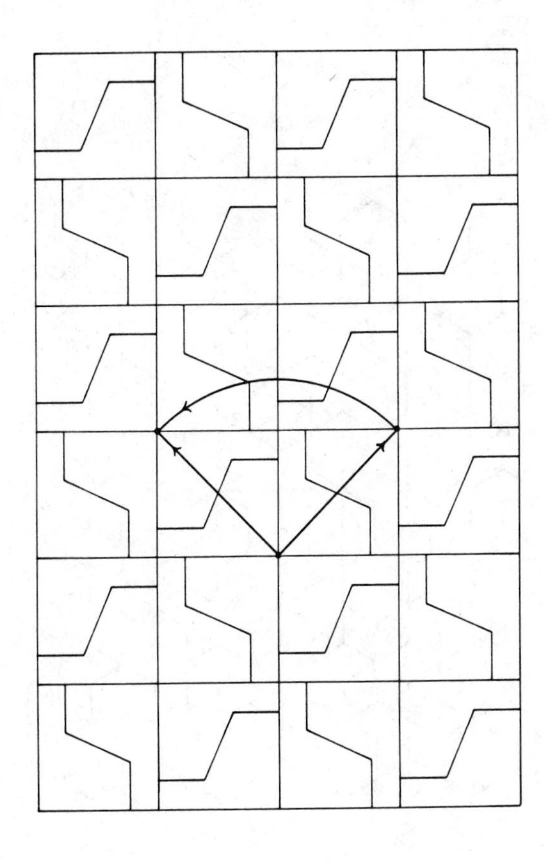

Figure 44

$\mathbb{Z}_4^q$ ⌊•—

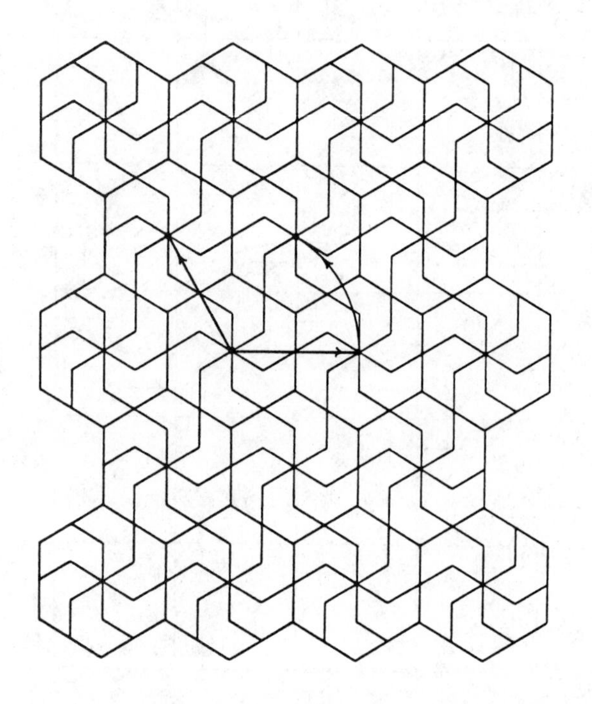

Figure 45

$\mathbb{Z}_6^h$

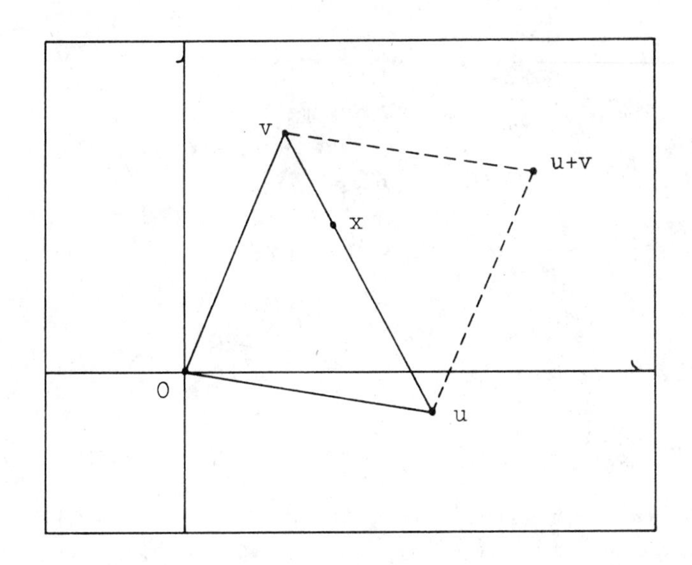

Figure 46

Hence:

$$[t,I][0,X][t,I]^{-1} = [t - X(t),X]$$
$$= [-u,I][x,X] .$$

Clearly, $J_{[t,I]}(\mathscr{F}_0)$ is a subgroup of $\mathscr{C}$ consonant with $(T, \mathscr{G})$. By the foregoing relation, it contains $[x,X]$. By Proposition 18, $J_{[t,I]}(\mathscr{F}_0)$ must equal $\mathscr{F}_x$. Therefore, $\mathscr{F}_0$ and $\mathscr{F}_x$ are conjugate in $\mathscr{A}$. We conclude that, for the arithmetic class $\mathbb{D}_1^r$, there is just one ornamental class which determines it, namely, the symmorphic case. We shall denote it by: $\mathbb{D}_1^r\llcorner$.

The calculations for the arithmetic class $\mathbb{D}_2^r$ are similar. Of course, the matrices for W and X would be replaced by the following:

$$W \longleftrightarrow \begin{bmatrix} -1 & 0 \\ 0 & -1 \end{bmatrix}, \qquad X \longleftrightarrow \begin{bmatrix} 0 & 1 \\ 1 & 0 \end{bmatrix}.$$

Condition (ξ) would again mean that $a + b$ is an integer; but condition (ρ) would now yield that $2.x$ is contained in T, so that $2a$ and $2b$ are integers. Hence, x must be either 0 or $(1/2).(u + v)$. Let us assume that it is the latter. We contend that $\mathscr{F}_0$ and $\mathscr{F}_x$ are conjugate in $\mathscr{A}$. Thus, let t stand for $(1/2).v$. We have:

$$[t,I][0,W][t,I]^{-1} = [2.t,-I]$$
$$= [v,-I],$$

and, as before:

$$[t,I][0,X][t,I]^{-1} = [-u,I][x,X].$$

By the first relation, $J_{[t,I]}(\mathscr{F}_0)$ is consonant with $(T, \mathscr{G})$. By the second, it contains $[x,X]$. Hence, $J_{[t,I]}(\mathscr{F}_0)$ must equal $\mathscr{F}_x$, and therefore $\mathscr{F}_0$ and $\mathscr{F}_x$ are conjugate in $\mathscr{A}$. It follows that the arithmetic class $\mathbb{D}_2^r$ yields only one ornamental class: the symmorphic case. We shall denote it by: $\mathbb{D}_2^r\llcorner$.

In Figures 47 and 48, we present plane ornaments illustrating the ornamental classes $\mathbb{D}_1^r\llcorner$ and $\mathbb{D}_2^r\llcorner$. In both cases, we have indicated generators for the symmetry groups.

Let us turn to the arithmetic classes $\mathbb{D}_1^o$ and $\mathbb{D}_2^o$. Let $(T, \mathscr{G})$ be an arithmetic group in $\mathbb{D}_1^o$, as described in Table 1. We may introduce generators W and X for $\mathscr{G}$ and a basis $\{u,v\}$ for T such that the matrices for W and X relative to $\{u,v\}$ are as follows:

$$W \longleftrightarrow \begin{bmatrix} 1 & 0 \\ 0 & 1 \end{bmatrix}, \qquad X \longleftrightarrow \begin{bmatrix} 1 & 0 \\ 0 & -1 \end{bmatrix}.$$

Let x be any member of R^2 of the form:

$$x = a.u + b.v,$$

where $0 \leq a < 1$ and $0 \leq b < 1$, and let us assume that x satisfies conditions (ρ) and (ξ). Condition (ρ) is in fact

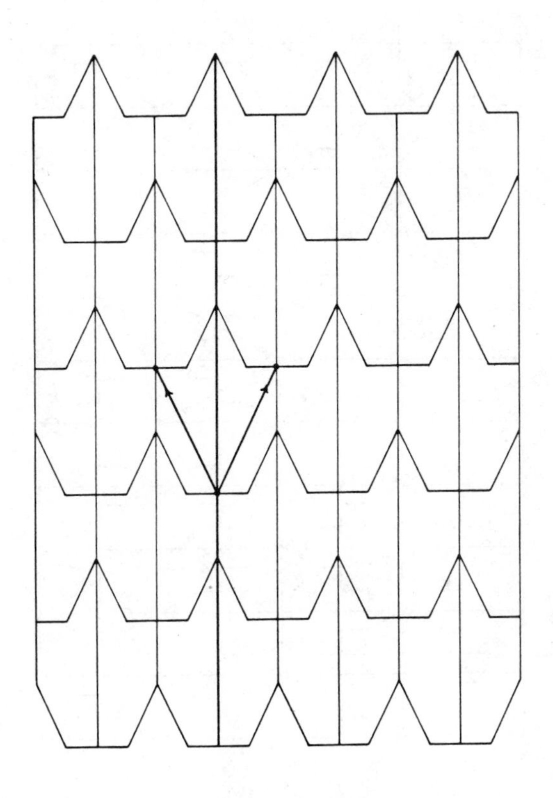

Figure 47

$\mathbb{D}_1^r$

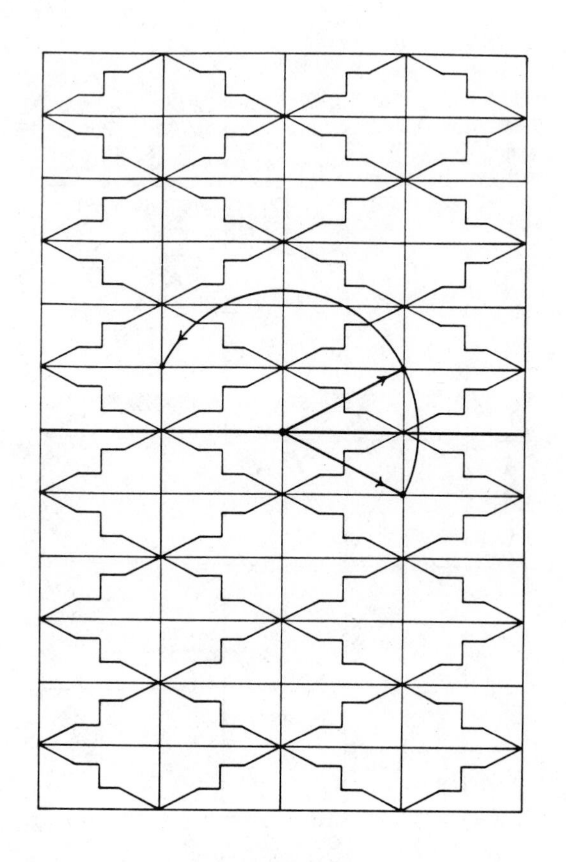

Figure 48

$\mathbb{D}_2^r$

neutral. However:

$$x + X(x) = 2a.u ,$$

so condition (ξ) means that $2a$ is an integer. It follows that either $a = 0$ or $a = 1/2$. Let y stand for $(1/2).u$. [See Figure 49.] Taking t to be $(-b/2).v$, we obtain:

$$\begin{aligned} t + x - X(t) &= (-b/2).v + a.u + b.v + (-b/2).v \\ &= a.u . \end{aligned}$$

Hence:

$$[t,I][x,X][t,I]^{-1} = [a.u, X] .$$

It follows that $J_{[t,I]}(\mathscr{F}_x)$ must equal either $\mathscr{F}_0$ or $\mathscr{F}_y$. Hence, $\mathscr{F}_x$ is conjugate in $\mathscr{A}$, either to $\mathscr{F}_0$ or to $\mathscr{F}_y$.

We contend now that $\mathscr{F}_0$ and $\mathscr{F}_y$ are not conjugate in

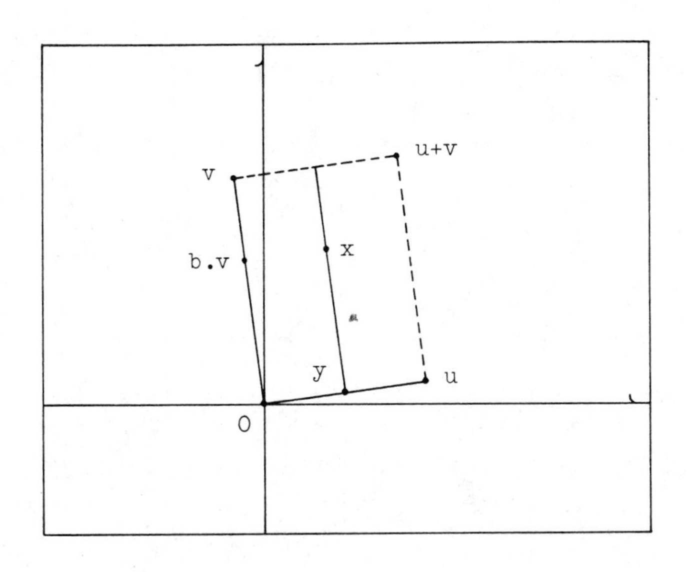

Figure 49

$\mathscr{A}$. Let us suppose to the contrary that there exists an affine transformation $[t,L]$ such that $J_{[t,L]}(\mathscr{F}_0) = \mathscr{F}_y$. We would have: $LXL^{-1} = X$, so that:

$$[t,L][0,X][t,L]^{-1} = [t - X(t), X] .$$

Hence:

$$t - X(t) = (1/2).u + s, \tag{1}$$

where s is a suitable member of T. However:

$$X(t - X(t)) = -(t - X(t)) ,$$

so that $t - X(t)$ must be a scalar multiple of v, contradicting relation (1). We infer that $\mathscr{F}_0$ and $\mathscr{F}_y$ are not conjugate in $\mathscr{A}$.

We conclude that, for the arithmetic class $\mathbb{D}_1^{\circ}$, there are two ornamental classes. We shall denote them by $\mathbb{D}_1^{\circ}\llcorner_{\bullet}$ and $\mathbb{D}_1^{\circ}\llcorner_{\bullet}$, the former containing $\mathscr{F}_0$ and the latter, $\mathscr{F}_y$. Of course, the former is the symmorphic case.

In Figures 50 and 51, one will find plane ornaments illustrating the foregoing ornamental classes. They are supplied with the usual annotations.

For the arithmetic class $\mathbb{D}_2^{\circ}$, we proceed in similar manner. Of course, the matrices for W and X would now be the following:

$$W \longleftrightarrow \begin{bmatrix} -1 & 0 \\ 0 & -1 \end{bmatrix}, \qquad X \longleftrightarrow \begin{bmatrix} 1 & 0 \\ 0 & -1 \end{bmatrix}.$$

Condition (ρ) is no longer neutral; it yields that $2.x$ is a member of T. Hence, x satisfies conditions (ρ) and (ξ) iff it is one of the following four members of R^2:

$$0, \quad w = \tfrac{1}{2}.(u+v), \quad y = \tfrac{1}{2}.u, \quad z = \tfrac{1}{2}.v.$$

[See Figure 52.]

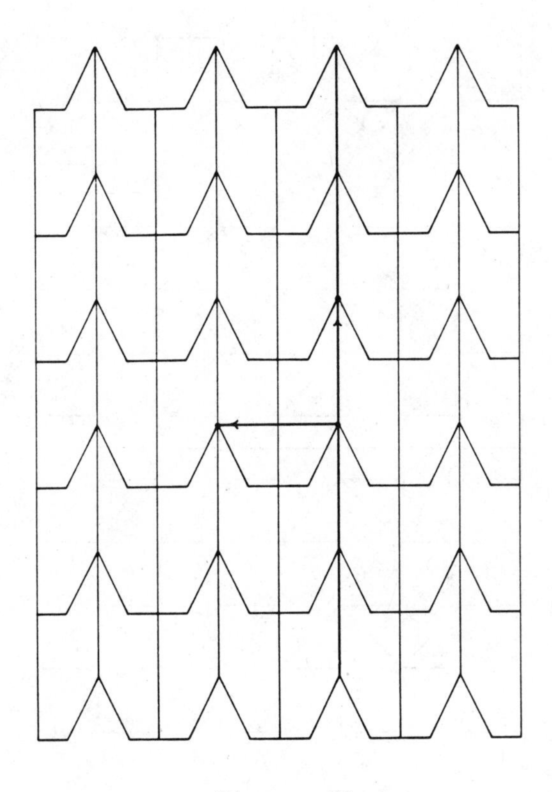

Figure 50

$\mathbb{D}_1^{\circ}$

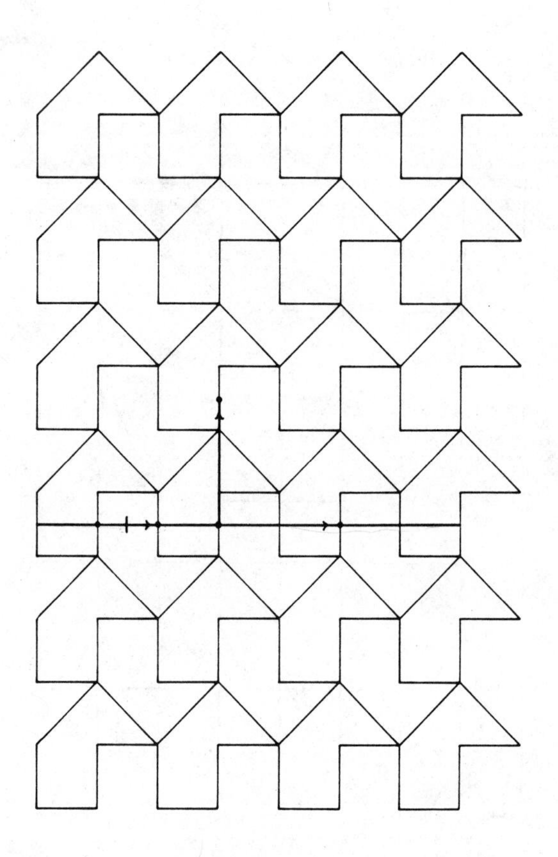

Figure 51

$\mathbb{D}_1^{\circ}$ L

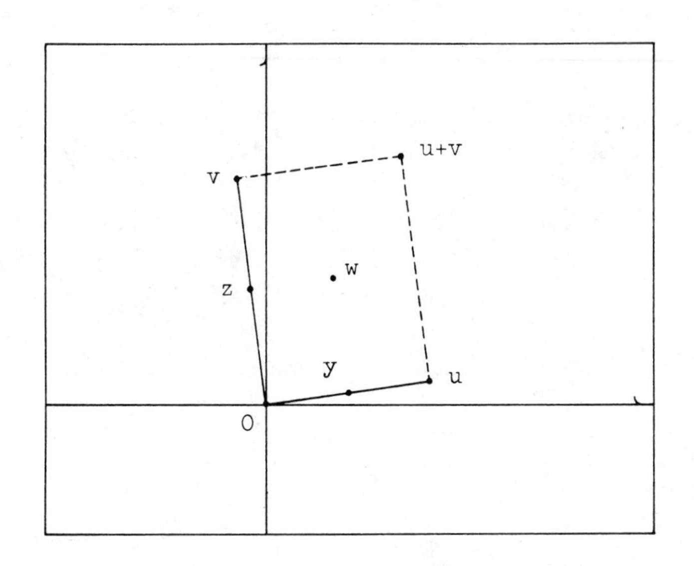

Figure 52

Now let $[t,L]$ be any affine transformation, and let us inquire whether $J_{[t,L]}$ transforms any one among $\mathscr{F}_0$, $\mathscr{F}_w$, $\mathscr{F}_y$, and $\mathscr{F}_z$ into another. Since:

$$[t,L][0,-I][t,L]^{-1} = [2.t,-I],$$

one would have:

$$2.t \in T.$$

Hence, there would be integers m and n such that $t = (m/2).u + (n/2).v$. Clearly, $t - X(t) = n.v$, so that:

$$t - X(t) \in T.$$

Let x be any one among 0, w, y, and z. Obviously:

$$[t,I][0,-I][t,I]^{-1} = [2.t,-I],$$

$$[t,I][x,X][t,I]^{-1} = [t - X(t), I][x,X].$$

The last four relations yield that $J_{[t,I]}(\mathscr{F}_x)$ is consonant with $(T, \mathscr{G})$ and contains $[x,X]$, hence that $J_{[t,I]}(\mathscr{F}_x) = \mathscr{F}_x$. It follows that, for the purpose of determining whether $J_{[t,L]}$ transforms one among $\mathscr{F}_0$, $\mathscr{F}_w$, $\mathscr{F}_y$, and $\mathscr{F}_z$ into another, one might as well assume that $t = 0$.

Of course, LXL^{-1} must equal either X or $-X$, which is to say that $L(M_X)$ must equal either M_X or M_{-X}. Hence, $L(y)$ must be a scalar multiple, either of u or of v. As a result, $L(y) - w$ cannot be a member of T. Since:

$$\begin{aligned} [0,L][y,X][0,L]^{-1} &= [L(y), \pm X] \\ &= [L(y) - w, I][w, \pm X], \end{aligned}$$

it follows that $J_{[0,L]}(\mathscr{F}_y)$ cannot equal $\mathscr{F}_w$.

Since:

$$[0,L][0,X][0,L]^{-1} = [0, \pm X],$$

$J_{[0,L]}(\mathscr{F}_0)$ could only equal $\mathscr{F}_0$.

Finally, let us take L to be the member of $\mathscr{G}$ for which $L(u) = v$ and $L(v) = u$. Clearly, $L(y) = z$. One can readily verify that $J_{[0,L]}(\mathscr{F}_y)$ is consonant with $(T, \mathscr{G})$. Since:

$$[0,L][y,X][0,L]^{-1} = [z,-X],$$

it follows that $J_{[0,L]}(\mathscr{F}_y) = \mathscr{F}_z$.

Altogether, we obtain that, for the arithmetic class $\mathbb{D}_2^{o}$, there are three ornamental classes, corresponding to $\mathscr{F}_0$, $\mathscr{F}_w$, and $\mathscr{F}_y$. The first is the symmorphic case. We shall denote them by: $\mathbb{D}_2^{o}{\bullet\!\!\llcorner}$, $\mathbb{D}_2^{o}\llcorner\!\bullet$, and $\mathbb{D}_2^{o}\llcorner\!\!\bullet$.

Figures 53 - 55 provide plane ornaments illustrating the foregoing ornamental classes.

Now let us consider the arithmetic classes $\mathbb{D}_6^{h}$, $\mathbb{D}_3'$, and $\mathbb{D}_3''$. Let $(T, \mathscr{G})$ be an arithmetic group drawn from $\mathbb{D}_6^{h}$. By Table 1, we are led to introduce a basis $\{u,v\}$

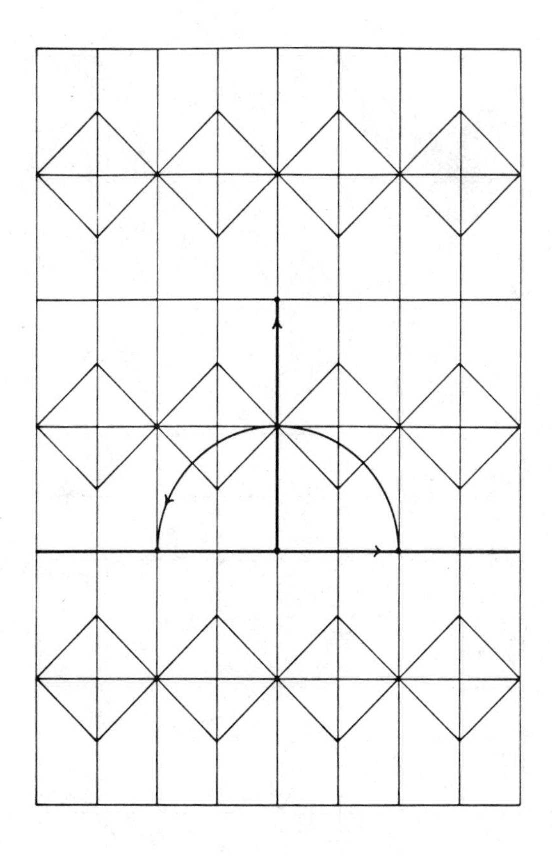

Figure 53

$\mathbb{D}_2^{\circ}$ └

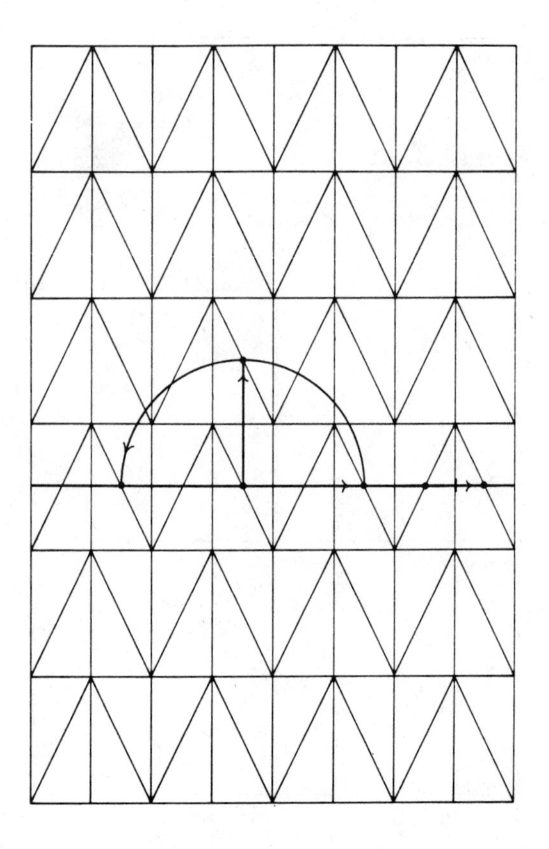

Figure 54

$\mathbb{D}_2^{\circ}$L•-

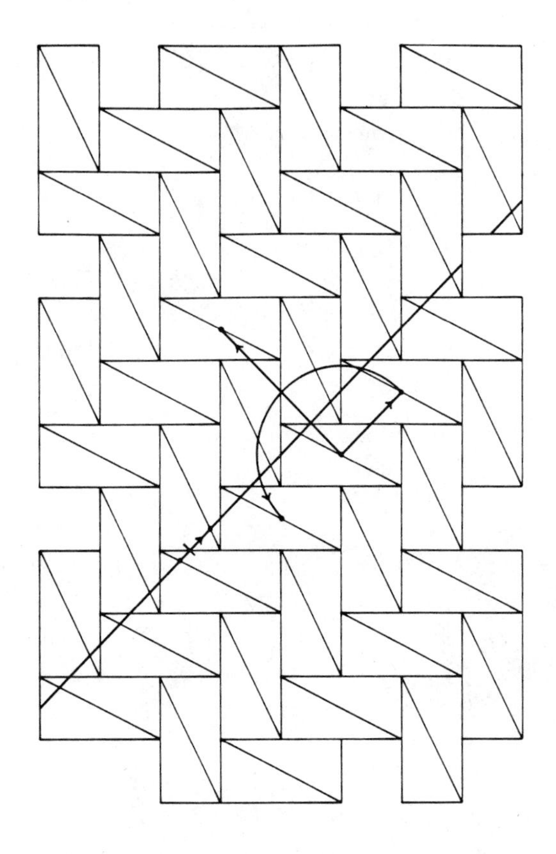

Figure 55

$\mathbb{D}_2^{\circ}$⌊•

for T and generators W and X for $\mathscr{G}$ such that the matrices for W and X relative to {u,v} are the following:

$$W \longleftrightarrow \begin{bmatrix} 1 & -1 \\ 1 & 0 \end{bmatrix}, \qquad X \longleftrightarrow \begin{bmatrix} 1 & -1 \\ 0 & -1 \end{bmatrix}.$$

We immediately obtain the matrix for I - W:

$$I - W \longleftrightarrow \begin{bmatrix} 0 & 1 \\ -1 & 1 \end{bmatrix}.$$

By problem 8 in 8, we may infer that (I - W)(T) = T. It follows that, among all members of R^2 of the form:

$$a.u + b.v,$$

where $0 \leq a < 1$ and $0 \leq b < 1$, there is only one which satisfies condition (ρ), namely, 0. Of course, 0 also satisfies condition (ξ). Hence, among all subgroups of $\mathscr{C}$, only $\mathscr{F}_0$ is consonant with (T, $\mathscr{G}$). We conlcude that the arithmetic class $\mathbb{D}_6^h$ yields just one ornamental class: the symmorphic case. It will be denoted by: $\mathbb{D}_6^h\llcorner$.

The plane ornament depicted in Figure 56 illustrates this ornamental class.

For the arithmetic class $\mathbb{D}_3'$, we must replace W by W^2, while retaining X. Relative to {u,v}, the matrix for W^2 is the following:

$$W^2 \longleftrightarrow \begin{bmatrix} 0 & -1 \\ 1 & -1 \end{bmatrix}.$$

Let x be any member of R^2 of the usual form. We have:

$$\begin{aligned} (I - W^2)(x) &= (I - W^2)(a.u + b.v) \\ &= (a.u + b.v) - (a.v + b.(-(u + v))) \\ &= (a + b).u + (2b - a).v . \end{aligned}$$

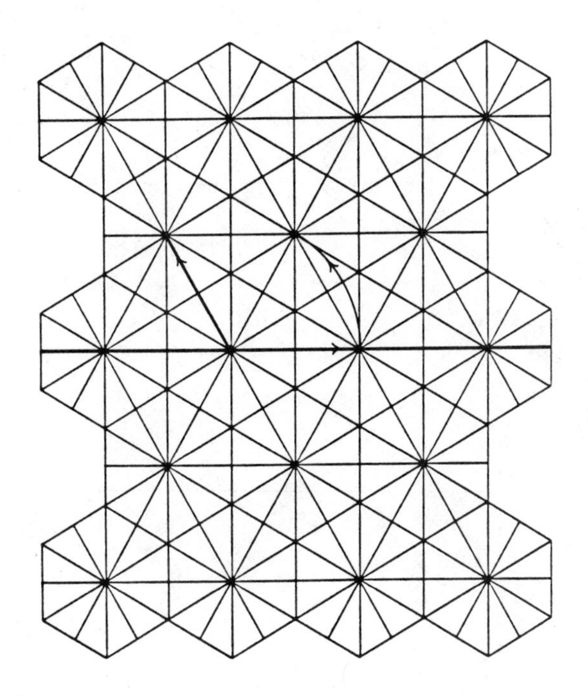

Figure 56

$\mathbb{D}_6^h$ ⌊

Hence, x satisfies condition (ρ) iff both a b and $2b-a$ are integers, which is to say that x is one of the following members of R^2:

$$0, \quad y = \frac{2}{3}.u + \frac{1}{3}.v, \quad z = \frac{1}{3}.u + \frac{2}{3}.v \,. \tag{2}$$

[See Figure 57.] These members also satisfy condition (ξ), because:

$$y + X(y) = u,$$
$$z + X(z) = 0.$$

We contend that both $\mathscr{F}_y$ and $\mathscr{F}_z$ are conjugate to $\mathscr{F}_0$ in $\mathscr{A}$. Indeed, we have:

$$[y,I][0,W^2][y,I]^{-1} = [u,W^2], \tag{3}$$
$$[z,I][0,W^2][z,I]^{-1} = [u+v,W^2], \tag{4}$$

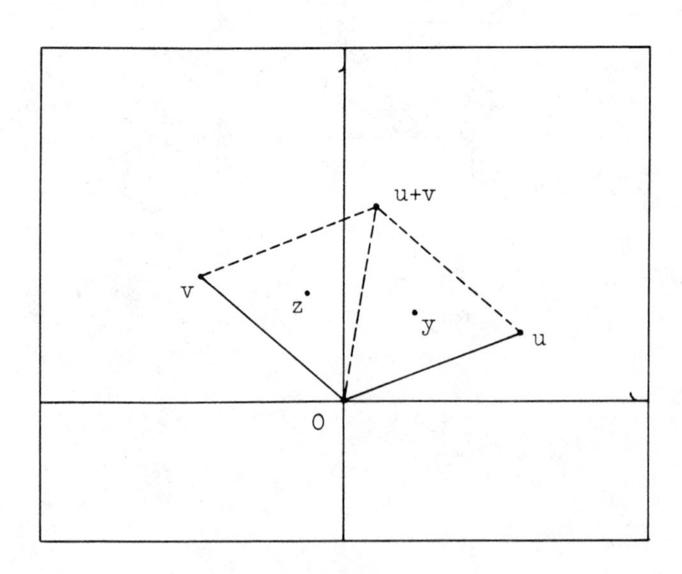

Figure 57

$$[y,I][0,X][y,I]^{-1} = [z,X], \tag{5}$$

$$[z,I][0,X][z,I]^{-1} = [v+y,X]. \tag{6}$$

Relations (3) and (5) entail that $J_{[y,I]}(\mathscr{F}_0)$ is consonant with $(T,\mathscr{G})$ and contains $[z,X]$, hence that $J_{[y,I]}(\mathscr{F}_0) = \mathscr{F}_z$. In the same way, relations (4) and (6) entail that $J_{[z,I]}(\mathscr{F}_0) = \mathscr{F}_y$.

We conclude that, for the arithmetic class $\mathbb{D}_3'$, the symmorphic case is the only corresponding ornamental class. It will be denoted by: $\mathbb{D}_3'\!\lfloor$.

For the arithmetic class $\mathbb{D}_3''$, we must replace not only W by W^2 but also X by -X. With respect to condition (ρ), we are again led to consider the members of R^2 listed in item (2). However:

$$y + (-X)(y) = z,$$

$$z + (-X)(z) = y + v,$$

so the only member of R^2 which satisfies both conditions (ρ) and (ξ), is 0.

Hence, for the arithmetic class $\mathbb{D}_3''$, we obtain only the symmorphic case: $\mathbb{D}_3''\!\lfloor$.

In Figures 58 and 59, one will find plane ornaments which illustrate, and which should help to discriminate the ornamental classes $\mathbb{D}_3'\!\lfloor$ and $\mathbb{D}_3''\!\lfloor$.

Finally, let us examine the arithmetic class $\mathbb{D}_4^q$. Let $(T,\mathscr{G})$ be an arithmetic group in $\mathbb{D}_4^q$. With reference to Table 1, we may introduce a basis $\{u,v\}$ for T and generators W and X for $\mathscr{G}$ such that the matrices for W and X relative to $\{u,v\}$ are as follows:

$$W \longleftrightarrow \begin{bmatrix} 0 & -1 \\ 1 & 0 \end{bmatrix}, \qquad X \longleftrightarrow \begin{bmatrix} 1 & 0 \\ 0 & -1 \end{bmatrix}.$$

Let x be any member of R^2 of the form:

$$x = a.u + b.v,$$

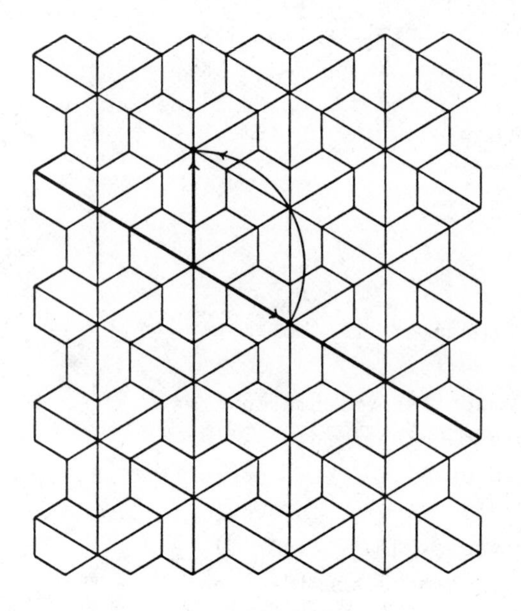

Figure 58

$\mathbb{D}_3'$

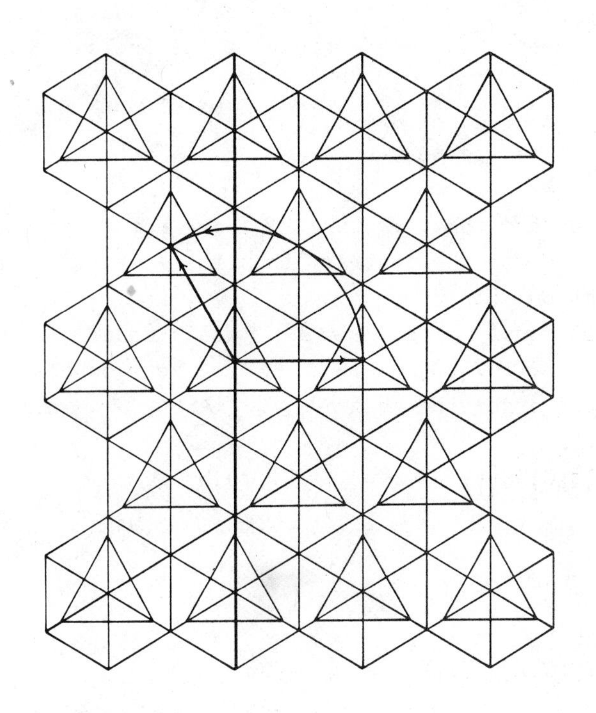

Figure 59

$\mathbb{D}_3''$

where $0 \leq a < 1$ and $0 \leq b < 1$, and let x satisfy conditions (ρ) and (ξ). Since:

$$x - W(x) = (a-b).u + (a+b).v ,$$

$$x + X(x) = 2a.u ,$$

it follows that either a and b both equal 0 or a and b both equal $1/2$. [In point of fact, condition (ξ) is redundant.] Hence, x must be one of the following members of R^2:

$$0, \quad w = \tfrac{1}{2}.u + \tfrac{1}{2}.v .$$

We contend that $\mathscr{F}_0$ and $\mathscr{F}_w$ are not conjugate in $\mathscr{A}$. Let us suppose to the contrary that there exists an affine transformation $[t,L]$ such that $J_{[t,L]}(\mathscr{F}_0) = \mathscr{F}_w$. Clearly, $L(T)$ would equal T and $L\mathscr{G}L^{-1}$ would equal $\mathscr{G}$. It would follow that $J_{[0,L]}(\mathscr{F}_0) = \mathscr{F}_0$. Hence, our supposition reduces to the assertion that $J_{[t,I]}(\mathscr{F}_0) = \mathscr{F}_w$. Since:

$$[t,I][0,W][t,I]^{-1} = [t - W(t),W]$$
$$= [t - W(t),I][0,W] ,$$

and:

$$[t,I][0,X][t,I]^{-1} = [t - X(t),X]$$
$$= [t - X(t) - w, I][w, X] ,$$

we would obtain:

$$t - W(t) \in T, \tag{7}$$
$$t - X(t) \in T. \tag{8}$$

Let c and d be the coordinates of t relative to $\{u,v\}$:

$$t = c.u + d.v.$$

By relation (7), $c + d$ and $c - d$ would have to be integers. By relation (8), $-\frac{1}{2} + 2d$ would have to be an integer. Obviously, these results are contradictory. We must conclude that that $\mathscr{F}_0$ and $\mathscr{F}_w$ are not conjugate in $\mathscr{A}$.

Hence, for the arithmetic class $\mathbb{D}_4^q$, there are two ornamental classes, corresponding to $\mathscr{F}_0$ and $\mathscr{F}_w$. The first is the symmorphic case. We shall denote them by: $\mathbb{D}_4^q$⌊• and $\mathbb{D}_4^q$⌊•⌋.

Figures 60 and 61 provide plane ornaments illustrating these ornamental classes.

7° In summary, the foregoing calculations yield that there are 17 ornamental classes. Of course, 13 of them are the symmorphic cases, corresponding to the 13 arithmetic classes. Of the four non-symmorphic cases, one corresponds to $\mathbb{D}_1^o$, two to $\mathbb{D}_2^o$, and one to $\mathbb{D}_4^q$.

The transitions from geometric to arithmetic, and from arithmetic to ornamental classes, are displayed in Table 2.

8° For each of the 17 ornamental classes, we have determined a representative group $\mathscr{F}$ (in cartesian form) generated by four members:

$$\underline{U} = [u, I],$$

$$\underline{V} = [v, I],$$

$$\underline{W} = [0, W],$$

$$\underline{X} = [x, X].$$

Of course, for the first five cases, the generator $\underline{X}$ (and all references to it) should be deleted.

The members u and v of R^2 compose a basis for the translational part T of $\mathscr{F}$, and the members W and X of $\mathscr{O}$ generate the linear part $\mathscr{G}$. The matrices for W and X relative to $\{u,v\}$ are those which appear in Table 1, under the arithmetic class which would correspond to $\mathscr{F}$. The member x of R^2 completes the determination of $\mathscr{F}$. For the symmorphic cases, x may be taken to be 0. For the cases $\mathbb{D}_1^o$⌊•

Figure 60

$\mathbb{D}_4^q$

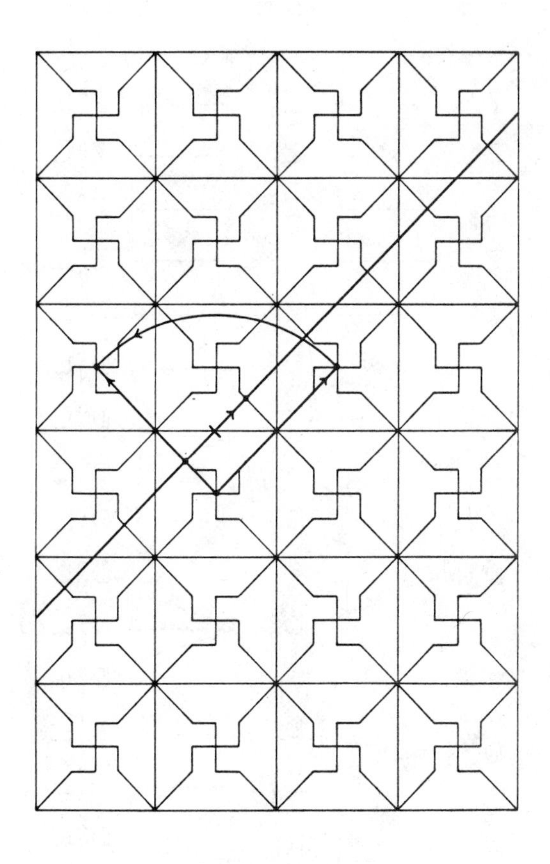

Figure 61

$\mathbb{D}_4^q \llcorner\!\bullet$

GEOMETRIC CLASS	ARITHMETIC CLASS	ORNAMENTAL CLASS
$\mathbb{Z}_1$	$\mathbb{Z}_1^a$	$\mathbb{Z}_1^a\llcorner_{\bullet}$
$\mathbb{Z}_2$	$\mathbb{Z}_2^a$	$\mathbb{Z}_2^a\llcorner_{\bullet}$
$\mathbb{Z}_3$	$\mathbb{Z}_3^h$	$\mathbb{Z}_3^h\llcorner_{\bullet}$
$\mathbb{Z}_4$	$\mathbb{Z}_4^q$	$\mathbb{Z}_4^q\llcorner_{\bullet}$
$\mathbb{Z}_6$	$\mathbb{Z}_6^h$	$\mathbb{Z}_6^h\llcorner_{\bullet}$
$\mathbb{D}_1$	$\mathbb{D}_1^r$	$\mathbb{D}_1^r\llcorner_{\bullet}$
	$\mathbb{D}_1^o$	$\mathbb{D}_1^o\llcorner_{\bullet}$
		$\mathbb{D}_1^o\llcorner\!\bullet$
$\mathbb{D}_2$	$\mathbb{D}_2^r$	$\mathbb{D}_2^r\llcorner_{\bullet}$
	$\mathbb{D}_2^o$	$\mathbb{D}_2^o\llcorner_{\bullet}$
		$\mathbb{D}_2^o\llcorner\!\bullet$
		$\mathbb{D}_2^o\llcorner^{\bullet}$
$\mathbb{D}_3$	$\mathbb{D}_3'$	$\mathbb{D}_3'\llcorner_{\bullet}$
	$\mathbb{D}_3''$	$\mathbb{D}_3''\llcorner_{\bullet}$
$\mathbb{D}_4$	$\mathbb{D}_4^q$	$\mathbb{D}_4^q\llcorner_{\bullet}$
		$\mathbb{D}_4^q\llcorner^{\bullet}$
$\mathbb{D}_6$	$\mathbb{D}_6^h$	$\mathbb{D}_6^h\llcorner_{\bullet}$

Table 2

and $\mathbb{D}_2^{o}\llcorner_{\bullet}$, x may be taken to be $\frac{1}{2}.u$, and, for the cases $\mathbb{D}_2^{o}\llcorner\bullet$ and $\mathbb{D}_4^{q}\llcorner\bullet$, to be $\frac{1}{2}.(u+v)$.

In Table 3, one will find a schematic display of the foregoing data. For each case, we have suggested that the generators $\underline{U}$ and $\underline{V}$ satisfy the condition:

$$|u| = |v|. \tag{1}$$

Clearly, one can arrange that this condition be satisfied, without straying from the given ornamental class. In fact, one can arrange that the lattice T be either quadratic or hexagonal. Whenever possible, we have portrayed T as quadratic, otherwise, as hexagonal.

Condition (1) will prove to be convenient in subsequent calculations. [See 3.4.]

For later reference, we have established a conventional serial ordering of the representative groups, numbering them from 01 through 17. We have also supplied the commonly used International Symbols for (the ornamental classes of) these groups. For example, the case of $\mathbb{D}_2^{r}\llcorner$ is numbered 09, and has International Symbol cmm.

The members of the group $\mathscr{F}$ may be presented in the form:

$$\underline{U}^a\underline{V}^b\underline{W}^c\underline{X}^d, \tag{2}$$

where a, b, c, and d are any integers. Of course, c and d may be constrained as follows:

$$0 \le c < k,$$
$$0 \le d < 2,$$

where k is the order of the rotation W. [For the first five cases, references to X and d should be deleted.] Subject to these constraints, the foregoing presentation of the members of $\mathscr{F}$ is unique.

Now one may expect to reconstruct the operation of multiplication on $\mathscr{F}$, in the form:

ORNAMENTAL GROUP	ORNAMENTAL CLASS		TRANSLATIONAL PART	LINEAR PART		EXTENSION GENERATOR
	TEXT SYMBOL	INT'L SYMBOL	u v	W	X	x
$\mathfrak{F}_{01}$	$\mathbb{Z}_1^a$	p1	v u	$\begin{bmatrix}1 & 0\\ 0 & 1\end{bmatrix}$	-	-
$\mathfrak{F}_{02}$	$\mathbb{Z}_2^a$	p2	v u	$\begin{bmatrix}-1 & 0\\ 0 & -1\end{bmatrix}$	-	-
$\mathfrak{F}_{03}$	$\mathbb{Z}_3^h$	p3	v u	$\begin{bmatrix}0 & -1\\ 1 & -1\end{bmatrix}$	-	-
$\mathfrak{F}_{04}$	$\mathbb{Z}_4^q$	p4	v u	$\begin{bmatrix}0 & -1\\ 1 & 0\end{bmatrix}$	-	-
$\mathfrak{F}_{05}$	$\mathbb{Z}_6^h$	p6	v u	$\begin{bmatrix}1 & -1\\ 1 & 0\end{bmatrix}$	-	-
$\mathfrak{F}_{06}$	$\mathbb{D}_1^r$	cm	v u	$\begin{bmatrix}1 & 0\\ 0 & 1\end{bmatrix}$	$\begin{bmatrix}0 & 1\\ 1 & 0\end{bmatrix}$	0
$\mathfrak{F}_{07}$	$\mathbb{D}_1^o$	pm	v u	$\begin{bmatrix}1 & 0\\ 0 & 1\end{bmatrix}$	$\begin{bmatrix}1 & 0\\ 0 & -1\end{bmatrix}$	0
$\mathfrak{F}_{08}$	$\mathbb{D}_1^o$	pg	v u	$\begin{bmatrix}1 & 0\\ 0 & 1\end{bmatrix}$	$\begin{bmatrix}1 & 0\\ 0 & -1\end{bmatrix}$	$\frac{1}{2}\cdot u$
$\mathfrak{F}_{09}$	$\mathbb{D}_2^r$	cmm	v u	$\begin{bmatrix}-1 & 0\\ 0 & -1\end{bmatrix}$	$\begin{bmatrix}0 & 1\\ 1 & 0\end{bmatrix}$	0
$\mathfrak{F}_{10}$	$\mathbb{D}_2^o$	pmm	v u	$\begin{bmatrix}-1 & 0\\ 0 & -1\end{bmatrix}$	$\begin{bmatrix}1 & 0\\ 0 & -1\end{bmatrix}$	0
$\mathfrak{F}_{11}$	$\mathbb{D}_2^o$	pmg	v u	$\begin{bmatrix}-1 & 0\\ 0 & -1\end{bmatrix}$	$\begin{bmatrix}1 & 0\\ 0 & -1\end{bmatrix}$	$\frac{1}{2}\cdot u$
$\mathfrak{F}_{12}$	$\mathbb{D}_2^o$	pgg	v u	$\begin{bmatrix}-1 & 0\\ 0 & -1\end{bmatrix}$	$\begin{bmatrix}1 & 0\\ 0 & -1\end{bmatrix}$	$\frac{1}{2}\cdot(u+v)$
$\mathfrak{F}_{13}$	$\mathbb{D}_3'$	p31m	v u	$\begin{bmatrix}0 & -1\\ 1 & -1\end{bmatrix}$	$\begin{bmatrix}1 & -1\\ 0 & -1\end{bmatrix}$	0
$\mathfrak{F}_{14}$	$\mathbb{D}_3''$	p3m1	v u	$\begin{bmatrix}0 & -1\\ 1 & -1\end{bmatrix}$	$\begin{bmatrix}-1 & 1\\ 0 & 1\end{bmatrix}$	0
$\mathfrak{F}_{15}$	$\mathbb{D}_4^q$	p4m	v u	$\begin{bmatrix}0 & -1\\ 1 & 0\end{bmatrix}$	$\begin{bmatrix}1 & 0\\ 0 & -1\end{bmatrix}$	0
$\mathfrak{F}_{16}$	$\mathbb{D}_4^q$	p4g	v u	$\begin{bmatrix}0 & -1\\ 1 & 0\end{bmatrix}$	$\begin{bmatrix}1 & 0\\ 0 & -1\end{bmatrix}$	$\frac{1}{2}\cdot(u+v)$
$\mathfrak{F}_{17}$	$\mathbb{D}_6^h$	p6m	v u	$\begin{bmatrix}1 & -1\\ 1 & 0\end{bmatrix}$	$\begin{bmatrix}1 & -1\\ 0 & -1\end{bmatrix}$	0

Table 3

$$(\underline{U}^{a'}\underline{V}^{b'}\underline{W}^{c'}\underline{X}^{d'})(\underline{U}^{a''}\underline{V}^{b''}\underline{W}^{c''}\underline{X}^{d''}) = \underline{U}^{a}\underline{V}^{b}\underline{W}^{c}\underline{X}^{d}, \quad (3)$$

where a, b, c, and d must be computed from a', b', c', d', a", b", c", and d". In order to carry out the computation, one must know sufficiently many relations governing $\underline{U}$, $\underline{V}$, $\underline{W}$, and $\underline{X}$ to support an effective regrouping of the terms on the left side of relation (3). Specifically, it would be sufficient to know the order k of $\underline{W}$:

$$\underline{W}^k = \underline{I},$$

where $\underline{I}$ stands for the identity, and to know expressions for the following members of $\mathscr{F}$, in the form of the foregoing item (2):

$$\begin{aligned}
\underline{U}\,\underline{V}\,\underline{U}^{-1} &= \quad , \\
\underline{W}\,\underline{U}\,\underline{W}^{-1} &= \quad , \\
\underline{W}\,\underline{V}\,\underline{W}^{-1} &= \quad , \\
\underline{X}^2 &= \quad , \\
\underline{X}\,\underline{U}\,\underline{X}^{-1} &= \quad , \\
\underline{X}\,\underline{V}\,\underline{X}^{-1} &= \quad , \\
\underline{X}\,\underline{W}\,\underline{X}^{-1} &= \quad .
\end{aligned}$$

In Table 4, we have presented this information. For each of the 17 ornamental classes, one should equate the members of the representative group $\mathscr{F}$ standing at the heads of the columns, with the members of $\mathscr{F}$ lying in the corresponding row. Thus, for the case of $\mathscr{F}_{15}$, we have:

$$\underline{X}\,\underline{W}\,\underline{X}^{-1} = \underline{W}^{-1},$$

and, for the case of $\mathscr{F}_{16}$:

$$\underline{X}\,\underline{W}\,\underline{X}^{-1} = \underline{V}\,\underline{W}^{-1}.$$

ORNAMENTAL GROUP	$\underline{V}\,\underline{U}\,\underline{V}^{-1}$	$\underline{W}^k = \underline{I}$	$\underline{W}\,\underline{U}\,\underline{W}^{-1}$	$\underline{W}\,\underline{V}\,\underline{W}^{-1}$	$\underline{X}^2$	$\underline{X}\,\underline{U}\,\underline{X}^{-1}$	$\underline{X}\,\underline{V}\,\underline{X}^{-1}$	$\underline{X}\,\underline{W}\,\underline{X}^{-1}$
$\mathfrak{F}_{01}$	$\underline{U}$	1	$\underline{U}$	$\underline{V}$	-	-	-	-
$\mathfrak{F}_{02}$	$\underline{U}$	2	$\underline{U}^{-1}$	$\underline{V}^{-1}$	-	-	-	-
$\mathfrak{F}_{03}$	$\underline{U}$	3	$\underline{V}$	$\underline{U}^{-1}\underline{V}^{-1}$	-	-	-	-
$\mathfrak{F}_{04}$	$\underline{U}$	4	$\underline{V}$	$\underline{U}^{-1}$	-	-	-	-
$\mathfrak{F}_{05}$	$\underline{U}$	6	$\underline{U}\,\underline{V}$	$\underline{U}^{-1}$	-	-	-	-
$\mathfrak{F}_{06}$	$\underline{U}$	1	$\underline{U}$	$\underline{V}$	$\underline{I}$	$\underline{V}$	$\underline{U}$	$\underline{I}$
$\mathfrak{F}_{07}$	$\underline{U}$	1	$\underline{U}$	$\underline{V}$	$\underline{I}$	$\underline{U}$	$\underline{V}^{-1}$	$\underline{I}$
$\mathfrak{F}_{08}$	$\underline{U}$	1	$\underline{U}$	$\underline{V}$	$\underline{U}$	$\underline{U}$	$\underline{V}^{-1}$	$\underline{I}$
$\mathfrak{F}_{09}$	$\underline{U}$	2	$\underline{U}^{-1}$	$\underline{V}^{-1}$	$\underline{I}$	$\underline{V}$	$\underline{U}$	$\underline{W}^{-1}$
$\mathfrak{F}_{10}$	$\underline{U}$	2	$\underline{U}^{-1}$	$\underline{V}^{-1}$	$\underline{I}$	$\underline{U}$	$\underline{V}^{-1}$	$\underline{W}^{-1}$
$\mathfrak{F}_{11}$	$\underline{U}$	2	$\underline{U}^{-1}$	$\underline{V}^{-1}$	$\underline{U}$	$\underline{U}$	$\underline{V}^{-1}$	$\underline{U}\,\underline{W}^{-1}$
$\mathfrak{F}_{12}$	$\underline{U}$	2	$\underline{U}^{-1}$	$\underline{V}^{-1}$	$\underline{U}$	$\underline{U}$	$\underline{V}^{-1}$	$\underline{U}\,\underline{V}\,\underline{W}^{-1}$
$\mathfrak{F}_{13}$	$\underline{U}$	3	$\underline{V}$	$\underline{U}^{-1}\underline{V}^{-1}$	$\underline{I}$	$\underline{U}$	$\underline{U}^{-1}\underline{V}^{-1}$	$\underline{W}^{-1}$
$\mathfrak{F}_{14}$	$\underline{U}$	3	$\underline{V}$	$\underline{U}^{-1}\underline{V}^{-1}$	$\underline{I}$	$\underline{U}^{-1}$	$\underline{U}\,\underline{V}$	$\underline{W}^{-1}$
$\mathfrak{F}_{15}$	$\underline{U}$	4	$\underline{V}$	$\underline{U}^{-1}$	$\underline{I}$	$\underline{U}$	$\underline{V}^{-1}$	$\underline{W}^{-1}$
$\mathfrak{F}_{16}$	$\underline{U}$	4	$\underline{V}$	$\underline{U}^{-1}$	$\underline{U}$	$\underline{U}$	$\underline{V}^{-1}$	$\underline{V}\,\underline{W}^{-1}$
$\mathfrak{F}_{17}$	$\underline{U}$	6	$\underline{U}\,\underline{V}$	$\underline{U}^{-1}$	$\underline{I}$	$\underline{U}$	$\underline{U}^{-1}\underline{V}^{-1}$	$\underline{W}^{-1}$

Table 4

The second column is exceptional, in that it presents the orders of $\underline{W}$.

We have now finished our description of the ornamental classes. In Table 2, one will find the schematic relations among geometric, arithmetic, and ornamental classes. From Tables 3 and 4, one may determine representative groups for each of the ornamental classes, the former table yielding concrete geometric descriptions, the latter, abstract descriptions in terms of generators and relations.

9° The calculations in 6° also yield that there are 17 classes of plane ornaments in E. In this connection, let us recall the assertion in 3.2°, that every ornamental subgroup of $\mathbf{E}$ is the symmetry group of some plane ornament in E. We are now in a position to prove that that is so.

Proposition 19. For each ornamental subgroup $\mathbf{F}$ of $\mathbf{E}$, there exists some plane ornament Ω in E such that $\mathbf{F} = \mathbf{E}_{\Omega}$.

Proof. Since the ornamental class containing $\mathbf{F}$ must be one of the 17 classes calculated in 6°, and since each of those classes has been illustrated by an appropriate plane ornament, we are justified in introducing a plane ornament Ω' in E such that $\mathbf{E}_{\Omega'}$ and $\mathbf{F}$ are isomorphic.

As a matter of fact, the illustrative plane ornaments which appear in 6° satisfy the following additional condition:

(*) for any affine transformation A on E, if $A.\Omega' = \Omega'$ then A is actually a euclidean transformation on E, so that A is a member of $\mathbf{E}_{\Omega'}$.

[See problem 14 in 8.] We may assume, then, that Ω' satisfies condition (*).

Applying THEOREM 1, we may introduce an affine transformation A on E such that $A\,\mathbf{E}_{\Omega'}A^{-1} = \mathbf{F}$. Let us denote by Ω the mosaic $A.\Omega'$ in E. We contend that:

$$\mathbf{E}_{\Omega} = A\,\mathbf{E}_{\Omega'}A^{-1},$$

so that

$$\mathbf{E}_{\Omega} = \mathbf{F}.$$

It will follow that Ω is a plane ornament in E. The proof of the proposition will then be complete.

Thus, for each member θ' of $\mathbf{E}_{\Omega'}$, $A\theta'A^{-1}$ is a member of $\mathbf{F}$, and hence must be a euclidean transformation on E. Obviously:

$$\begin{aligned}(A\theta'A^{-1}).\Omega &= (A\theta'A^{-1}A).\Omega' \\ &= \Omega,\end{aligned}$$

so that $A\theta'A^{-1}$ is a member of $\mathbf{E}_{\Omega}$. We infer that:

$$A\,\mathbf{E}_{\Omega'}A^{-1} \subseteq \mathbf{E}_{\Omega}. \qquad (1)$$

For each member θ of $\mathbf{E}_{\Omega}$, $A^{-1}\theta A$ is an affine transformation on E, and:

$$\begin{aligned}(A^{-1}\theta A).\Omega' &= (A^{-1}\theta).\Omega \\ &= \Omega'.\end{aligned}$$

By condition (*), $A^{-1}\theta A$ must be a member of $\mathbf{E}_{\Omega'}$. We infer that:

$$A^{-1}\mathbf{E}_{\Omega}A \subseteq \mathbf{E}_{\Omega'},$$

so that:

$$\mathbf{E}_{\Omega} \subseteq A\,\mathbf{E}_{\Omega'}A^{-1}. \qquad (2)$$

Relations (1) and (2) yield that $\mathbf{E}_{\Omega}$ and $A\,\mathbf{E}_{\Omega'}A^{-1}$ are the same. ///

2.6 PLANE CRYSTALS

1° In this section, we shall consider the relation between plane crystals in E and ornamental subgroups of $\mathbf{E}$. Our objective is simply to offer illustrations of the 17 ornamental classes, alternative to those provided by plane ornaments.

2° Let Δ be any subset of E. One says that Δ is (uniformly) <u>discrete</u> iff there exists a positive real number a such that, for any members α and β of Δ, if $\delta(\alpha,\beta) < a$ then $\alpha = \beta$. Of course, this usage is consistent with that for subgroups of R^2. [See 1.6°.]

Let Δ be any discrete subset of E. By a <u>symmetry</u> of Δ, we shall mean any euclidean transformation θ on E under which Δ is invariant: $\theta(\Delta) = \Delta$. Clearly, the family of all symmetries of Δ is a subgroup of $\mathbf{E}$. We shall refer to it as the <u>symmetry group</u> of Δ, and denote it by $\mathbf{E}_\Delta$.

By imitating the argument supporting Proposition 10, one can prove the following proposition.

<u>Proposition 20</u>. For each (nonempty) subset Δ of E and for any cartesian coordinate mapping K for E, the translational part T^K_Δ of $\mathbf{E}^K_\Delta$ is a discrete subgroup of R^2.

We shall refer to any (nonempty) discrete subset Δ of E as a <u>plane crystal</u> provided that it satisfies the condition of two-dimensional periodicity, which is to say that, for some (and hence for any) cartesian coordinate mapping K, T^K_Δ is a lattice in R^2. Obviously, Δ is a plane crystal in E iff $\mathbf{E}_\Delta$ is an ornamental subgroup of $\mathbf{E}$.

In Figures 62 - 64, one will find examples of plane crystals. The symmetry groups of these crystals are representatives of the ornamental classes $\mathbb{D}^q_4\llcorner_\bullet$, $\mathbb{D}^q_4\llcorner\!\bullet$, and $\mathbb{D}^h_6\llcorner_\bullet$, respectively.

p4m

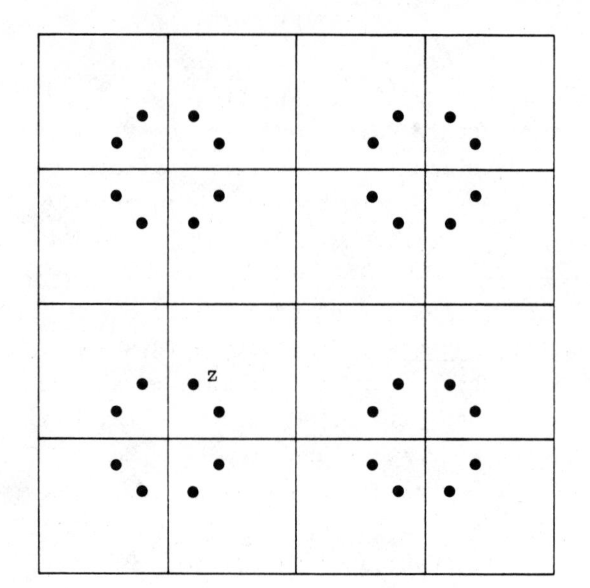

Figure 62

p4g

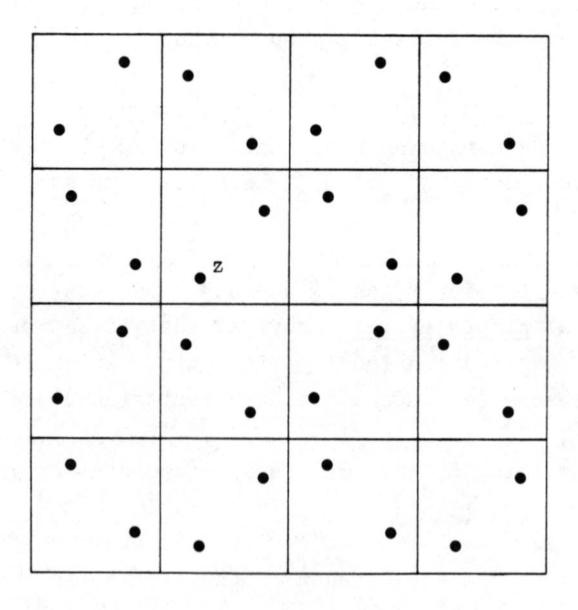

Figure 63

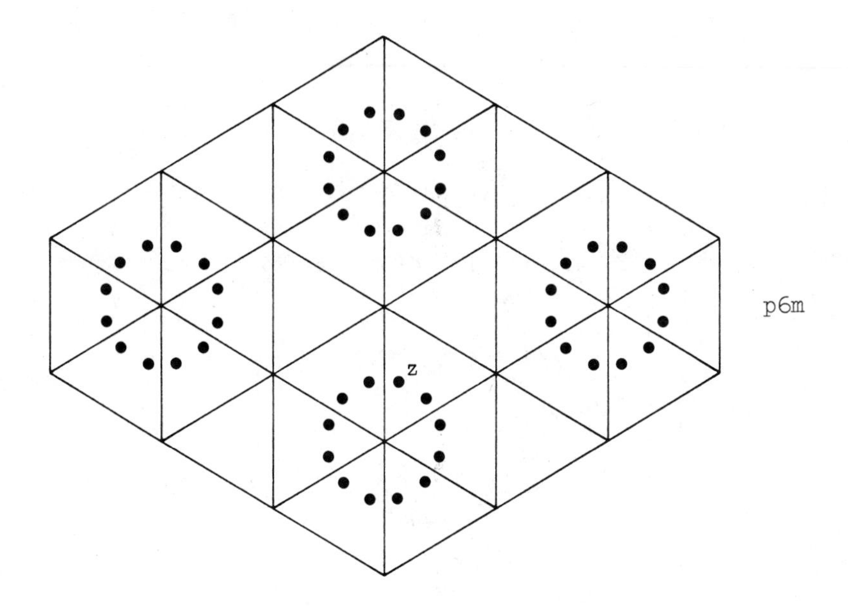

Figure 64

The quadratic and hexagonal grids underlying the figures are intended to suggest the lattice structure for the crystals.

3° Let us regard two plane crystals Δ' and Δ'' in E as <u>equivalent</u> iff their symmetry groups $\mathbf{E}_{\Delta'}$ and $\mathbf{E}_{\Delta''}$ are isomorphic. With respect to this relation, there can be at most 17 equivalence classes of plane crystals. We plan to prove that, for any ornamental subgroup $\mathbf{F}$ of $\mathbf{E}$, there exists some plane crystal Δ in E such that $\mathbf{F} = \mathbf{E}_{\Delta}$. From this result, one will be able to conclude that there are in fact 17 equivalence classes of plane crystals.

4° Let $\mathbf{F}$ be an ornamental subgroup of $\mathbf{E}$. Given any point α in E, we shall denote the orbit of α in E under $\mathbf{F}$ by $\mathbf{F}^{\alpha}$:

$$\mathbf{F}^{\alpha} = \{\theta(\alpha) : \theta \in \mathbf{F}\}.$$

Proposition 21. For any plane crystal Δ in E, there exists a finite subset J of Δ such that:

$$\Delta = \bigcup_{\alpha \in J} \mathbf{E}^{\alpha}_{\Delta}.$$

Proof. Let Δ be any plane crystal in E, and let K be any cartesian coordinate mapping for E. With respect to K, let us make the following notational substitutions:

D for $K(\Delta)$,
T for T^{K}_{Δ},
and $\mathscr{U}$ for $q(T)$.

We shall prove the proposition, by showing that there exists a finite subset $\mathring{J}$ of D such that:

$$D = \bigcup_{z \in \mathring{J}} \mathscr{U}^{z}. \tag{1}$$

Of course, by definition:

$$\mathscr{U}^{z} = \{t + z : \quad t \in T\}.$$

Let $\{u,v\}$ be a basis for the lattice T in R^2, and let P stand for the subset of R^2 consisting of all members of the form:

$$a.u + b.v,$$

where $0 \leq a < 1$ and $0 \leq b < 1$. Since D is a (uniformly) discrete subset of R^2, the intersection $D \cap P$ must be finite. Let x be any member of D, and let $\mathscr{r}$ and $\mathscr{s}$ be the real numbers for which:

$$x = \mathscr{r}.u + \mathscr{s}.v.$$

Let t and z be defined as follows:

$$t = [\mathscr{r}].u + [\mathscr{s}].v,$$

$$z = (\mathscr{r}).u + (\mathscr{s}).v.$$

Clearly, t is contained in T. Since D is invariant under $\mathscr{U}$, it follows that $x - t$ is a member of D, hence that z is a member of $D \cap P$. As a result:

$$x \in \mathscr{U}^z.$$

Taking $\mathring{J}$ to be $D \cap P$, we obtain relation (1). ///

Proposition 22. For any ornamental subgroup $\mathbf{F}$ of $\mathbf{E}$ and for any finite subset J of E, the subset

$$\Delta = \cup_{a \in J} \mathbf{F}^a$$

of E is a plane crystal. Moreover, there exists a finite subset J of E such that

$$\mathbf{F} = \mathbf{E}_\Delta,$$

where Δ is the plane crystal in E constructed from J, as just described.

Proof. Let $\mathbf{F}$ be any ornamental subgroup of $\mathbf{E}$. Let K be any cartesian coordinate mapping for E, and let us substitute:

T for T^K,
$\mathscr{U}$ for $q(T)$,
$\mathscr{F}$ for $\mathbf{F}^K$,
and $\mathscr{G}$ for $\mathscr{G}^K$.

For each member V of $\mathscr{G}$, let us introduce a member c_V of R^2 such that $[c_V, V]$ is contained in $\mathscr{F}$.

Now let J be a finite subset of E. For each member z of $K(J)$:

$$\begin{aligned}\mathscr{F}^z &= \{t + c_V + V(z) : \ t \in T, \ V \in \mathscr{G}\} \\ &= \cup_{V \in \mathscr{G}} (c_V + V(z)) + T.\end{aligned}$$

Hence, the set

$$D = \cup_{z \varepsilon K(J)} \mathscr{F}^z$$

is a finite union of cosets of T in R^2. By problem 10 in 8, the set

$$\Delta = \cup_{\alpha \varepsilon J} \mathbf{F}^{\alpha}$$

is a discrete subset of E.

By Proposition 20, T^K_Δ must be a discrete subgroup of R^2. Moreover, $\mathbf{F}$ is a subgroup of $\mathbf{E}_\Delta$, so T is a subgroup of T^K_Δ. We infer that T^K_Δ is a lattice in R^2, hence that Δ is a plane crystal in E.

In general, $\mathbf{F} \neq \mathbf{E}_\Delta$. For example, $\mathscr{F}$ might equal $\mathscr{U}$, and T might be quadratic. Moreover, K(J) might be {0}, so that:

$$\mathbf{E}^K_\Delta = T \rtimes \mathscr{O}_T.$$

In this situation, $\mathbf{F}$ would be contained in the ornamental class $\mathbb{Z}^a_1$⊥, while $\mathbf{E}_\Delta$ would be contained in $\mathbb{D}^q_4$⊥. Obviously, then, $\mathbf{F}$ and $\mathbf{E}_\Delta$ would be distinct.

However, by appropriate design of the (finite) set J, one can obtain: $\mathbf{F} = \mathbf{E}_\Delta$. To justify this assertion, we will show that:

(a) when $\mathscr{G}$ contains a rotation other than I, then there exists some member z of R^2 such that the only members of $\mathscr{C}$ under which $\mathscr{F}^z$ is invariant are the members of $\mathscr{F}$;

(b) when $\mathscr{G}$ contains no rotation other than I, then, for each member z of R^2, there exist members of $\mathscr{C}$ other than those contained in $\mathscr{F}$ under which $\mathscr{F}^z$ is invariant;

(b)' under case (b), when $\mathscr{G} = \mathscr{D}_1(X)$ for a suitable reflection X, then there exist members z' and z" of R^2 such that the only members of $\mathscr{C}$ under which $\mathscr{F}^{z'} \cup \mathscr{F}^{z''}$ is invariant are the members of $\mathscr{F}$;

(b)" under case (b), when $\mathscr{G} = \mathscr{Z}_1$, then, for any members z' and z" of R^2, there exist members of $\mathscr{C}$ other than those contained in $\mathscr{C}$ under which $\mathscr{F}^{z'} \cup \mathscr{F}^{z''}$ is invariant; but one can select members z', z^*, and z'' of R^2 such that the only members of $\mathscr{C}$ under which $\mathscr{F}^{z'} \cup \mathscr{F}^{z^*} \cup \mathscr{F}^{z''}$ is invariant are the members of $\mathscr{F}$.

From the foregoing results, one will be able to conclude that $\mathbf{F} = \mathbf{E}_{\triangle}$, where $\triangle$ is a plane crystal in E composed of as few as three orbits in E under **F.**

By Proposition 18, we may choose the cartesian coordinate mapping K so that $\mathscr{F}$ is consonant with $(T, \mathscr{G})$. Hence, we may arrange that, for every member W of $\mathscr{G}^+$, $c_W = 0$, and that, for any two members X' and X" of $\mathscr{G}^-$, $c_{X'} = c_{X''}$. [As usual, for the cases in which $\mathscr{G}$ contains no reflections, one should delete all references to reflections in $\mathscr{G}$.] Let x stand for the common value of c_X, where X is any member of $\mathscr{G}^-$. By the computation in 5.6°, we may arrange that 2.x is a member of T.

For the present, let us assume that $x = 0$, that is, that **F** has been drawn from one of the symmorphic ornamental classes.

Since T is (uniformly) discrete, we may introduce a positive real number a such that, for any members t' and t" of T, it $t' \neq t''$ then $a \leq |t' - t''|$. In turn, let us introduce a member z of R^2 satisfying the following condition:

$$(\sigma) \qquad 0 < |z| < a/4.$$

Clearly, $\mathscr{F}^z$ may be presented in the form:

$$\mathscr{F}^z = \cup_{t \in T} D_t ,$$

where, for each member t of T, D_t stands for $t + D_0$, and where D_0 stands for the (finite) set:

$$D_0 = \{V(z) : V \in \mathscr{G}\}.$$

For each member t of T, let b_t denote the "barycenter" of D_t:

$$b_t = (1/k).\Sigma_{y \in D_t} y ,$$

where k is the number of members of D_t. Clearly:

$$b_t = t + b_0 . \qquad (2)$$

Now let $\mathscr{F}^*$ denote the family of all cartesian transformations under which $\mathscr{F}^z$ is invariant. Of course, $\mathscr{F}^*$ is a subgroup of $\mathscr{C}$. Let $[s,U]$ be any member of $\mathscr{F}^*$. We contend that, for each member t' of T, there exists a member t'' of T such that $[s,U](D_{t'}) = D_{t''}$. Thus, for any members V_1' and V_2' of $\mathscr{G}$, there exist members $[t_1'',V_1'']$ and $[t_2'',V_2'']$ of $\mathscr{F}$ such that:

$$[s,U](t' + V_1'(z)) = t_1'' + V_1''(z) ,$$
$$[s,U](t' + V_2'(z)) = t_2'' + V_2''(z) .$$

By condition (σ):

$$\begin{aligned} &|t_2'' - t_1''| \\ &\quad = |(UV_2'(z) - V_2''(z)) - (UV_1'(z) - V_1''(z))| \\ &\quad \leq 4|z| \\ &\quad < a . \end{aligned}$$

Hence, $t_1'' = t_2''$. It follows that there exists a member t'' of T for which $[s,U](D_{t'}) \subseteq D_{t''}$. Since $D_{t'}$ and $D_{t''}$ have the same number of members, we may conclude that $[s,U](D_{t'}) = D_{t''}$.

We obtain:

$$\begin{aligned} [s,U](b_{t'}) &= (1/k).\Sigma_{y' \in D_{t'}} [s,U](y') \qquad (3) \\ &= (1/k).\Sigma_{y'' \in D_{t''}} y'' \\ &= b_{t''} . \end{aligned}$$

At this point, let us assume that $\mathscr{G}$ contains a rotation W, distinct from I. Moreover, let us refine our selection of z, by imposing the condition:

(τ) for each member V of $\mathscr{O}_T$, if $V \neq I$ then $V(z) \neq z$.

We plan to prove that [s,U] must be contained in $\mathscr{F}$. One will then be able to infer that $\mathscr{F}^* = \mathscr{F}$, as desired.

We have:

$$\begin{aligned} W(b_0) &= (1/k).\Sigma_{V' \varepsilon \mathscr{G}} WV'(z) \\ &= (1/k).\Sigma_{V'' \varepsilon \mathscr{G}} V''(z) \\ &= b_0. \end{aligned}$$

It follows that:

$$b_0 = 0. \tag{4}$$

Let t be the member of T for which $[s,U](D_0) = D_t$. By relations (2), (3), and (4), we have:

$$\begin{aligned} s &= [s,U](0) \\ &= [s,U](b_0) \\ &= b_t \\ &= t + b_0 \\ &= t . \end{aligned}$$

Hence, s is contained in T. In particular, it follows that the translational part of $\mathscr{F}^*$ coincides with T. By Proposition 1, U must be contained in $\mathscr{O}_T$.

Clearly, [0,U] is a member of $\mathscr{F}^*$, and $[0,U](D_0) = D_0$. Hence, we may introduce a member V of $\mathscr{G}$ such that $U(z) = V(z)$, so that $V^{-1}U(z) = z$. By condition (τ), U = V. Therefore, [s,U] is contained in $\mathscr{F}$.

In the foregoing Figures 62 and 64, and in the following Figures 65 - 72, one will find examples of plane crystals in E representing the cases considered so far. The annotations for these figures show the "seed point" z.

When $\mathscr{G}$ contains no rotation other than I, then the foregoing line of argument fails, because one cannot prove that $b_0 = 0$. The anomalous cases are those for which $\mathbf{F}$ lies in $\mathbb{Z}_1^a\llcorner$, $\mathbb{D}_1^r\llcorner$, or $\mathbb{D}_1^o\llcorner$. For the latter two cases, $\mathscr{G}$ would be $\mathscr{D}_1(X)$, where X is a suitable reflection. In these cases, one can readily verify that, for any member z of R^2, $\mathscr{F}^z$ is invariant under $[z + X(z), -X]$. Of course, $[z + X(z), -X]$ is not contained in $\mathscr{F}$. However, we may take z' to be a member of R^2 satisfying conditions (σ) and (τ), then take z'' to be $-(z' + X(z'))$. It would follow that $|z''| < |z'|$. Defining D_0 to be the set $\{z', X(z'), z''\}$, we obtain:

$$\mathscr{F}^{z'} \cup \mathscr{F}^{z''} = \cup_{t \in T} D_t .$$

Moreover, the barycenter b_0 of D_0 is 0. The foregoing argument may now be repeated, with minor adjustments, to show that the only cartesian transformations $[s,U]$ under which $\mathscr{F}^{z'} \cup \mathscr{F}^{z''}$ is invariant are the members of $\mathscr{F}$. [When showing that U must be contained in $\mathscr{G}$, one would use the fact that $U(z')$ must equal either z' or $X(z')$, because $|z''| < |z'|$.]

Figures 73 and 74 illustrate these cases.

The case of $\mathbb{Z}_1^a\llcorner$ is special. For any members z' and z'' of R^2, one can show that $\mathscr{F}^{z'} \cup \mathscr{F}^{z''}$ is invariant under $[z' + z'', -I]$. Hence, $\mathscr{F}^{z'} \cup \mathscr{F}^{z''}$ is invariant under cartesian transformations other than those contained in $\mathscr{F}$. However, one may select three members z', z^*, and z'' of R^2 satisfying conditions (σ) and (τ), having mutually distinct norms, and meeting the additional condition that $z' + z'' = -z^*$. One may then take D_0 to be the set $\{z', z^*, z''\}$, obtaining:

$$\mathscr{F}^{z'} \cup \mathscr{F}^{z^*} \cup \mathscr{F}^{z''} = \cup_{t \in T} D_t .$$

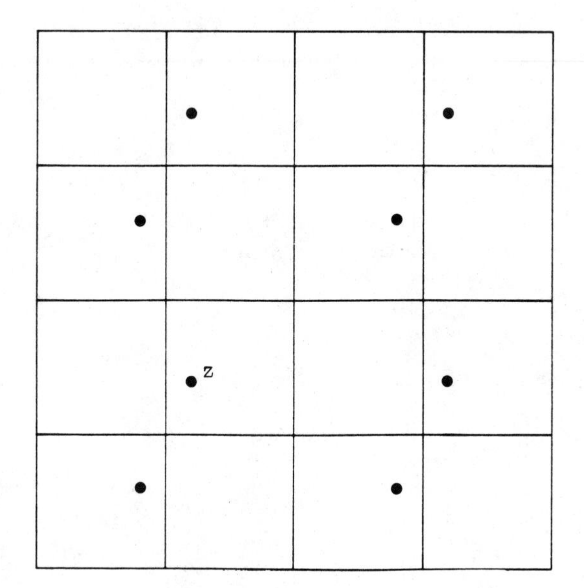

Figure 65

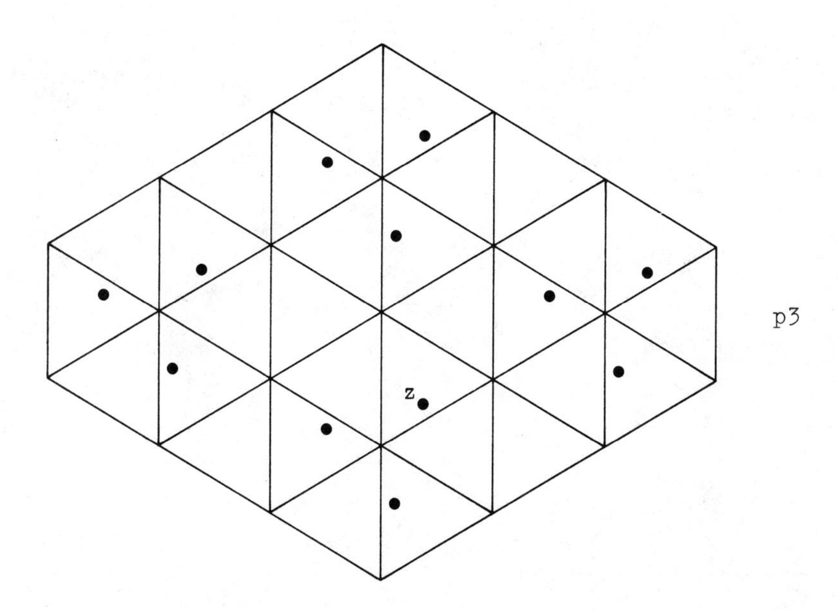

Figure 66

p4

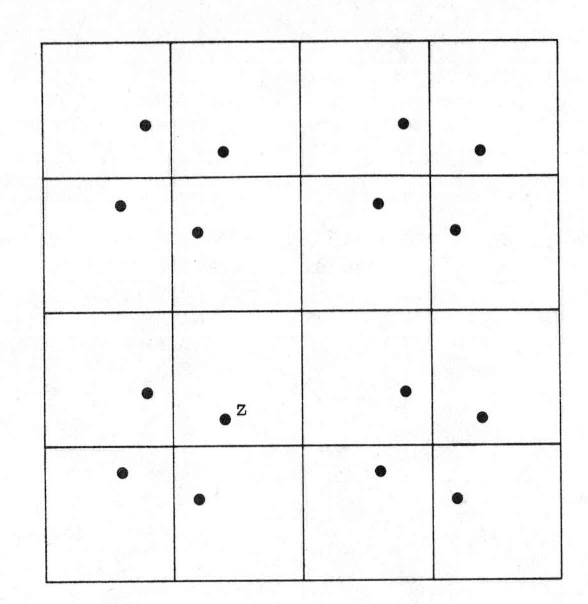

Figure 67

p6

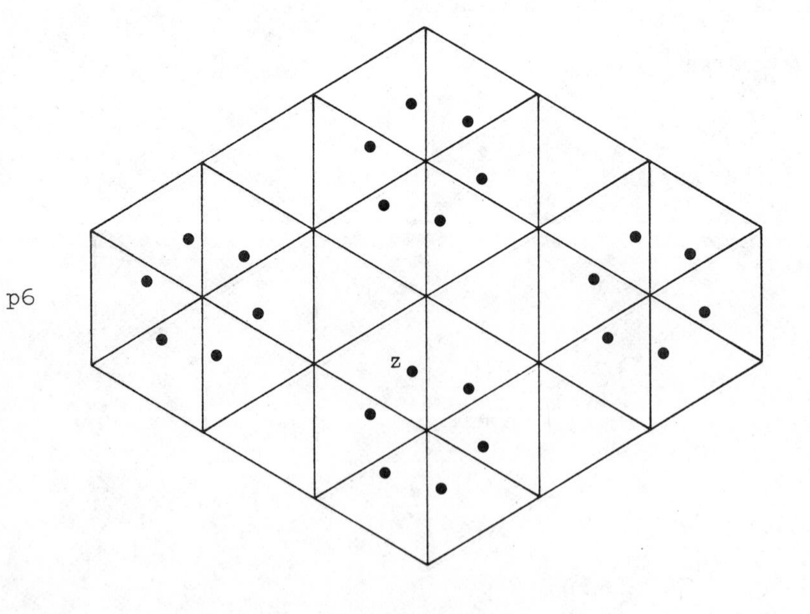

Figure 68

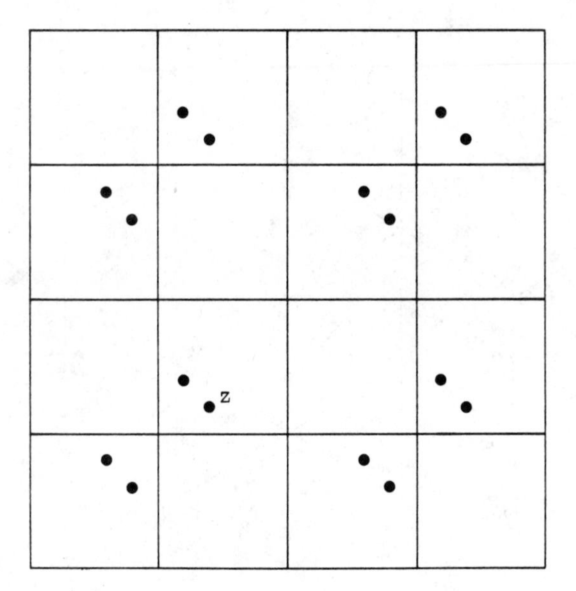

Figure 69

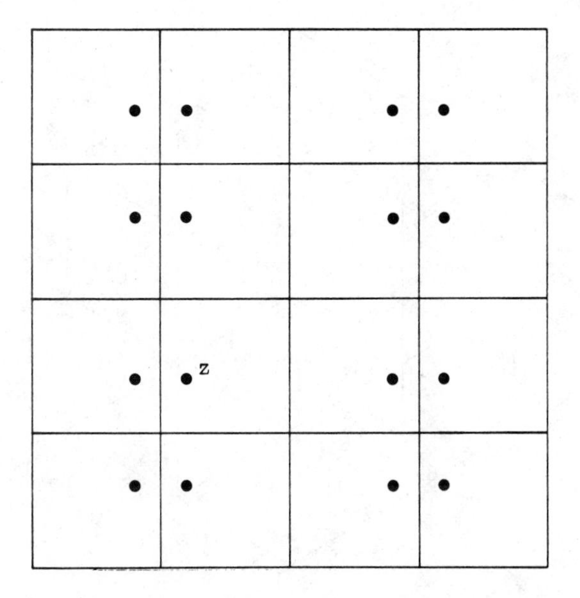

Figure 70

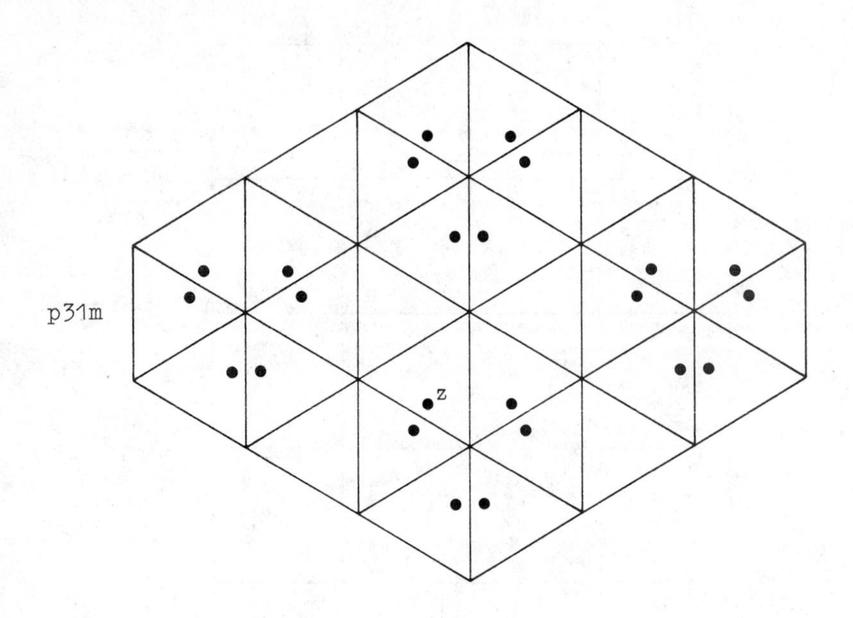

Figure 71

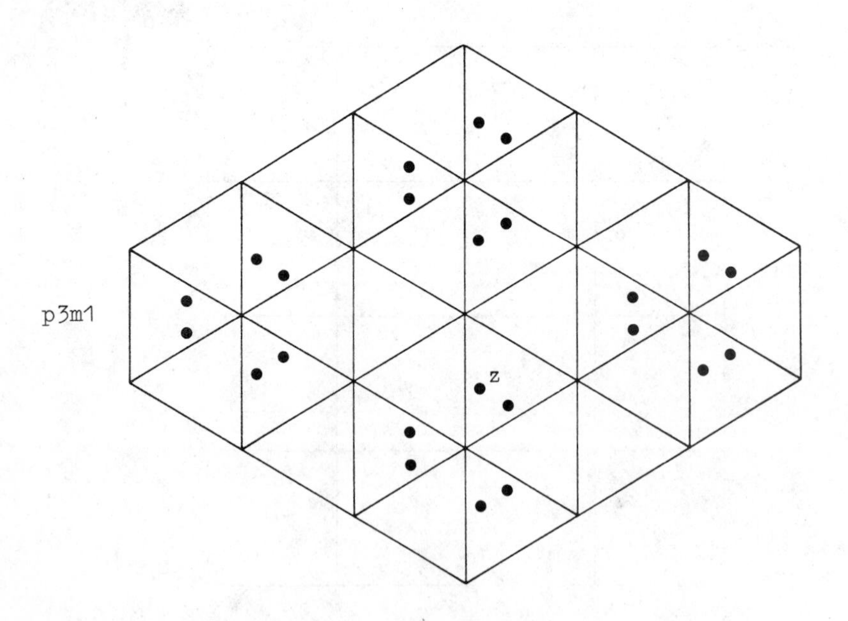

Figure 72

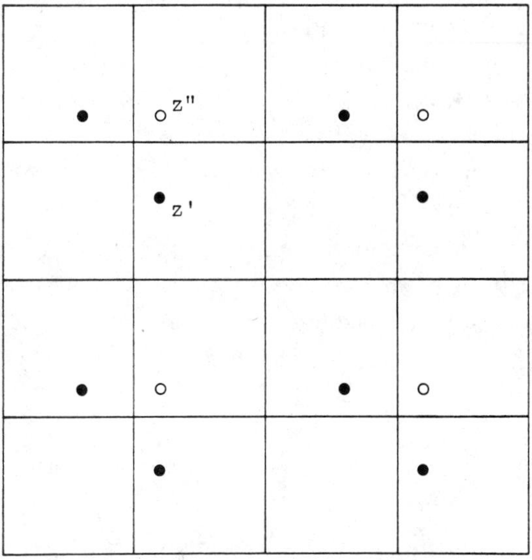

z"
z'

cm

Figure 73

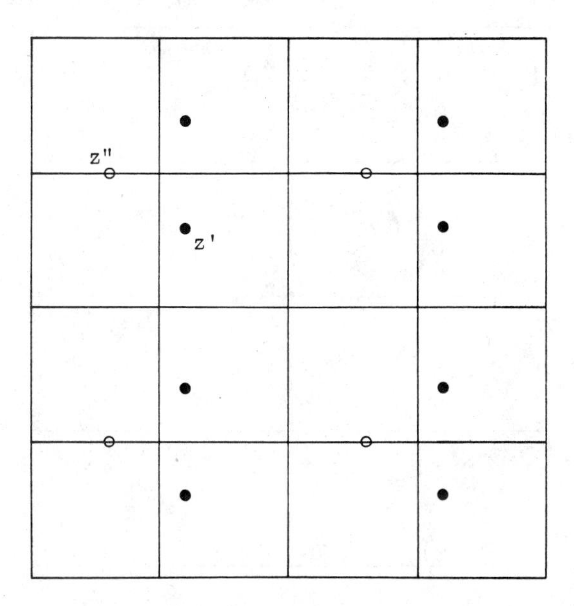

pm

Figure 74

Again, the barycenter b_0 of D_0 is 0. The foregoing argument is now effective, yielding that the only cartesian transformations under which $\mathscr{F}^{z'} \cup \mathscr{F}^{z^*} \cup \mathscr{F}^{z''}$ is invariant are the members of $\mathscr{F}$.

Figure 75 illustrates this case.

Let us turn to the nonsymmorphic cases, characterized by the condition that $x \neq 0$.

Let T^x stand for the union of T and $x + T$. By problem 10 in 8, we may introduce a positive real number a such that, for any member u' and u'' of T^x, if $u' \neq u''$ then $a \leq |u'' - u'|$. For each member z of R^2, we have:

$$^{z} = \cup_{u \in T^x} D_u ,$$

where, for each member u of T^x, D_u stands for:

p1

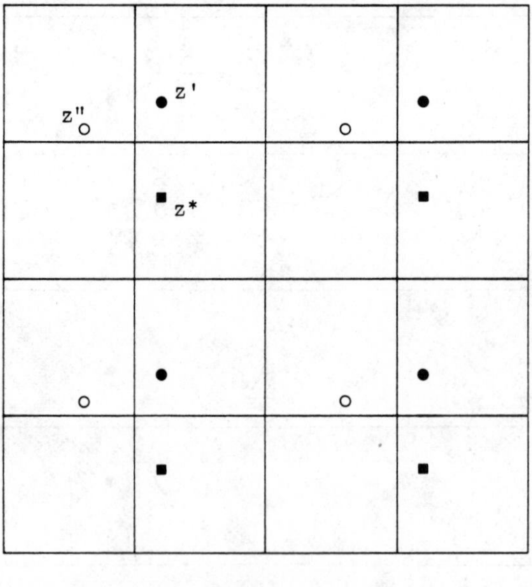

Figure 75

$u + D_0^+$ if $u \varepsilon T$,

$u + D_0^-$ if $u \varepsilon x + T$,

and where D_0^+ and D_0^- stand for the sets:

$$D_0^+ = \{W(z): \quad W \varepsilon \mathcal{G}^+\},$$

$$D_0^- = \{X(z): \quad X \varepsilon \mathcal{G}^-\}.$$

When $\mathcal{G}$ contains a rotation other than I, and when z satisfies conditions (σ) and (τ), then one can proceed as before to show that the only cartesian transformations [s,U] under which $\mathcal{F}^z$ is invariant are the members of $\mathcal{F}$. In particular, by relations (2), (3), and (4), one would find either that s is contained in T, in which case $U(D_0^+) = D_0^+$, or that s is contained in $x + T$, in which case $U(D_0^+) = D_0^-$. In the former case, U would lie in $\mathcal{G}^+$, in the latter, in $\mathcal{G}^-$; in either case, [s,U] would be contained in $\mathcal{F}$.

The foregoing Figure 63 and the following Figures 76 and 77 illustrate the nonsymmorphic cases considered so far.

The anomalous case is that for which **F** represents $\mathbb{D}_1^\circ$L. For this case, $\mathcal{G}$ would be $\mathcal{D}_1(X)$, where X is a suitable reflection. Moreover, $X(x) = x$. For each member z of R^2, $\mathcal{F}^z$ is invariant under $[z + X(z), -X]$, which is not contained in $\mathcal{F}$. However, one can select members z' and z" of R^2 such that the only members of $\mathcal{C}$ under which $\mathcal{F}^{z'} \cup \mathcal{F}^{z''}$ is invariant are the members of $\mathcal{F}$. In Figure 78, we have indicated (ad hoc) how such choices might be made.

One may also treat this case under the foregoing line of argument, that is, under the "barycenter argument," but, to do so, one must introduce three orbits in R^2 under $\mathcal{F}$. ///

pmg

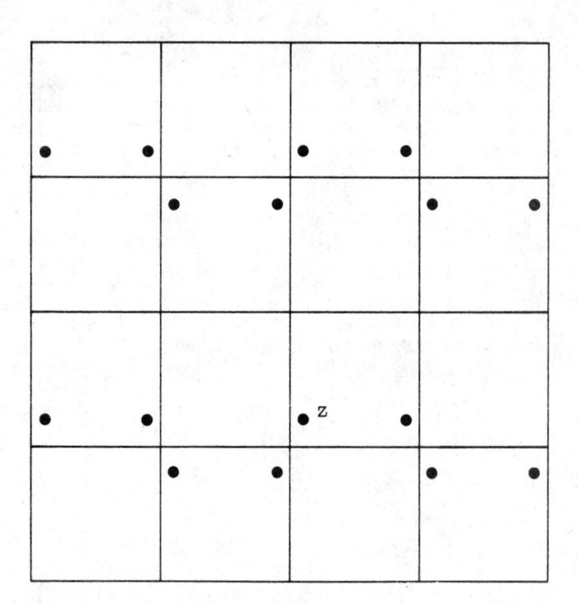

Figure 76

pgg

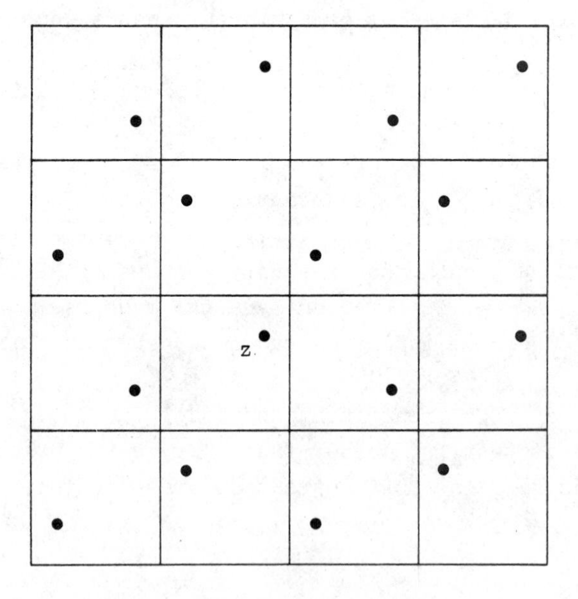

Figure 77

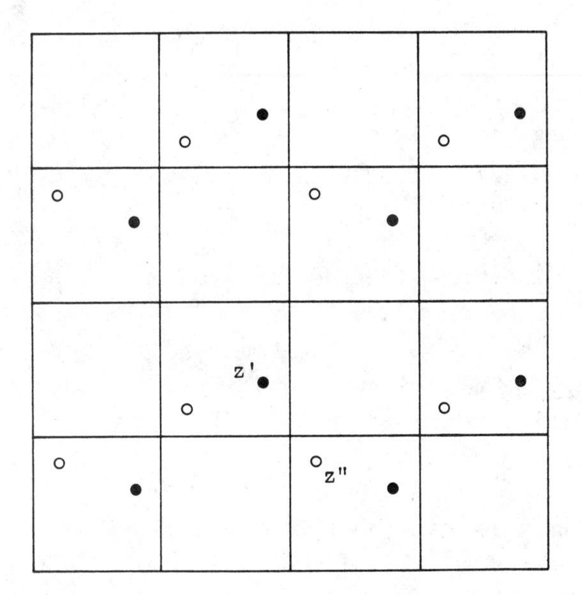

pg

Figure 78

2.7 ORNAMENTAL GROUPS

1° The object of this section is to characterize ornamental subgroups of **E** in abstract terms.

Let **F** be an ornamental subgroup of **E**, and let **U** stand for the subgroup $\mathbf{T} \cap \mathbf{F}$ of **F**, consisting of all translations in **F**. By the results of the preceding sections of this chapter, the following facts are evident:

(a) **U** is a free abelian group having rank two;
(b) **U** is a normal subgroup of **F**;
(c) the quotient group **F/U** is finite;
(d) the centralizer of **U** in **F** is **U** itself.

One obtains condition (c) by noting that, for any cartesian coordinate mapping K, the quotient group $\mathbf{F}^K/T^K$ is isomorphic to $\mathcal{G}$.

With respect to conditions (b) and (d), the subgroup

$\mathbf{U}$ of $\mathbf{F}$ is unique. [See Proposition 15.]

Now let $\mathbf{F}$ be an arbitrary group. By a trellis in $\mathbf{F}$, we shall mean any subgroup $\mathbf{U}$ of $\mathbf{F}$ which satisfies the foregoing conditions (a), (b), (c), and (d). We shall say that $\mathbf{F}$ is an ornamental group provided that there exists a trellis in $\mathbf{F}$.

2° The following theorem, due to H. Zassenhaus, characterizes ornamental subgroups of $\mathbf{E}$ in abstract terms.

THEOREM 2. For any group $\mathbf{F}$, $\mathbf{F}$ is isomorphic to an ornamental subgroup of $\mathbf{E}$ iff $\mathbf{F}$ is ornamental.

Proof. Obviously, when $\mathbf{F}$ is isomorphic to an ornamental subgroup of $\mathbf{E}$, then it is ornamental.

Let us assume that $\mathbf{F}$ is ornamental. We must prove that there exist an ornamental subgroup $\mathbf{F}_o$ of $\mathbf{E}$ and an isomorphism carrying $\mathbf{F}$ to $\mathbf{F}_o$. In fact, we shall prove that:

(*) there exist an arithmetic group $(T, \mathscr{G})$ and an injective homomorphism ω carrying $\mathbf{F}$ to $\mathscr{C}$ such that the range of ω is compatible with $(T, \mathscr{G})$.

Clearly, this will be sufficient.

Let $\mathbf{U}$ be a trellis in $\mathbf{F}$, and let us consider the sequence:

$$\mathbf{U} \xrightarrow{\iota} \mathbf{F} \xrightarrow{\pi} \mathbf{G}, \tag{1}$$

where ι is the natural injection mapping, where $\mathbf{G}$ stands for the quotient group $\mathbf{F}/\mathbf{U}$, and where π is the natural projection mapping. In this context, ι is injective, π is surjective, and the range of ι equals the kernel of π. Moreover, $\mathbf{U}$ is abelian. Using these facts, one can show that there is an implicitly defined action of $\mathbf{G}$ by automorphisms on $\mathbf{U}$. To this end, let g be any member of $\mathbf{G}$ and let h be any member of $\mathbf{U}$. For any members f' and f''

of $\mathbf{F}$, if $\pi(f') = g$ and $\pi(f'') = g$ then:

$$\begin{aligned} f''\iota(h)f''^{-1} &= f'f'^{-1}f''\iota(h)f''^{-1}f'f'^{-1} \\ &= f'\iota(h)f'^{-1}, \end{aligned}$$

because $f'^{-1}f''$ is contained in $\iota(\mathbf{U})$, and $\mathbf{U}$ is abelian. Hence, we may unambiguously define a member $g.h$ of $\mathbf{U}$, by the following relation:

$$\iota(g.h) = f\iota(h)f^{-1},$$

where f is any member of $\mathbf{F}$ for which $\pi(f) = g$. Clearly, for each member g of $\mathbf{G}$, the mapping

$$\alpha_g(h \longrightarrow g.h)$$

carrying $\mathbf{U}$ to itself, is an automorphism of $\mathbf{U}$. Moreover, the mapping

$$\alpha(g \longrightarrow \alpha_g)$$

carrying $\mathbf{G}$ to the automorphism group of $\mathbf{U}$, is a homomorphism. This homomorphism presents the action of $\mathbf{G}$ by automorphisms on $\mathbf{U}$. By applying condition (d), one can readily verify that α is injective.

Now let us focus upon the initial assertion (*).

We shall first show that the ordered pair $(\mathbf{U},\mathbf{G})$ can be interpreted as an arithmetic group $(T, \mathscr{G})$, in such a way that the implicitly defined action of $\mathbf{G}$ on $\mathbf{U}$ corresponds to the natural action of $\mathscr{G}$ on T. Then we shall construct an injective homomorphism ω carrying $\mathbf{F}$ to $\mathscr{C}$ for which $\omega(\mathbf{F})$ is compatible with $(T, \mathscr{G})$.

By condition (a), we may introduce a lattice T^o in R^2 and an isomorphism τ^o carrying $\mathbf{U}$ to T^o. For each member g of $\mathbf{G}$, $\tau^o\alpha_g\tau^{o-1}$ is an automorphism of T^o; hence, there exists a member L^o_g of $\mathscr{L}$ such that, for any member t^o of T^o:

$$L^o_g(t^o) = (\tau^o a_g \tau^{o-1})(t^o). \tag{2}$$

Clearly, the set

$$\mathscr{G}^o = \{L^o_g : g \in \mathbf{G}\}$$

is a finite subgroup of the group $\mathscr{L}$. By problem 4 in 1.4, we may introduce a member L of $\mathscr{L}$ such that, for each member g of $\mathbf{G}$, $LL^o_gL^{-1}$ is orthogonal. Let L_g stand for $LL^o_gL^{-1}$. Let T stand for $L(T^o)$, and let τ be the isomorphism

$$\tau(h \longrightarrow L(\tau^o(h)))$$

carrying $\mathbf{U}$ to T. Let ϕ be the homomorphism

$$\phi(g \longrightarrow L_g)$$

carrying $\mathbf{G}$ to $\mathscr{O}$, and let $\mathscr{G}$ stand for the range of ϕ. Of course, since a is injective, it follows that ϕ is injective.

Clearly, T is a lattice in R^2, and $\mathscr{G}$ is a finite subgroup of $\mathscr{O}$. Moreover, by relation (2):

$$L_g(t) = (\tau a_g \tau^{-1})(t), \tag{3}$$

where g is any member of $\mathbf{G}$ and where t is any member of T. As a result, $L_g(T) = T$. Hence, $(T, \mathscr{G})$ is an arithmetic group.

Now let us replace the sequence (1) by the following:

$$T \xrightarrow{\ q^*\ } \mathbf{F} \xrightarrow{\ p^*\ } \mathscr{G}, \tag{4}$$

where q^* stands for $\iota\tau^{-1}$ and p^*, for $\phi\pi$. Applying relation (3), we find that the action of $\mathscr{G}$ by automorphisms on T, defined implicitly by the sequence (4), coincides with the natural action. That is:

$$fq^*(t)f^{-1} = q^*(V(t)), \tag{5}$$

where V is any member of $\mathcal{G}$, where f is any member of for which $p^*(f) = V$, and where t is any member of T.

The first step of the proof is now complete. Let us advance to the construction of ω.

Since p^* is surjective, we may introduce a "cross-section" of p^*, that is, a mapping c carrying $\mathcal{G}$ to $\mathbf{F}$ such that p^*c is the identity mapping on $\mathcal{G}$. We may (and shall) assume that $c(I)$ is the identity element in $\mathbf{F}$. For each member f of $\mathbf{F}$, let V stand for $p^*(f)$. Clearly, $p^*(fc(V)^{-1}) = I$, so there exists a (unique) member t of T such that $q^*(t) = fc(V)^{-1}$. Hence:

$$f = q^*(t)c(V).$$

We shall refer to t and V as the "translational" and "linear" parts of f, relative, of course, to c.

Now let k be any mapping carrying $\mathcal{G}$ to R^2, for which:

$$k(I) = 0. \tag{6}$$

Combining this mapping with the decomposition of the members of $\mathbf{F}$ into their translational and linear parts, one may define the mapping

$$\omega(f \longrightarrow [t + k(V), V])$$

carrying $\mathbf{F}$ to $\mathcal{C}$. Clearly, ω is injective. We contend that, by proper design of the mapping k, ω will prove to be a homomorphism. From relation (6), it will then follow that $\omega(\mathbf{F})$ is compatible with $(T, \mathcal{G})$.

Let us develop conditions upon k, sufficient to insure that ω is a homomorphism.

Thus, let t' and t'' be any members of T, and let V' and V'' be any members of $\mathcal{G}$. Let f' stand for $q^*(t')c(V')$ and f'', for $q^*(t'')c(V'')$. We have:

$$q^*(t')c(V')q^*(t'')c(V'') \tag{7}$$
$$= q^*(t')c(V')q^*(t'')c(V')^{-1}c(V')c(V'')c(V'V'')^{-1}c(V'V'').$$

Clearly:

$$p^*(c(V')c(V'')c(V'V'')^{-1}) = I.$$

Let $m(V',V'')$ be the (unique) member of T such that:

$$q^*(m(V',V'')) = c(V')c(V'')c(V'V'')^{-1}.$$

Now relations (5) and (7) yield:

$$f'f'' = q^*(t' + V'(t'') + m(V',V''))c(V'V''). \tag{8}$$

The last relation presents the operation of multiplication on **F**, in a form congenial to our subsequent calculations.

In order that ω be a homomorphism, one must have:

$$\omega(f'f'') = [t' + k(V'), V'][t'' + k(V''), V''].$$

By relation (8), the foregoing condition is equivalent to the following:

$$m(V',V'') + k(V'V'') = k(V') + V'(k(V'')), \tag{9}$$

where V' and V'' are any members of $\mathcal{G}$.

Thus, we shall be able to complete the proof of assertion (*), by constructing a mapping k carrying $\mathcal{G}$ to R^2 which meets condition (9). [Incidentally, by taking both V' and V'' to be I, one can readily see that condition (9) implies condition (6).]

Let V', V'', and V^o be any members of $\mathcal{G}$. We have:

$$c(V')c(V'')c(V^o)c(V'V''V^o)^{-1}$$
$$= c(V')c(V'')c(V'V'')^{-1}c(V'V'')c(V^o)c(V'V''V^o)^{-1}$$
$$= q^*(m(V',V''))q^*(m(V'V'',V^o)) ;$$

and:

$$\begin{aligned} &c(V')c(V'')c(V^o)c(V'V''V^o)^{-1} \\ &\quad = c(V')c(V'')c(V^o)c(V''V^o)^{-1}c(V''V^o)c(V'V''V^o)^{-1} \\ &\quad = c(V')q^*(m(V'',V^o))c(V')^{-1}c(V')c(V''V^o)c(V'V''V^o)^{-1} \\ &\quad = q^*(V'(m(V'',V^o)))q^*(m(V',V''V^o)) \; . \end{aligned}$$

Hence:

$$\begin{aligned} &m(V',V'') + m(V'V'',V^o) \\ &\quad = V'(m(V'',V^o)) + m(V',V''V^o) \; . \end{aligned}$$

Summing over V^o, we find that:

$$\begin{aligned} &n.m(V',V'') + \Sigma_{V^o \varepsilon \mathcal{G}} m(V'V'',V^o) \\ &\quad = V'(\Sigma_{V^o \varepsilon \mathcal{G}} m(V'',V^o)) + \Sigma_{V^o \varepsilon \mathcal{G}} m(V',V''V^o) \\ &\quad = \Sigma_{V^o \varepsilon \mathcal{G}} m(V',V^o) + V'(\Sigma_{V^o \varepsilon \mathcal{G}} m(V'',V^o)), \end{aligned}$$

where n is the order of $\mathcal{G}$. With reference to condition (9), we are led to define k as follows:

$$k(V) = (1/n).\Sigma_{V^o \varepsilon \mathcal{G}} m(V,V^o) \; ,$$

for any member V of $\mathcal{G}$. By design, k satisfies condition (9). ///

2.8 PROBLEMS

1. In several of the following problems, we shall make use of the concept of polygon in E. We shall treat polygons as a special kind of tile; hence, we intend that a polygon contain not only the points of a certain simple closed (continuous) curve, composed of a finite number of line segments, but also the points interior to that curve. We shall make use of the corresponding

measures of perimeter and area as well, and of the idea of convexity. One should have no difficulty devising formal definitions of these terms, in a manner compatible with the foregoing development of euclidean geometry.

2. Let τ' and τ'' be any tiles in E. One says that τ' and τ'' are <u>congruent</u> iff there exists a euclidean transformation θ on E such that $\tau'' = \theta(\tau')$.
Now let Ω be any mosaic in E. One says that Ω is <u>monohedral</u> iff, for any tiles τ' and τ'' in Ω, τ' and τ'' are congruent. Given a tile τ and a monohedral mosaic Ω in E, one says that τ is a <u>prototile</u> for Ω iff τ is congruent to some (and hence to any) tile in Ω.
Let τ be any tile in E. One may inquire whether:

(*) there exists a monohedral mosaic Ω for which τ is a prototile.

Not much is known about the general question. However, for the case in which τ is a polygon, significant results are quite accessible. In this problem, and in several of the following, we shall indicate some of these results. The treatise by Grünbaum and Shephard provides a substantial discussion of the matter.
Prove that if τ is either a triangle or a quadrilateral then statement (*) is true.
Find examples of a (convex) pentagon and of a (convex) hexagon for which statement (*) is false. Try to describe all convex pentagons and all convex hexagons for which the statement is true. For this effort, one may wish to consult the review articles by M. Gardner and D. Schattschneider. The results for hexagons are known, and are "relatively easy" to obtain, while the results for pentagons are only partially known, and are already quite complicated.
The next problem bears upon the case in which τ has seven or more edges.

3. Let Ω be a mosaic in E, composed entirely of polygons. Prove that at least one of the following four statements must be false:

(a) each tile in Ω has seven or more edges;
(b) each tile in Ω is convex;
(c) there exists a positive real number a such that, for each tile τ in Ω, the area of τ is greater than a;
(d) there exists a positive real number b such that, for each tile τ in Ω, the perimeter of τ is less than b.

Conversely, for any three of the foregoing statements, construct an example of a mosaic Ω for which the three are true.
To develop a proof that at least one of the statements is false, one should consult the article by I. Niven.

4. To set the present problem in context, one should refer to the review article by J. Milnor, dealing with the eighteenth problem of Hilbert.
Let $\mathbf{F}$ be any subgroup of $\mathbf{E}$. One says that a subset D of E is a <u>fundamental domain</u> for $\mathbf{F}$ iff:

(i) for each point α in E, there exists a member θ of $\mathbf{F}$ such that α is contained in $\theta(D)$;
(j) for each member θ of $\mathbf{F}$ (other than the identity), $\text{int}(D) \cap \text{int}(\theta(D))$ is empty.

Prove that $\mathbf{F}$ is an ornamental subgroup of $\mathbf{E}$ iff there exists a fundamental domain D for $\mathbf{F}$ which is compact and which has nonempty interior.
In this situation, the family

$$\Omega = \{\theta(D) : \theta \in \mathbf{F}\}$$

would be a plane ornament in E. In general, however,

the symmetry group of Ω will be larger than **F**.
With reference to problem 2, it is interesting to note that there exist polygons in E which can serve as prototiles for monohedral plane ornaments, but which cannot serve as fundamental domains for ornamental subgroups of **E**. In Figure 79, one will find an example of such a polygon, designed by H. Heesch.

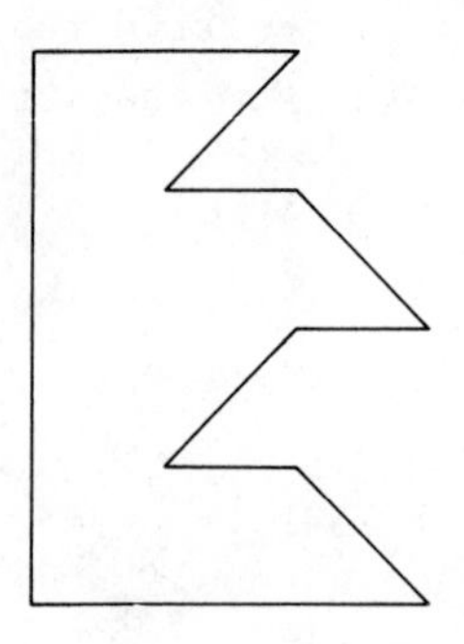

Figure 79

5. Let Ω be a plane ornament in E. Let a be a point in E, and let k be a positive integer. One says that a is a <u>k-rotocenter</u> for Ω iff the (finite) subgroup of **E** consisting of all rotations in $\mathbf{E}_{\Omega}$ having center a, has order k. [Of course, k would have to be 1, 2, 3, 4, or 6.]
For each of the illustrative plane ornaments presented in 5.6°, locate the k-rotocenters.
Show that a given plane ornament Ω in E admits a 6-rotocenter iff it admits a 2-rotocenter and a 3-rotocenter. Show that Ω cannot admit both 3- and 4-rotocenters.

6. Let **F** be an ornamental subgroup of **E**, contained in one of the nonsymmorphic ornamental classes. Show that if **F** is contained in either $\mathbb{D}^{\circ}_{1}\llcorner_{\bullet}$ or $\mathbb{D}^{\circ}_{2}\llcorner_{\bullet}$ then

there are no reflections in $\mathbf{F}$, though there are glide-reflections. Show that if $\mathbf{F}$ is contained in either $\mathbb{D}_2^o$L• or $\mathbb{D}_4^q$L• then there are both reflections and glide-reflections in $\mathbf{F}$. For the latter two cases, locate the reflections, relative to the appropriate plane ornaments depicted in 5.6^o.

Glide-reflections are by definition the composition of a translation and a reflection, the axis of the given reflection being invariant under the given translation. [See $1.3.5^o$.] When one calls attention to the fact that $\mathbf{F}$ contains a glide-reflection, one intends to say that the component translation and reflection are not themselves contained in $\mathbf{F}$. Otherwise, there would be little reason to take special note of it.

7. Let τ be a polygon in E, having m edges. Let

$$v_1, \ v_2, \ \dots, \ v_m, \ v_{m+1}$$

be an enumeration of the vertices of τ, where $v_{m+1} = v_1$. For each integer i $(1 \leq i \leq m)$, let J_i denote the edge of τ containing the vertices v_i and v_{i+1}. Let X_i denote the reflection on E, the axis of which includes J_i. Finally, let $\mathbf{F}$ be the subgroup of $\mathbf{E}$ generated by the reflections

$$X_1, \ X_2, \ \dots, \ X_m.$$

Let us say that the polygon τ is <u>reflective</u> iff τ is a fundamental domain for $\mathbf{F}$. [See problem 4.] Of course, when τ is reflective, then the family

$$\Omega = \{\theta(\tau) : \theta \in \mathbf{F}\}$$

of tiles, is a plane ornament in E.

The various members of $\mathbf{F}$ have the form:

$Y_{\langle 1,2,\ldots,n-1,n\rangle} = (Y_1Y_2\cdots Y_{n-1})Y_n(Y_{\langle 1,2,\ldots,n-1\rangle}\cdots Y_{\langle 1,2\rangle}Y_{\langle 1\rangle})^{-1}$

$Y_1Y_2 \ldots Y_n$,

where, for each integer j $(1 \leq j \leq n)$, there exists some integer $i(j)$ $(1 \leq i(j) \leq m)$ for which $X_{i(j)} = Y_j$. Prove that $Y_1Y_2 \ldots Y_n$ can be characterized as a "moving product" of reflections, determined by the successive positions of certain edges of τ:

$$Y_1Y_2 \ldots Y_n = Y_{<1,2,\ldots,n>} \ldots Y_{<1,2>} Y_{<1>},$$

where, for each integer j $(1 \leq j \leq n)$, the reflection $Y_{<1,2,\ldots,j>}$ on E is inductively defined, as follows:

$j = 1$: $Y_{<1>} = Y_1$;

$j > 1$: the axis of $Y_{<1,2,\ldots,j>}$ includes the line segment $Y_{<1,2,\ldots,j-1>}(J_{i(j)})$.

As a result, one may say, informally, that the polygon τ is reflective iff the polygons which arise by repeated reflections of (the template) τ in its sides, compose a mosaic, in fact, a plane ornament in E. For an example, see Figure 56.
Prove that if τ is reflective then it must be convex.
Now let us assume that τ is reflective. For each integer i $(1 \leq i \leq m)$, let μ_i be the measure of the angle at the vertex v_i of τ, and let c_i stand for $2\pi/\mu_i$. Of course:

(a) $\Sigma_{i=1}^{m} (1/c_i) = (m-2)/2$.

Show that:

(b) for each integer i $(1 \leq i \leq m)$, c_i is an integer; in particular, $3 \leq c_i$.

Under condition (b), find all solutions of relation

(a). By applying the crystallographic restriction and the results of problem 5, show that certain of the solutions cannot arise from a reflective polygon. The remaining solutions are the following:

m	c_1	c_2	c_3	c_4	c_5	c_6
3	6	6	6			
3	4	8	8			
3	4	6	12			
3	3	12	12			
4	4	4	4	4		
4	3	3	6	6		
6	3	3	3	3	3	3

By interpreting this table, describe all reflective polygons in E. The following condition will be useful:

(c) for each integer i $(1 \leq i \leq m)$, if c_i is odd then the two edges of τ containing v_i must have the same length.

There are in fact essentially seven reflective polygons, not simply because there are seven surviving solutions of relation (a) but because each such solution yields (within scale transformation) one and only one reflective polygon.
For the sixth case, there are two interpretations, as suggested in Figure 80. However, one can easily see that the second is untenable. The fifth case is exceptional, in that it represents all rectangles.

8. Let T' and T'' be lattices in R^2, and let L be a linear mapping carrying R^2 to itself. Let $\{u',v'\}$ and $\{u'',v''\}$ be bases for T' and T'', respectively.

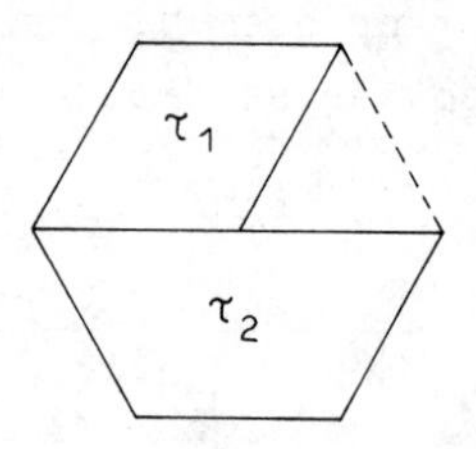

Figure 80

Let a, b, c, and d be the real numbers for which:

$L(u') = a.u'' + b.v''$,

$L(v') = c.u'' + d.v''$.

Prove that the following statements are mutually equivalent:

(1) $L(T') = T''$;

(2) $\{L(u'), L(v')\}$ is a basis for T'';

(3) a, b, c, and d are integers, and $ad - bc = \pm 1$.

9. Let T be a lattice in R^2, and let T_o be a subgroup of T. Prove that T_o is a lattice in R^2 iff the index of T_o in T is finite.

10. Let T be a lattice in R^2, and let J be a finite subset of R^2. Let U stand for the subset

$$U = \{x + t : x \in J, t \in T\}$$

of R^2. Prove that U is (uniformly) discrete. [See 6.2^o.]

11. Let Z^2 be realized as the <u>standard lattice</u> in R^2:

$Z^2 = \{m.r + n.s : m,n \in Z\}$.

Let $\mathcal{Z}$ stand for the subgroup of $\mathcal{L}$ consisting of all members L for which $L(Z^2) = Z^2$. By problem 8, the members of $\mathcal{Z}$ can be presented in matrix form, relative to $\{r,s\}$, as follows:

$$\begin{bmatrix} a & c \\ b & d \end{bmatrix},$$

where a, b, c, and d are integers and where $ad - bc = \pm 1$.

Let us use the term <u>bieberbach group</u> to refer to any subgroup $\mathcal{B}$ of the affine group $\mathcal{A}$ on R^2, for which the translational part equals Z^2 and the linear part $\mathcal{H}$ is finite. Of course, $\mathcal{H}$ must be included in $\mathcal{Z}$.

Let K be any cartesian coordinate mapping for E. Prove that, for every ornamental subgroup **F** of **E**, there exists an affine transformation A on R^2 such that $A\,\mathbf{F}^K A^{-1}$ is a bieberbach group. Prove that, for every bieberbach group $\mathcal{B}$, there exist an ornamental subgroup **F** of **E** and an affine transformation A on R^2 such that $A\,\mathbf{F}^K A^{-1}$ equals $\mathcal{B}$. [Use problem 4 in 1.4, or else THEOREM 2.] Prove that, for any two bieberbach groups $\mathcal{B}'$ and $\mathcal{B}''$, $\mathcal{B}'$ and $\mathcal{B}''$ are isomorphic iff they are conjugate in $\mathcal{A}$. [Use THEOREM 1.] Of course, every affine transformation $[t,L]$ for which $\mathcal{B}'' = [t,L]\mathcal{B}'[t,L]^{-1}$, must have the property that L is contained in $\mathcal{Z}$.

Conclude that the 17 ornamental classes derived in section 5 can be identified as the equivalence classes of bieberbach groups, under the relation of conjugacy in $\mathcal{A}$.

In the same manner, show that the 13 arithmetic classes derived in section 4 can be identified as the conjugacy classes of finite subgroups of $\mathcal{Z}$.

12. Let A be an affine transformation on E. Prove that

if A is a similarity transformation then:

(*) for each mosaic Ω in E, $\mathbf{E}_{A.\Omega} = A\,\mathbf{E}_{\Omega}\,A^{-1}$.

[See problem 5 in 1.4.] Conversely, prove that if A satisfies condition (*) then it must be a similarity transformation.

13. Let Ω be any mosaic in E. We shall say that Ω is <u>transitive</u> provided that $\mathbf{E}_{\Omega}$ acts transitively on Ω:

(t) for any tiles τ' and τ'' in Ω, there exists some member θ of $\mathbf{E}_{\Omega}$ such that $\theta(\tau') = \tau''$.

We shall say that Ω is <u>free</u> provided that $\mathbf{E}_{\Omega}$ acts freely on Ω:

(f) for any tile τ in Ω and for any member θ of $\mathbf{E}_{\Omega}$, if $\theta(\tau) = \tau$ then θ is the identity transformation.

Show that each of the 17 plane ornaments exhibited in 5.6° is free. Verify that 14 of them are transitive while three are not.
With respect to the condition of transitivity, the exceptional cases are the plane ornaments illustrating the ornamental classes $\mathbb{D}_2^{\circ}\!\llcorner$, $\mathbb{D}_3^{''}\!\llcorner$, and $\mathbb{D}_4^{q}\!\llcorner$. For each of these classes, construct some other representative plane ornament which does satisfy the condition of transitivity.
Now let Ω be a plane ornament for which $\mathbf{E}_{\Omega}$ lies in one of the foregoing three classes, and let Ω have the property that, for each tile τ in Ω, $\mathrm{int}(\tau)$ is connected. Prove that Ω cannot be both transitive and free.

14. For each mosaic Ω in E, let $\mathbf{A}_{\Omega}$ denote the subgroup of $\mathbf{A}$ consisting of all affine symmetries of Ω:

$$\mathbf{A}_\Omega = \{A \in \mathbf{A}: \quad A.\Omega = \Omega\}.$$

To distinguish $\mathbf{A}_\Omega$ and $\mathbf{E}_\Omega$, we shall refer to the former as the <u>affine</u>, and to the latter as the <u>euclidean</u> symmetry group of Ω.
Obviously, $\mathbf{E}_\Omega$ is a subgroup of $\mathbf{A}_\Omega$. Design a plane ornament Ω in E for which $\mathbf{E}_\Omega \neq \mathbf{A}_\Omega$.
Verify that, for each of the 17 plane ornaments depicted in 5.6°, the affine and euclidean symmetry groups are in fact the same.

15. Show that, for two mosaics in general, the corresponding symmetry groups may be isomorphic, but not conjugate in $\mathbf{A}$. [See Figures 28 and 29.]

16. Let A be any affine transformation on E. For each cartesian coordinate mapping K, let t^K and V^K stand for the translational and linear parts of A^K, so that $A^K = [t^K, V^K]$. Prove that, for any cartesian coordinate mappings K' and K":

$$0 < \det(V^{K'}) \quad \text{iff} \quad 0 < \det(V^{K''}).$$

Hence, one may say that the affine transformation A is <u>positive</u> provided that, for some (and hence for any) cartesian coordinate mapping K, $\det(V^K)$ is positive.
Let $\mathbf{A}^+$ be the family of all positive affine transformations on E. Clearly, $\mathbf{A}^+$ is a subgroup of $\mathbf{A}$ having index 2 in $\mathbf{A}$.
Now let $\mathbf{F}'$ and $\mathbf{F}''$ be any ornamental subgroups of $\mathbf{E}$. Prove that $\mathbf{F}'$ and $\mathbf{F}''$ are conjugate in $\mathbf{A}$ iff they are conjugate in $\mathbf{A}^+$. [Simply review, and revise where necessary the proof of THEOREM 1.] Conclude that the classification of ornamental subgroups of $\mathbf{E}$ by the relation of conjugacy in $\mathbf{A}^+$ coincides with the classification by the relation of isomorphism.

17. Let Ω' and Ω'' be any plane ornaments in E. Among all conditions appropriate to the classification of plane ornaments, the relation of isomorphism between the corresponding symmetry groups is surely the weakest. In contrast, the following condition is surely the strongest:

(s) there exists a similarity transformation S on E such that $S.\Omega'$ and Ω'' are the same.

Under this condition, two plane ornaments would be equivalent iff the one could be obtained from the other by suitable transformations of position, orientation, and scale. However, as a principle of classification, condition (s) is probably too restrictive: a description of all equivalence classes following that condition would hardly be distinguishable from a description of all plane ornaments.
The following natural modification of condition (s) proves to be useful:

(s)' there exists a similarity transformation S on E such that the symmetry groups of $S.\Omega'$ and Ω'' are equal.

By problem 12, this condition is equivalent to the condition that:

(s)" $\mathbf{E}_{\Omega'}$ and $\mathbf{E}_{\Omega''}$ are conjugate in the similarity group **S**.

Find examples of plane ornaments in E which are equivalent under condition (s)" but not under condition (s).
Motivated by the foregoing discussion, determine the equivalence classes of ornamental subgroups of **E** following the relation of conjugacy in **S**. [One may proceed by refining the classification following the rela-

tion of conjugacy in **A**, obtained in 5.6°.]

18. Let us say that two plane ornaments Ω' and Ω'' in E are <u>strongly</u> equivalent provided that:

(+) there exists an isomorphism Φ carrying $\mathbf{E}_{\Omega'}$ to $\mathbf{E}_{\Omega''}$ and a bijective mapping H carrying Ω' to Ω'' such that, for each member θ of $\mathbf{E}_{\Omega'}$ and for any tile τ in Ω', $\Phi(\theta).H(\tau) = H(\theta.\tau)$.

To satisfy condition (+), not only must $\mathbf{E}_{\Omega'}$ and $\mathbf{E}_{\Omega''}$ be isomorphic but they must also act in equivalent manner on Ω' and Ω'', respectively.
Try to describe the classification of plane ornaments in E, under condition (+).

19. For the plane ornaments depicted in Figures 81 - 84, find the corresponding ornamental class.

20. Ornamental subgroups of **E** were defined in terms of the concept of lattice in R^2. In similar manner, one may define "marginal" subgroups of **E**, using the concept of chain in R^2. [See 2.1°.] Thus, let us say that a subgroup **M** of **E** is <u>marginal</u> provided that, for some (and hence for any) cartesian coordinate mapping K, the translational part T^K of $\mathbf{M}^K$ is a chain in R^2.
Motivate the study of marginal subgroups of **E**, by defining an appropriate analogue of plane ornament, namely, <u>border ornament</u> in E. Describe the equivalence classes of marginal subgroups of **E**, under the relation of conjugacy in **A**. In particular, show that there are seven such classes, represented (in cartesian form) by the following groups. In this context, H denotes the <u>standard chain</u> in R^2:

$$H = \{m.r : m \in Z\},$$

Figure 81

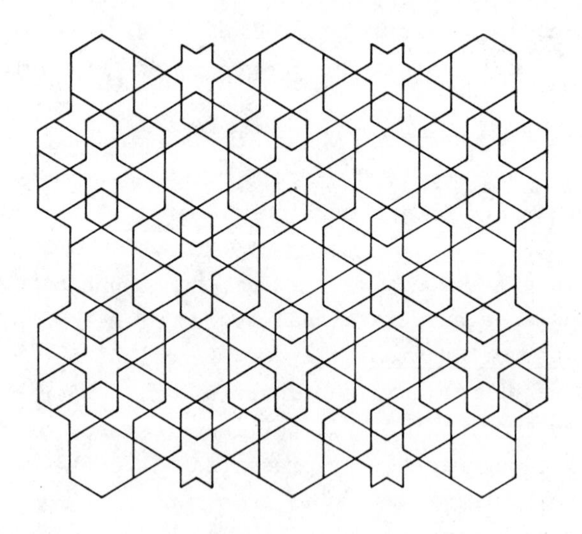

Figure 82

Figure 83

Figure 84

and X and Y denote the reflections on R^2 defined by the conditions: $X(r) = r$, $Y(s) = s$. Moreover, c_I and c_{-I} equal 0, and c_X and c_Y equal $(1/2).r$.

$\mathbb{Z}^a_{1\bullet-}$: $\{[t,I][0\,,\,I] : \; t \in H\}$

$\mathbb{Z}^a_{2\bullet-}$: $\{[t,I][0\,,\,V] : \; t \in H, \;\; V \in \mathcal{Z}_2\}$

$\mathbb{D}^h_{1\bullet-}$: $\{[t,I][0\,,\,V] : \; t \in H, \;\; V \in \mathcal{D}_1(X)\}$

$\mathbb{D}^h_{1-\bullet-}$: $\{[t,I][c_V, V] : \; t \in H, \;\; V \in \mathcal{D}_1(X)\}$

$\mathbb{D}^v_{1\bullet-}$: $\{[t,I][0\,,\,V] : \; t \in H, \;\; V \in \mathcal{D}_1(Y)\}$

$\mathbb{D}^o_{2\bullet-}$: $\{[t,I][0\,,\,V] : \; t \in H, \;\; V \in \mathcal{D}_2(X)\}$

$\mathbb{D}^o_{2-\bullet-}$: $\{[t,I][c_V, V] : \; t \in H, \;\; V \in \mathcal{D}_2(X)\}$.

In Figure 85, we have presented (starkly simple) examples of border ornaments illustrating the seven classes. One can find more interesting examples, based upon African textiles, in the book by C. Zaslavsky. Prove that the groups representing the types $\mathbb{Z}^a_{1\bullet-}$ and $\mathbb{D}^h_{1-\bullet-}$ are isomorphic, and that the groups representing the types $\mathbb{Z}^a_{2\bullet-}$, $\mathbb{D}^v_{1\bullet-}$, and $\mathbb{D}^o_{2-\bullet-}$ are isomorphic. Conclude that there are only four equivalence classes of marginal subgroups of **E**, under the relation of isomorphism. As a result, the analogue of THEOREM 1 for marginal subgroups of **E** is definitely false.

Show that the equivalence classes of marginal subgroups of **E** under the relation of conjugacy in **S**, may be obtained simply by refining the foregoing classification, to take account of scale transformations.

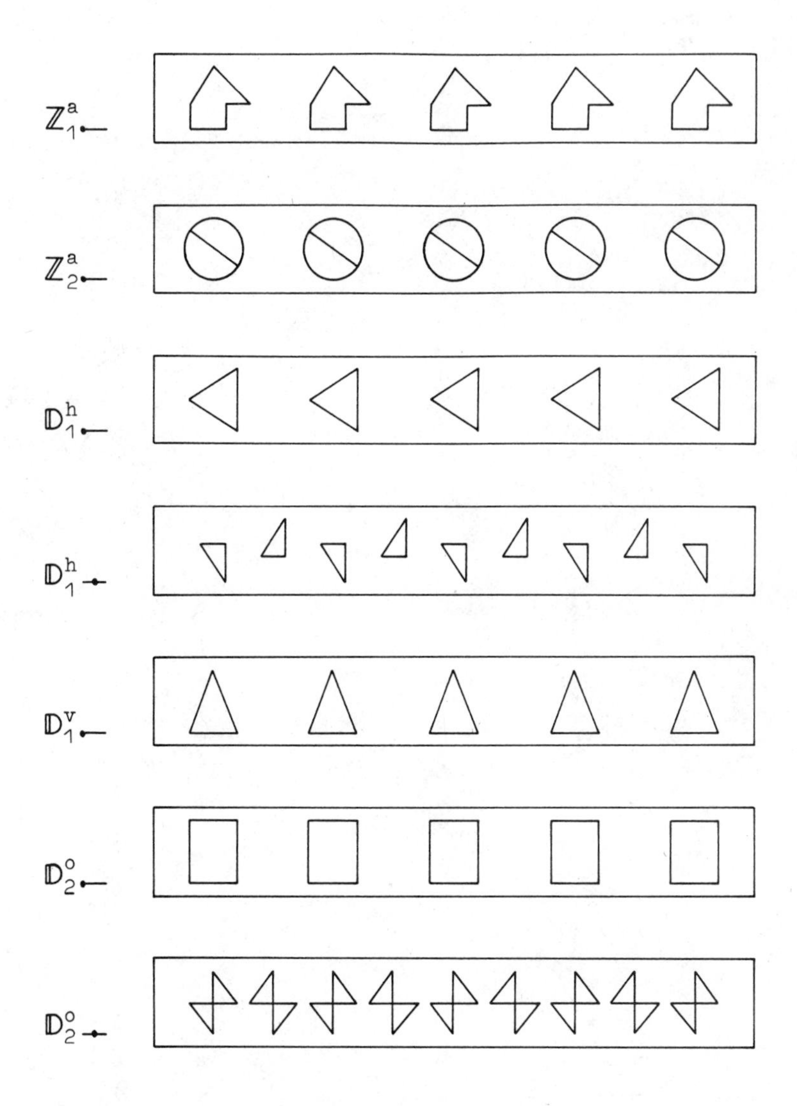
$\mathbb{Z}_1^a$
$\mathbb{Z}_2^a$
$\mathbb{D}_1^h$
$\mathbb{D}_1^h$
$\mathbb{D}_1^v$
$\mathbb{D}_2^o$
$\mathbb{D}_2^o$

Figure 85

Chapter 3
CHROMATIC PLANE ORNAMENTS

3.0 INTRODUCTION

The object of this chapter is to present the theory of colorations of plane ornaments. We shall first define chromatic plane ornaments, and describe a principle of classification of such ornaments compatible with that by which (neutral) plane ornaments were classified in the preceding chapter. We shall then consider in detail the problems involved in enumerating the equivalence classes of chromatic plane ornaments. In due course, we will be led to design algorithms suitable for machine computation, by which not only counts of the equivalence classes can be obtained but also specific instructions for executing representative colorings.

We have implemented such algorithms on a PDP 11/70, equipped with a UNIX operating system. We will report the raw counts of equivalence classes for a range of colors from 1 through 60, the coloring instructions, for a range from 1 through 8. Examples based on the coloring instructions will be displayed in a folio, for four colors.

3.1 CHROMATIC MOSAICS

1° Let n be any positive integer, and let N be the set:

$$N = \{1,2,3,\ldots,n\} .$$

One should regard the members of N as symbols for colors.

By an n-chromatic mosaic in E, we shall mean any ordered pair (Ω,k), where Ω is a mosaic in E and where k is a surjective mapping carrying Ω to N. In this context, one may refer to Ω as the "underlying mosaic," to k as the "color mapping," and to n as the "number of colors." Moreover, for each tile τ in Ω, one should interpret $k(\tau)$ as the "color" assigned to τ.

We shall use the term chromatic mosaic, to refer to n-chromatic mosaics in general.

One will find simple examples of chromatic mosaics in Figures 86 and 87. In both cases, the underlying mosaic is the standard quadratic mosaic, and the number of colors is two. For the first case, we have constructed the color mapping by imitating the array of light and dark squares in a chessboard. For the second, we have arranged the colors as a system of light and dark stripes. Of course, one must interpret "light" as 1 and "dark" as 2, or conversely.

There are many other examples in the folio composing section 8.

2° Now let S_N denote the group of all permutations on N, and, as usual, let $\mathbf{E}$ denote the group of all euclidean transformations on E. Let $\mathbf{E}\times S_N$ stand for the direct product of $\mathbf{E}$ and S_N.

Let (Ω,k) be any n-chromatic mosaic in E. Given a member θ of $\mathbf{E}_\Omega$, let us denote by $\dot{\theta}$ the mapping

$$(\tau \longmapsto \theta.\tau)$$

carrying Ω to $\theta.\Omega$. For every member (θ,σ) of $\mathbf{E}\times S_N$, the mapping $\sigma k \dot{\theta}^{-1}$ carrying $\theta.\Omega$ to N is surjective. Hence, one may form the n-chromatic mosaic $(\theta.\Omega,\sigma k\dot{\theta}^{-1})$. We shall refer to it as the transform of (Ω,k) under (θ,σ), and denote it by $(\theta,\sigma).(\Omega,k)$:

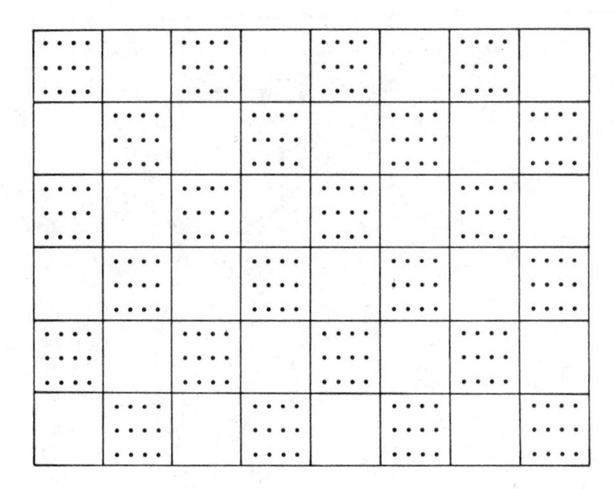

Figure 86

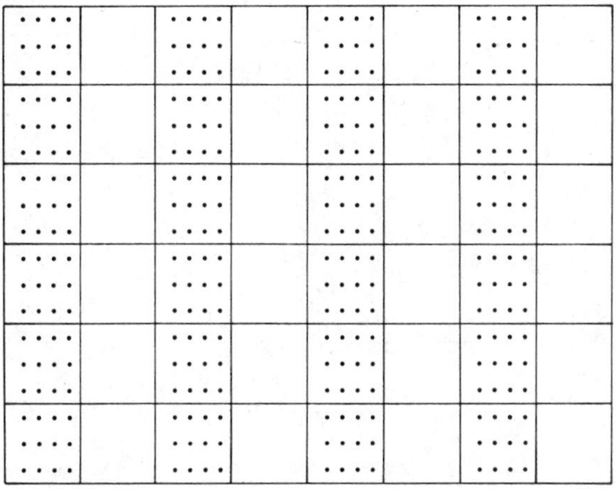

Figure 87

$$(\theta,\sigma).(\Omega,\boldsymbol{k}) \;=\; (\theta.\Omega,\sigma\boldsymbol{k}\overset{\bullet}{\theta}^{-1}).$$

The foregoing definition yields an action of the group $\mathbf{E}\times S_N$ on the family of all n-chromatic mosaics in E. In fact, for any members (θ',σ') and (θ'',σ'') of $\mathbf{E}\times S_N$ and for each n-chromatic mosaic $(\Omega,\boldsymbol{k})$, we have:

$$\begin{aligned}
((\theta',\sigma')(\theta'',\sigma'')).(\Omega,\boldsymbol{k}) \\
&= (\theta'\theta'',\sigma'\sigma'').(\Omega,\boldsymbol{k}) \\
&= ((\theta'\theta'').\Omega,(\sigma'\sigma'')\boldsymbol{k}(\overset{\bullet}{\theta'\theta''})^{-1}) \\
&= (\theta'.(\theta''.\Omega),\sigma'(\sigma''\boldsymbol{k}\overset{\bullet}{\theta}''^{-1})\overset{\bullet}{\theta}'^{-1}) \\
&= (\theta',\sigma').(\theta''.\Omega,\sigma''\boldsymbol{k}\overset{\bullet}{\theta}''^{-1}) \\
&= (\theta',\sigma').((\theta'',\sigma'').(\Omega,\boldsymbol{k})).
\end{aligned}$$

Moreover:

$$(\iota,\varepsilon).(\Omega,\boldsymbol{k}) \;=\; (\Omega,\boldsymbol{k}),$$

where ι and ε stand for the identities in $\mathbf{E}$ and S_N, respectively.

3° Let $(\Omega,\boldsymbol{k})$ be any n-chromatic mosaic in E. By a <u>symmetry</u> of $(\Omega,\boldsymbol{k})$, we shall mean any member (θ,σ) of $\mathbf{E}\times S_N$ under which $(\Omega,\boldsymbol{k})$ is invariant:

$$(\theta,\sigma).(\Omega,\boldsymbol{k}) \;=\; (\Omega,\boldsymbol{k}).$$

Thus, (θ,σ) is a symmetry of $(\Omega,\boldsymbol{k})$ iff:

$$\theta.\Omega \;=\; \Omega\;; \tag{1)'}$$

$$\boldsymbol{k}\overset{\bullet}{\theta} \;=\; \sigma\boldsymbol{k}. \tag{1)''}$$

The first of these conditions requires that θ be a symmetry of Ω. The second requires that, for any tile τ in Ω, the application of σ to the color assigned to τ yields the color assigned to the transform of τ under θ.

To make verbal distinction between symmetries of Ω and symmetries of (Ω, k), we shall sometimes refer to the latter as "chromatic symmetries."

The family of all symmetries of (Ω, k) equals the stabilizer of (Ω, k) under the action of $\mathbf{E} \times S_N$. Hence, it is a subgroup of $\mathbf{E} \times S_N$. We shall refer to it as the <u>symmetry group</u> of (Ω, k), and denote it by:

$$(\mathbf{E} \times S_N)_{(\Omega, k)} .$$

Thus:

$$(\mathbf{E} \times S_N)_{(\Omega, k)} = \{(\theta, \sigma) \in \mathbf{E} \times S_N : (\theta, \sigma).(\Omega, k) = (\Omega, k)\}.$$

For examples of chromatic symmetries, one should review Figures 86 and 87. In Figure 86, every rotation having order four and leaving fixed the center of a square, may be paired with the identity permutation to obtain a chromatic symmetry. Every rotation having order four and leaving fixed one corner of a square, may be paired with the permutation which interchanges 1 and 2. Indeed, every symmetry of the underlying quadratic mosaic yields a chromatic symmetry, upon being paired with the appropriate permutation.

In Figure 87, no rotational symmetry having order four will yield a chromatic symmetry. The indicated coloration is incompatible with the fourfold symmetric structure of the mosaic. However, certain rotations having order two will yield chromatic symmetries. Thus, one may pair any rotation having order two and leaving fixed either the center of a square or the center of one horizontal edge of a square, with the identity permutation; and one may pair any rotation having order two and leaving fixed either one corner of a square or the center of one vertical edge of a square, with the permutation which interchanges 1 and 2.

The other surviving symmetries of the quadratic mosaic can be readily identified.

The second example shows that, in general, the family of those symmetries of the underlying mosaic with which one can pair permutations to obtain chromatic symmetries, may be strictly smaller than the full symmetry group of the mosaic. Indeed, by means of a suitably whimsical coloring, one can reduce the surviving symmetries to the identity alone.

We shall say that an n-chromatic mosaic (Ω, k) is <u>complete</u> iff, for each member θ of $\mathbf{E}_\Omega$, there exists some permutation σ on N such that (θ, σ) is a symmetry of (Ω, k).

4° Let (Ω, k) be any n-chromatic mosaic in E, and let (θ, σ) be a symmetry of (Ω, k). We shall refer to θ as the <u>geometric</u> part, and to σ as the <u>combinatorial</u> part of (θ, σ). The following proposition asserts that the latter is uniquely determined by the former.

<u>Proposition</u> <u>23</u>. For any symmetries (θ', σ') and (θ'', σ'') of (Ω, k), if $\theta' = \theta''$ then $\sigma' = \sigma''$.

Proof. One need only review relation (1)" in 3°. If θ' and θ'' are equal then $\sigma' k$ and $\sigma'' k$ are equal. Since k is surjective, it would follow that σ' and σ'' are equal. ///

By the <u>geometric</u> part of $(\mathbf{E} \times S_N)_{(\Omega, k)}$, we shall mean the subgroup of $\mathbf{E}_\Omega$ consisting of all members θ for which there exists some permutation σ in S_N such that (θ, σ) is a symmetry of (Ω, k). Let us denote this subgroup by $\mathbf{E}_{(\Omega, k)}$. By the <u>combinatorial</u> part, we shall mean the subgroup of S_N consisting of all permutations σ for which there exists some member θ of $\mathbf{E}_\Omega$ such that (θ, σ) is a symmetry of (Ω, k).

From the foregoing proposition, we infer that, for each member θ of $\mathbf{E}_{(\Omega, k)}$, there exists precisely one member σ of S_N such that (θ, σ) lies in $(\mathbf{E} \times S_N)_{(\Omega, k)}$.

Let h_θ stand for σ. We obtain, then, the mapping

$$h(\theta \longmapsto h_\theta)$$

carrying $\mathbf{E}_{(\Omega,k)}$ to S_N. Obviously, the domain of h is the geometric part of $(\mathbf{E}\times S_N)_{(\Omega,k)}$, and the range is the combinatorial part. By design, the graph of h coincides with the symmetry group of (Ω,k):

$$(\mathbf{E}\times S_N)_{(\Omega,k)} = \{(\theta,h_\theta): \quad \theta \in \mathbf{E}_{(\Omega,k)}\}.$$

Moreover, since $(\mathbf{E}\times S_N)_{(\Omega,k)}$ is a subgroup of $\mathbf{E}\times S_N$, it is plain that h is a homomorphism. We shall refer to h as the <u>structural homomorphism</u> for (Ω,k).

The concept of structural homomorphism will provide the means for distinguishing chromatic plane ornaments among chromatic mosaics, for defining a natural equivalence relation on the family of all such ornaments, and for calculating the resulting equivalence classes.

3.2 CHROMATIC PLANE ORNAMENTS

1° Let n be any positive integer.

By an <u>n-chromatic plane ornament</u>, we shall mean any n-chromatic mosaic (Ω,k) meeting the following conditions:

(o) Ω is a plane ornament;
(g) the geometric part of $(\mathbf{E}\times S_N)_{(\Omega,k)}$ equals $\mathbf{E}_\Omega$;
(c) the combinatorial part acts transitively on N.

We shall use the term <u>chromatic plane ornament</u>, to refer to n-chromatic plane ornaments in general.

In this context, condition (g) is simply the requirement that (Ω,k) be complete. It demands compatibility between the color mapping k and the symmetry structure of Ω. Condition (c) requires that the colors assigned by k to the various tiles in Ω be "well distributed" with respect to the action of $\mathbf{E}_\Omega$ on Ω, which is to say that it prevents the tiles from falling into two nonempty disjoint invariant families, having no colors in common.

While condition (o) could hardly be different, conditions (g) and (c) are subject to discussion. For example, one could weaken condition (g) by requiring only that $\mathbf{E}_{(\Omega,k)}$ be an ornamental subgroup of $\mathbf{E}$, whether or not equal to $\mathbf{E}_\Omega$. Moreover, one could interpret condition (c) as provisional, with the intent of reducing the general case to the transitive. [See problem 15 in 9.] However, we submit that conditions (o), (g), and (c), as they stand, are reasonable.

2° By an <u>n-chromatic homomorphism</u>, we shall mean any homomorphism f such that:

(d) the domain of f is an ornamental subgroup of $\mathbf{E}$;

(r) the codomain of f is S_N, and the range of f acts transitively on N.

Let us use the term <u>chromatic homomorphism</u>, to refer to n-chromatic homomorphisms in general.

Of course, this definition is motivated by the relation between chromatic mosaics and structural homomorphisms. Thus, for each n-chromatic plane ornament (Ω,k), the corresponding structural homomorphism h is an n-chromatic homomorphism. In the next proposition, we establish the converse.

Let f be any n-chromatic homomorphism, and let Ω be a plane ornament for which $\mathbf{E}_\Omega$ equals the domain of f.

<u>Proposition 24</u>. The following statements are equivalent:

(a) there exists a color mapping k for Ω such that the structural homomorphism for the n-chromatic mosaic (Ω,k) equals f;

(b) for each tile τ in Ω, there exists some member j of N such that, for each member θ of $\mathbf{E}_\Omega$, if $\theta.\tau = \tau$ then $f(\theta)(j) = j$.

Proof. When condition (a) is satisfied, then, for every tile τ in Ω and for every member θ of $\mathbf{E}_\Omega$:

$$k(\theta.\tau) = f(\theta)(k(\tau)).$$

[See relation (1)" in 1.3°.] Clearly, by taking j to be $k(\tau)$, one can obtain condition (b).

Now let condition (b) be satisfied. Let Δ be a subset of Ω which contains exactly one tile from each orbit in Ω under $\mathbf{E}_\Omega$. For each tile υ in Δ, let j_υ be a member of N such that, for each member θ of $\mathbf{E}_\Omega$, if $\theta.\upsilon = \upsilon$ then $f(\theta)(j_\upsilon) = j_\upsilon$. With these ingredients, we may define a color mapping for Ω. Let us first note that, for any tiles υ' and υ'' in Δ and for any members θ' and θ'' of $\mathbf{E}_\Omega$, if $\theta'.\upsilon' = \theta''.\upsilon''$ then $\theta''^{-1}\theta'.\upsilon' = \upsilon''$, so $\upsilon' = \upsilon''$; hence, $f(\theta''^{-1}\theta')(j_{\upsilon'}) = j_{\upsilon''}$, so $f(\theta')(j_{\upsilon'}) = f(\theta'')(j_{\upsilon''})$. Now we obtain a well defined mapping

$$k(\tau \longmapsto f(\theta)(j_\upsilon))$$

carrying Ω to N, where, for each tile τ in Ω, θ and υ are members of $\mathbf{E}_\Omega$ and Δ, respectively, for which $\tau = \theta.\upsilon$. Since the range of f is a transitive subgroup of S_N, the mapping k is surjective. Hence, one may form the n-chromatic mosaic (Ω, k). We contend that the structural homomorphism for (Ω, k) equals f. [It would follow, then, that (Ω, k) is an n-chromatic plane ornament.]

Thus, let θ be any member of $\mathbf{E}_\Omega$ and let τ be any tile in Ω. Let θ' and υ be members of $\mathbf{E}_\Omega$ and Δ, respectively, for which $\theta'.\upsilon = \tau$. Clearly, $\theta.\tau = \theta\theta'.\upsilon$, and hence:

$$\begin{aligned} k(\theta.\tau) &= f(\theta\theta')(j_\upsilon) \\ &= f(\theta)(f(\theta')(j_\upsilon)) \\ &= f(\theta)(k(\tau)). \end{aligned}$$

Therefore:

$$\mathscr{k}\dot{\theta} = f(\theta)\mathscr{k}.$$

By relation (1)" in 1.3°, we conclude that the structural homomorphism for $(\Omega, \mathscr{k})$ equals f. ///

The various plane ornaments Ω exhibited in $2.5.6^{\circ}$, all have the property that $\mathbf{E}_{\Omega}$ acts freely on Ω. [See problem 13 in 2.8.] Obviously, these plane ornaments meet condition (b). It follows that every n-chromatic homomorphism is the structural homomorphism for some n-chromatic plane ornament.

3.3 THE PROBLEM OF CLASSIFICATION

1° Let us now introduce the operative equivalence relation in terms of which the subsequent classification of chromatic plane ornaments will be developed.

At the outset, we must note that it is neither natural nor interesting to classify chromatic plane ornaments by the relation of isomorphism between the corresponding (chromatic) symmetry groups. Indeed, for each chromatic mosaic $(\Omega, \mathscr{k})$, the projection homomorphism

$$((\theta, \sigma) \longmapsto \theta)$$

carrying the symmetry group of $(\Omega, \mathscr{k})$ to its geometric part, is an isomorphism. [See Proposition 23.] Hence, the stark relation of isomorphism would in effect (observe the geometric but) ignore the combinatorial parts of the corresponding symmetry groups. Moreover, for chromatic plane ornaments, the relation of isomorphism would yield nothing but the 17 ornamental classes derived in the preceding chapter.

The appropriate condition for the equivalence of chromatic plane ornaments is the condition demanding that the corresponding structural homomorphisms be equivalent. Thus, let n be any positive integer, and let $(\Omega', \mathscr{k}')$ and $(\Omega'', \mathscr{k}'')$ be any two n-chromatic plane ornaments. Let h' and

h'' be the corresponding structural homomorphisms, carrying $\mathbf{E}_{\Omega'}$ and $\mathbf{E}_{\Omega''}$ to S_N. We shall regard (Ω', k') and (Ω'', k'') as <u>equivalent</u> provided that h' and h'' are equivalent, in the sense that there exist an isomorphism Φ carrying $\mathbf{E}_{\Omega'}$ to $\mathbf{E}_{\Omega''}$ and a permutation ρ on N such that:

$$h''\Phi = J_\rho h',$$

where J_ρ is the inner automorphism on S_N determined by ρ. In diagrammatic terms, this relation appears as follows:

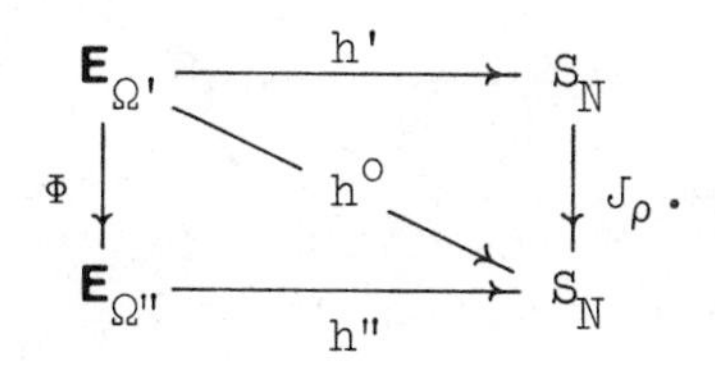

The intermediary homomorphism h° (which is the structural homomorphism for the n-chromatic plane ornament $(\Omega', \rho k')$) may help to give intuitive content to the equivalence relation just defined. Thus, the relation: $h^{\circ} = J_\rho h'$, asserts that the symmetry groups for (Ω', k') and $(\Omega', \rho k')$ are the same, once the permutation of colors effected by ρ is taken into account. The relation: $h''\Phi = h^{\circ}$, asserts not only that the symmetry groups for (Ω'', k'') and $(\Omega', \rho k')$ are isomorphic, but in fact that there exists an isomorphism for which the combinatorial parts of corresponding (chromatic) symmetries are equal.

2° In the light of Proposition 24, it is now evident that the problem of classifying chromatic plane ornaments may be replaced by the problem of classifying chromatic homomorphisms. At this point, the chromatic plane ornaments themselves will slip into the background.

Let n be any positive integer. Let f' and f'' be any two n-chromatic homomorphisms, and let $\mathbf{F}'$ and $\mathbf{F}''$ stand for the domains of f' and f'', respectively. With reference to the preceding article, we shall say that f' and f'' are <u>equivalent</u> iff there exist an isomorphism Φ carrying **F'** to **F"** and a permutation ρ on N such that:

$$f''\Phi = J_\rho f' .$$

In diagrammatic form, we would have:

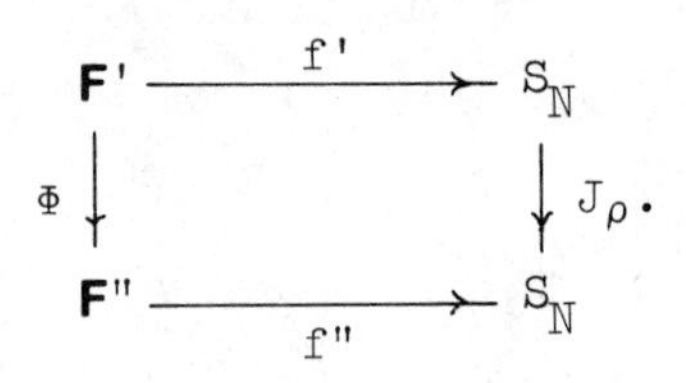

We shall refer to the resulting equivalence classes as <u>n-chromatic ornamental classes</u>.

For general reference, we shall use the term <u>chromatic ornamental class</u>.

3° For the purpose of describing the n-chromatic ornamental classes, one need not consider every n-chromatic homomorphism. Rather, one may select a representative group from each of the ornamental classes obtained in the preceding chapter, thus generating a list of 17 groups, and one may then restrict attention to those n-chromatic homomorphisms the domains of which lie in the list. Of course, two such homomorphisms can be equivalent only if their domains are the same. Hence, the problem of classification breaks into 17 subproblems.

4° Now let us adopt once for all the list of 17 ornamental groups (in cartesian form), presented in Table 3, together with the indicated serial notation:

$$\mathscr{F}_m , \qquad 1 \leq m \leq 17,$$

and let us introduce the normalizers of those groups in the affine group $\mathscr{A}$ on R^2:

$$\mathscr{F}_m^{\triangle}, \qquad 1 \le m \le 17.$$

By (the proof of) THEOREM 1, we know that every isomorphism Φ carrying $\mathscr{F}_m$ to itself, can be implemented by an inner automorphism J_A of $\mathscr{A}$. Obviously, the affine transformation A would have to be contained in $\mathscr{F}_m^{\triangle}$. Hence, for any two n-chromatic homomorphisms f' and f'' having common domain $\mathscr{F}_m$, f' and f'' are equivalent iff there exist an affine transformation A in $\mathscr{F}_m^{\triangle}$ and a permutation ρ on N such that:

$$f''J_A = J_\rho f' . \tag{1}$$

Let $\mathscr{F}_m^{\triangle} \times S_N$ stand for the direct product of the groups $\mathscr{F}_m^{\triangle}$ and S_N, and let $\mathbb{F}_m^n$ stand for the set of all n-chromatic homomorphisms having domain $\mathscr{F}_m$. From the foregoing relation, we are led to define an action of $\mathscr{F}_m^{\triangle} \times S_N$ on $\mathbb{F}_m^n$. Thus:

$$(A, \rho).f = J_\rho f J_A^{-1}, \tag{2}$$

where (A,ρ) is any member of $\mathscr{F}_m^{\triangle} \times S_N$ and where f is any member of $\mathbb{F}_m^n$. Obviously, two homomorphisms f' and f'' in $\mathbb{F}_m^n$ lie in the same orbit under $\mathscr{F}_m^{\triangle} \times S_N$ iff they are equivalent, in the sense of relation (1).

At this point, we may conclude that, for each positive integer n, the n-chromatic ornamental classes are identifiable as the orbits in the sets

$$\mathbb{F}_m^n, \qquad 1 \le m \le 17,$$

under the actions of the groups

$$\mathscr{F}_m^{\triangle} \times S_N, \qquad 1 \leq m \leq 17,$$

as defined by relation (2).

5° Our formulation of the problem of classification is now complete. In sections 5 and 6, we shall describe algorithms by which one may enumerate all chromatic ornamental classes, and, for each such class, exhibit a representative chromatic homomorphism. In sections 7 and 8, we shall report data and present examples.

3.4 NORMALIZERS

1° To support the subsequent computations, let us develop a procedure for calculating the normalizers in $\mathscr{A}$ of the given ornamental groups $\mathscr{F}_m$ $(1 \leq m \leq 17)$.

Thus, let $\mathscr{F}$ stand for any one of the given groups, and let $\mathscr{F}^{\triangle}$ be its normalizer in $\mathscr{A}$. Let T denote the translational part of $\mathscr{F}$, and $\mathscr{G}$, the linear part. By $2.5.8^\circ$, we may introduce generators of $\mathscr{F}$, of the form:

$$\underline{U} = [u, I],$$
$$\underline{V} = [v, I],$$
$$\underline{W} = [0, W],$$
$$\underline{X} = [x, X].$$

Of course, $\{u,v\}$ is a basis for T, and W and X generate $\mathscr{G}$. For convenience, we shall assume that: $|u| = |v|$. [See $2.5.8^\circ$.]

When $\mathscr{F}$ represents one of the first five ornamental classes, then the generator $\underline{X}$ should be deleted.

Let $T^{\triangle}$ denote the translational part of $\mathscr{F}^{\triangle}$, and $\mathscr{G}^{\triangle}$, the linear part. One should take care not to confuse $T^{\triangle}$ with the normalizer of T in R^2, which would of course be R^2 itself, nor to confuse $\mathscr{G}^{\triangle}$ with the normalizer of $\mathscr{G}$ in $\mathscr{L}$, though in some cases the two are the same.

Now let A be an arbitrary member of $\mathscr{A}$: $A = [t,L]$. If A is contained in $\mathscr{F}^{\Delta}$, then, by relations (5) and (6) in 1.1.2°, we would have:

$$(\Delta 1) \quad L(T) = T;$$

$$(\Delta 2) \quad L\mathscr{G}L^{-1} = \mathscr{G}.$$

Moreover, since:

$$[t,L][0,L^{-1}WL][t,L]^{-1} = [t - W(t), W],$$

we would also have:

$$(\Delta 3) \quad t - W(t) \in T.$$

Finally, by Proposition 18, we know that $[x,L^{-1}XL]$ is a member of $\mathscr{F}$; since:

$$[t,L][x,L^{-1}XL][t,L]^{-1} = [t + L(x) - X(t), X],$$

we would obtain:

$$(\Delta 4) \quad (t - X(t)) - (x - L(x)) \in T.$$

Proposition 25. For each affine transformation $[t,L]$, $[t,L]$ is a member of $\mathscr{F}^{\Delta}$ iff it satisfies conditions (Δ1), (Δ2), (Δ3), and (Δ4).

Proof. One can prove this proposition by inductive argument, from the relations:

$$t - W^{i+1}(t) = (t - W^{i}(t)) + W^{i}(t - W(t)),$$

$$t - (W^{j+1}X)(t) = (t - W^{j+1}(t)) + W^{j+1}(t - X(t)),$$

where i and j are any nonnegative integers. We omit the details. ///

Incidentally, when $\mathscr{F}$ represents one of the symmorphic ornamental classes, then x would be 0. In such cases, condition (Δ4) would reduce to the following:

$$(\Delta 4)' \quad t - X(t) \in T.$$

For the symmorphic cases, it is now clear that conditions (Δ1) and (Δ2) determine $\mathscr{G}^{\Delta}$, and conditions (Δ3) and (Δ4)' determine T^{Δ}. Hence:

$$\mathscr{F}^{\Delta} = T^{\Delta} \rtimes \mathscr{G}^{\Delta}.$$

As a matter of fact, the same relation holds in the nonsymmorphic cases as well, but, in view of the hybrid nature of condition (Δ4), it must be verified by computation.

2° With reference to condition (Δ1), let us denote by $\mathscr{L}_T$ the subgroup of $\mathscr{L}$ consisting of all members L for which $L(T) = T$.

3° By applying conditions (Δ1), (Δ2), (Δ3), and (Δ4), one can calculate the normalizers in $\mathscr{A}$ of the given ornamental groups. Let us carry out the calculations for a few cases, then assemble the complete results in a table, Table 5.

Let us consider first the group $\mathscr{F}_1$. For this case, $\mathscr{G}$ equals $\mathscr{Z}_1$. Clearly, then, only condition (Δ1) need be applied, so that the affine transformation $[t,L]$ lies in $\mathscr{F}_1^{\Delta}$ iff L is contained in $\mathscr{L}_T$. Hence, one may describe $\mathscr{F}_1^{\Delta}$ as the semi-direct product of R^2 and $\mathscr{L}_T$:

$$\mathscr{F}_1^{\Delta} = R^2 \rtimes \mathscr{L}_T.$$

Now let us consider the group $\mathscr{F}_6$. In this case, $\mathscr{G}$ equals $\mathscr{D}_1(X)$, for a suitable reflection X. With respect to $\{u,v\}$, the matrix for X may be taken to be:

$$X \longleftrightarrow \begin{bmatrix} 0 & 1 \\ 1 & 0 \end{bmatrix}.$$

Under condition ($\Delta 1$), L would have to lie in $\mathscr{L}_T$. By problem 8 in 2.8, the matrix for L relative to {u,v} would be of the form:

$$L \longleftrightarrow \begin{bmatrix} a & c \\ b & d \end{bmatrix},$$

where a, b, c, and d are integers and where $ad - bc = \pm 1$. Under condition ($\Delta 2$), one would have $LXL^{-1} = X$; that is:

$$\begin{bmatrix} a & c \\ b & d \end{bmatrix}\begin{bmatrix} 0 & 1 \\ 1 & 0 \end{bmatrix} = \begin{bmatrix} 0 & 1 \\ 1 & 0 \end{bmatrix}\begin{bmatrix} a & c \\ b & d \end{bmatrix},$$

so that $a = d$ and $b = c$. Since $ad - bc = \pm 1$, one would then obtain $(a - b)(a + b) = \pm 1$. Hence, L satisfies both conditions ($\Delta 1$) and ($\Delta 2$) iff the matrix for L relative to {u,v} is one of the following:

$$\begin{bmatrix} 1 & 0 \\ 0 & 1 \end{bmatrix}, \quad \begin{bmatrix} 0 & 1 \\ 1 & 0 \end{bmatrix}, \quad \begin{bmatrix} -1 & 0 \\ 0 & -1 \end{bmatrix}, \quad \begin{bmatrix} 0 & -1 \\ -1 & 0 \end{bmatrix},$$

which is to say that L is a member of $\mathscr{D}_2(X)$.

Let t be presented in terms of {u,v}:

$$t = r.u + s.v,$$

where r and s are suitable real numbers. Under condition ($\Delta 4$)', $r - s$ would have to be an integer. Accordingly, we are led to introduce the subgroup T/ of R^2, consisting of all members of the form:

$$k.u + s.(u + v),$$

where k is any integer and where s is any real number. Now we obtain:

$$\mathscr{F}_6^{\Delta} = T / \rtimes \mathscr{D}_2(X) .$$

Let us continue the calculations, by considering the group $\mathscr{F}_{11}$. For this case, $\mathscr{G}$ equals $\mathscr{D}_2(X)$, where X is a suitable reflection. Relative to $\{u,v\}$, the matrix for X may be taken to be:

$$X \longleftrightarrow \begin{bmatrix} 1 & 0 \\ 0 & -1 \end{bmatrix} .$$

Arguing as in the preceding case, one can show that L satisfies conditions ($\Delta 1$) and ($\Delta 2$) iff the matrix for L relative to $\{u,v\}$ is one of the following:

$$\begin{bmatrix} 1 & 0 \\ 0 & 1 \end{bmatrix}, \quad \begin{bmatrix} 1 & 0 \\ 0 & -1 \end{bmatrix}, \quad \begin{bmatrix} 0 & -1 \\ 1 & 0 \end{bmatrix}, \quad \begin{bmatrix} 0 & 1 \\ 1 & 0 \end{bmatrix},$$

$$\begin{bmatrix} -1 & 0 \\ 0 & -1 \end{bmatrix}, \quad \begin{bmatrix} -1 & 0 \\ 0 & 1 \end{bmatrix}, \quad \begin{bmatrix} 0 & 1 \\ -1 & 0 \end{bmatrix}, \quad \begin{bmatrix} 0 & -1 \\ -1 & 0 \end{bmatrix} .$$

By our special assumption that $|u| = |v|$, the foregoing family can be identified as $\mathscr{D}_4(X)$.

Let t be presented in the form:

$$t = r.u \quad s.v ,$$

where r and s are suitable real numbers. Under condition ($\Delta 3$), both $2r$ and $2s$ would have to be integers. Hence, t must lie in the lattice $\frac{1}{2}.T$, consisting of all members w of R^2 for which $2.w$ is contained in T. However, when t is a member of $\frac{1}{2}.T$ and L is a member of $\mathscr{D}_4(X)$, then condition ($\Delta 4$) can be satisfied only when in fact L lies in $\mathscr{D}_2(X)$. We conclude that:

$$\mathscr{F}_{11}^{\Delta} = \frac{1}{2}.T \rtimes \mathscr{D}_2(X) .$$

In similar manner, one can show that:

$$\mathscr{F}_{10}^{\Delta} = \tfrac{1}{2}.T \ltimes \mathscr{D}_4(X) ,$$

$$\mathscr{F}_{12}^{\Delta} = \tfrac{1}{2}.T \ltimes \mathscr{D}_4(X) .$$

For these cases, the foregoing constraint involving solutions of condition ($\Delta 4$) does not arise.

Finally, let us briefly consider the groups $\mathscr{F}_{13}$ and $\mathscr{F}_{14}$. In both cases, $\mathscr{G}$ equals $\mathscr{D}_3(X)$, where X is an appropriate reflection. For $\mathscr{F}_{13}$, one may take the matrix for X relative to $\{u,v\}$ to be:

$$X \longleftrightarrow \begin{bmatrix} 1 & -1 \\ 0 & -1 \end{bmatrix},$$

and for $\mathscr{F}_{14}$, to be:

$$X \longleftrightarrow \begin{bmatrix} -1 & 1 \\ 0 & 1 \end{bmatrix}.$$

In each case, $\mathscr{G}^{\Delta}$ proves to be $\mathscr{D}_6(X)$.

For both $\mathscr{F}_{13}$ and $\mathscr{F}_{14}$, condition ($\Delta 3$) entails that t be of the form:

$$i.(\tfrac{2}{3}.u + \tfrac{1}{3}.v) + j.(\tfrac{1}{3}.u + \tfrac{2}{3}.v) ,$$

where i and j are any integers. Let T^h stand for the lattice in R^2 consisting of all such members.

For $\mathscr{F}_{14}$, the members t of T^h satisfy not only condition ($\Delta 3$) but also condition ($\Delta 4$)'. Hence:

$$\mathscr{F}_{14}^{\Delta} = T^h \ltimes \mathscr{D}_6(X) .$$

However, for $\mathscr{F}_{13}$, the only members t of T^h which satisfy condition ($\Delta 4$)' are those contained in T. Hence:

$$\mathscr{F}_{13}^{\Delta} = T \ltimes \mathscr{D}_6(X) .$$

4° Now let us return to the context of 1°.

By the preceding calculations, it is plain that $\mathscr{F}^{\triangle}$ need not itself be an ornamental group. Indeed, $T^{\triangle}$ need not be a lattice in R^2, and $\mathscr{G}^{\triangle}$ need not be a geometric group. However, in each of the cases considered, $\mathscr{F}^{\triangle}$ did prove to be the semi-direct product of $T^{\triangle}$ and $\mathscr{G}^{\triangle}$, even when $\mathscr{F}$ itself represented a nonsymmorphic case. As stated in 1°, this condition is valid for all 17 groups.

In Table 5, we have assembled descriptions of the normalizers. For each case, we have displayed four members:

$$\underline{U}^{\triangle} = [u^{\triangle}, I],$$
$$\underline{V}^{\triangle} = [v^{\triangle}, I],$$
$$\underline{W}^{\triangle} = [0, W^{\triangle}],$$
$$\underline{X}^{\triangle} = [0, X^{\triangle}],$$

of $\mathscr{F}^{\triangle}$, in terms of the given basis $\{u,v\}$ for T. The members $W^{\triangle}$ and $X^{\triangle}$ of $\mathscr{L}$ generate $\mathscr{G}^{\triangle}$, and the members $u^{\triangle}$ and $v^{\triangle}$ of R^2 determine $T^{\triangle}$, as indicated in the table.

To clarify the cases of $\mathscr{F}_1$ and $\mathscr{F}_2$, one should consult problem 3 in 9. The object of that problem is to show that the group $\mathscr{L}_T$ is finitely generated, and, in particular, to justify the corresponding entries in Table 5.

5° While $\mathscr{F}^{\triangle}$ is sometimes finitely generated and sometimes not, the automorphism group of $\mathscr{F}$ (implemented by $\mathscr{F}^{\triangle}$) proves always to be finitely generated. In fact, by straightforward computation, one can verify that, for each of the 17 cases, the automorphisms of $\mathscr{F}$ implemented by the members $\underline{U}^{\triangle}$, $\underline{V}^{\triangle}$, $\underline{W}^{\triangle}$, and $\underline{X}^{\triangle}$ of $\mathscr{F}^{\triangle}$ displayed in Table 5, are sufficient to generate the automorphism group of $\mathscr{F}$.

One may characterize the automorphisms of $\mathscr{F}$ implemented by $\underline{U}^{\triangle}$, $\underline{V}^{\triangle}$, $\underline{W}^{\triangle}$, and $\underline{X}^{\triangle}$, in the following manner:

$$\underline{U}^{\Delta}\,\underline{U}\,\underline{U}^{\Delta-1} = \underline{U}^{e111}\underline{V}^{e112}\underline{W}^{e113}\underline{X}^{e114} ,$$
$$\underline{U}^{\Delta}\,\underline{V}\,\underline{U}^{\Delta-1} = \underline{U}^{e121}\underline{V}^{e122}\underline{W}^{e123}\underline{X}^{e124} ,$$
$$\underline{U}^{\Delta}\,\underline{W}\,\underline{U}^{\Delta-1} = \underline{U}^{e131}\underline{V}^{e132}\underline{W}^{e133}\underline{X}^{e134} ,$$
$$\underline{U}^{\Delta}\,\underline{X}\,\underline{U}^{\Delta-1} = \underline{U}^{e141}\underline{V}^{e142}\underline{W}^{e143}\underline{X}^{e144} ,$$
$$\underline{V}^{\Delta}\,\underline{U}\,\underline{V}^{\Delta-1} = \underline{U}^{e211}\underline{V}^{e212}\underline{W}^{e213}\underline{X}^{e214} ,$$
$$\underline{V}^{\Delta}\,\underline{V}\,\underline{V}^{\Delta-1} = \underline{U}^{e221}\underline{V}^{e222}\underline{W}^{e223}\underline{X}^{e224} ,$$
$$\underline{V}^{\Delta}\,\underline{W}\,\underline{V}^{\Delta-1} = \underline{U}^{e231}\underline{V}^{e232}\underline{W}^{e233}\underline{X}^{e234} ,$$
$$\underline{V}^{\Delta}\,\underline{X}\,\underline{V}^{\Delta-1} = \underline{U}^{e241}\underline{V}^{e242}\underline{W}^{e243}\underline{X}^{e244} ,$$
$$\underline{W}^{\Delta}\,\underline{U}\,\underline{W}^{\Delta-1} = \underline{U}^{e311}\underline{V}^{e312}\underline{W}^{e313}\underline{X}^{e314} ,$$
$$\underline{W}^{\Delta}\,\underline{V}\,\underline{W}^{\Delta-1} = \underline{U}^{e321}\underline{V}^{e322}\underline{W}^{e323}\underline{X}^{e324} ,$$
$$\underline{W}^{\Delta}\,\underline{W}\,\underline{W}^{\Delta-1} = \underline{U}^{e331}\underline{V}^{e332}\underline{W}^{e333}\underline{X}^{e334} ,$$
$$\underline{W}^{\Delta}\,\underline{X}\,\underline{W}^{\Delta-1} = \underline{U}^{e341}\underline{V}^{e342}\underline{W}^{e343}\underline{X}^{e344} ,$$
$$\underline{X}^{\Delta}\,\underline{U}\,\underline{X}^{\Delta-1} = \underline{U}^{e411}\underline{V}^{e412}\underline{W}^{e413}\underline{X}^{e414} ,$$
$$\underline{X}^{\Delta}\,\underline{V}\,\underline{X}^{\Delta-1} = \underline{U}^{e421}\underline{V}^{e422}\underline{W}^{e423}\underline{X}^{e424} ,$$
$$\underline{X}^{\Delta}\,\underline{W}\,\underline{X}^{\Delta-1} = \underline{U}^{e431}\underline{V}^{e432}\underline{W}^{e433}\underline{X}^{e434} ,$$
$$\underline{X}^{\Delta}\,\underline{X}\,\underline{X}^{\Delta-1} = \underline{U}^{e441}\underline{V}^{e442}\underline{W}^{e443}\underline{X}^{e444} ,$$

where the eijk $(1 \le i,j,k \le 4)$ are appropriate integers. [See 2.5.7°.] One will find these integers stored in Table 6. The format is self-evident.

NORMALIZER	TRANSLATIONAL PART			LINEAR PART		
	$u^{\triangle}$	$v^{\triangle}$	$T^{\triangle}$	$W^{\triangle}$	$X^{\triangle}$	$\mathcal{G}^{\triangle}$
$\mathfrak{F}_{01}^{\triangle}$	u	v	R^2	$\begin{bmatrix} 1 & 1 \\ 0 & 1 \end{bmatrix}$	$\begin{bmatrix} 0 & 1 \\ 1 & 0 \end{bmatrix}$	$\mathcal{L}_T$
$\mathfrak{F}_{02}^{\triangle}$	$\frac{1}{2}.u$	$\frac{1}{2}.v$	$Z.u^{\triangle} + Z.v^{\triangle}$	$\begin{bmatrix} 1 & 1 \\ 0 & 1 \end{bmatrix}$	$\begin{bmatrix} 0 & 1 \\ 1 & 0 \end{bmatrix}$	$\mathcal{L}_T$
$\mathfrak{F}_{03}^{\triangle}$	$\frac{2}{3}.u + \frac{1}{3}.v$	$\frac{1}{3}.u + \frac{2}{3}.v$	$Z.u^{\triangle} + Z.v^{\triangle}$	$\begin{bmatrix} 1 & -1 \\ 1 & 0 \end{bmatrix}$	$\begin{bmatrix} 1 & -1 \\ 0 & -1 \end{bmatrix}$	$\mathcal{D}_6(X^{\triangle})$
$\mathfrak{F}_{04}^{\triangle}$	$\frac{1}{2}.u + \frac{1}{2}.v$	$-\frac{1}{2}.u + \frac{1}{2}.v$	$Z.u^{\triangle} + Z.v^{\triangle}$	$\begin{bmatrix} 0 & -1 \\ 1 & 0 \end{bmatrix}$	$\begin{bmatrix} 1 & 0 \\ 0 & -1 \end{bmatrix}$	$\mathcal{D}_4(X^{\triangle})$
$\mathfrak{F}_{05}^{\triangle}$	u	v	$Z.u^{\triangle} + Z.v^{\triangle}$	$\begin{bmatrix} 1 & -1 \\ 1 & 0 \end{bmatrix}$	$\begin{bmatrix} 1 & -1 \\ 0 & -1 \end{bmatrix}$	$\mathcal{D}_6(X^{\triangle})$
$\mathfrak{F}_{06}^{\triangle}$	u	$u + v$	$Z.u^{\triangle} + R.v^{\triangle}$	$\begin{bmatrix} -1 & 0 \\ 0 & -1 \end{bmatrix}$	$\begin{bmatrix} 0 & 1 \\ 1 & 0 \end{bmatrix}$	$\mathcal{D}_2(X^{\triangle})$
$\mathfrak{F}_{07}^{\triangle}$	u	$\frac{1}{2}.v$	$R.u^{\triangle} + Z.v^{\triangle}$	$\begin{bmatrix} -1 & 0 \\ 0 & -1 \end{bmatrix}$	$\begin{bmatrix} 1 & 0 \\ 0 & -1 \end{bmatrix}$	$\mathcal{D}_2(X^{\triangle})$
$\mathfrak{F}_{08}^{\triangle}$	u	$\frac{1}{2}.v$	$R.u^{\triangle} + Z.v^{\triangle}$	$\begin{bmatrix} -1 & 0 \\ 0 & -1 \end{bmatrix}$	$\begin{bmatrix} 1 & 0 \\ 0 & -1 \end{bmatrix}$	$\mathcal{D}_2(X^{\triangle})$
$\mathfrak{F}_{09}^{\triangle}$	$\frac{1}{2}.u + \frac{1}{2}.v$	$-\frac{1}{2}.u + \frac{1}{2}.v$	$Z.u^{\triangle} + Z.v^{\triangle}$	$\begin{bmatrix} 0 & -1 \\ 1 & 0 \end{bmatrix}$	$\begin{bmatrix} 1 & 0 \\ 0 & -1 \end{bmatrix}$	$\mathcal{D}_4(X^{\triangle})$
$\mathfrak{F}_{10}^{\triangle}$	$\frac{1}{2}.u$	$\frac{1}{2}.v$	$Z.u^{\triangle} + Z.v^{\triangle}$	$\begin{bmatrix} 0 & -1 \\ 1 & 0 \end{bmatrix}$	$\begin{bmatrix} 1 & 0 \\ 0 & -1 \end{bmatrix}$	$\mathcal{D}_4(X^{\triangle})$
$\mathfrak{F}_{11}^{\triangle}$	$\frac{1}{2}.u$	$\frac{1}{2}.v$	$Z.u^{\triangle} + Z.v^{\triangle}$	$\begin{bmatrix} -1 & 0 \\ 0 & -1 \end{bmatrix}$	$\begin{bmatrix} 1 & 0 \\ 0 & -1 \end{bmatrix}$	$\mathcal{D}_2(X^{\triangle})$
$\mathfrak{F}_{12}^{\triangle}$	$\frac{1}{2}.u$	$\frac{1}{2}.v$	$Z.u^{\triangle} + Z.v^{\triangle}$	$\begin{bmatrix} 0 & -1 \\ 1 & 0 \end{bmatrix}$	$\begin{bmatrix} 1 & 0 \\ 0 & -1 \end{bmatrix}$	$\mathcal{D}_4(X^{\triangle})$
$\mathfrak{F}_{13}^{\triangle}$	u	v	$Z.u^{\triangle} + Z.v^{\triangle}$	$\begin{bmatrix} 1 & -1 \\ 1 & 0 \end{bmatrix}$	$\begin{bmatrix} 1 & -1 \\ 0 & -1 \end{bmatrix}$	$\mathcal{D}_6(X^{\triangle})$
$\mathfrak{F}_{14}^{\triangle}$	$\frac{2}{3}.u + \frac{1}{3}.v$	$\frac{1}{3}.u + \frac{2}{3}.v$	$Z.u^{\triangle} + Z.v^{\triangle}$	$\begin{bmatrix} 1 & -1 \\ 1 & 0 \end{bmatrix}$	$\begin{bmatrix} 1 & -1 \\ 0 & -1 \end{bmatrix}$	$\mathcal{D}_6(X^{\triangle})$
$\mathfrak{F}_{15}^{\triangle}$	$\frac{1}{2}.u + \frac{1}{2}.v$	$-\frac{1}{2}.u + \frac{1}{2}.v$	$Z.u^{\triangle} + Z.v^{\triangle}$	$\begin{bmatrix} 0 & -1 \\ 1 & 0 \end{bmatrix}$	$\begin{bmatrix} 1 & 0 \\ 0 & -1 \end{bmatrix}$	$\mathcal{D}_4(X^{\triangle})$
$\mathfrak{F}_{16}^{\triangle}$	$\frac{1}{2}.u + \frac{1}{2}.v$	$-\frac{1}{2}.u + \frac{1}{2}.v$	$Z.u^{\triangle} + Z.v^{\triangle}$	$\begin{bmatrix} 0 & -1 \\ 1 & 0 \end{bmatrix}$	$\begin{bmatrix} 1 & 0 \\ 0 & -1 \end{bmatrix}$	$\mathcal{D}_4(X^{\triangle})$
$\mathfrak{F}_{17}^{\triangle}$	u	v	$Z.u^{\triangle} + Z.v^{\triangle}$	$\begin{bmatrix} 1 & -1 \\ 1 & 0 \end{bmatrix}$	$\begin{bmatrix} 1 & -1 \\ 0 & -1 \end{bmatrix}$	$\mathcal{D}_6(X^{\triangle})$

Table 5

ORNAMENTAL GROUP	eijk																
	i	1				2				3				4			
	j \ k	1	2	3	4	1	2	3	4	1	2	3	4	1	2	3	4
$\mathfrak{F}_{01}$	1	1	0	0	0	1	0	0	0	1	0	0	0	0	1	0	0
	2	0	1	0	0	0	1	0	0	1	1	0	0	1	0	0	0
	3	0	0	1	0	0	0	1	0	0	0	1	0	0	0	1	0
	4	0	0	0	0	0	0	0	0	0	0	0	0	0	0	0	0
$\mathfrak{F}_{02}$	1	1	0	0	0	1	0	0	0	1	0	0	0	0	1	0	0
	2	0	1	0	0	0	1	0	0	1	1	0	0	1	0	0	0
	3	1	0	1	0	0	1	1	0	0	0	1	0	0	0	1	0
	4	0	0	0	0	0	0	0	0	0	0	0	0	0	0	0	0
$\mathfrak{F}_{03}$	1	1	0	0	0	1	0	0	0	1	1	0	0	1	0	0	0
	2	0	1	0	0	0	1	0	0	-1	0	0	0	-1	-1	0	0
	3	1	0	1	0	1	1	1	0	0	0	1	0	0	0	-1	0
	4	0	0	0	0	0	0	0	0	0	0	0	0	0	0	0	0
$\mathfrak{F}_{04}$	1	1	0	0	0	1	0	0	0	0	1	0	0	1	0	0	0
	2	0	1	0	0	0	1	0	0	-1	0	0	0	0	-1	0	0
	3	1	0	1	0	0	1	1	0	0	0	1	0	0	0	-1	0
	4	0	0	0	0	0	0	0	0	0	0	0	0	0	0	0	0
$\mathfrak{F}_{05}$	1	1	0	0	0	1	0	0	0	1	1	0	0	1	0	0	0
	2	0	1	0	0	0	1	0	0	-1	0	0	0	-1	-1	0	0
	3	0	-1	1	0	1	1	1	0	0	0	1	0	0	0	-1	0
	4	0	0	0	0	0	0	0	0	0	0	0	0	0	0	0	0
$\mathfrak{F}_{06}$	1	1	0	0	0	1	0	0	0	-1	0	0	0	0	1	0	0
	2	0	1	0	0	0	1	0	0	0	-1	0	0	1	0	0	0
	3	0	0	1	0	0	0	1	0	0	0	1	0	0	0	1	0
	4	1	-1	0	1	0	0	0	1	0	0	0	1	0	0	0	1
$\mathfrak{F}_{07}$	1	1	0	0	0	1	0	0	0	-1	0	0	0	1	0	0	0
	2	0	1	0	0	0	1	0	0	0	-1	0	0	0	-1	0	0
	3	0	0	1	0	0	0	1	0	0	0	1	0	0	0	1	0
	4	0	0	0	1	0	1	0	1	0	0	0	1	0	0	0	1
$\mathfrak{F}_{08}$	1	1	0	0	0	1	0	0	0	-1	0	0	0	1	0	0	0
	2	0	1	0	0	0	1	0	0	0	-1	0	0	0	-1	0	0
	3	0	0	1	0	0	0	1	0	0	0	1	0	0	0	1	0
	4	0	0	0	1	0	1	0	1	-1	0	0	1	0	0	0	1
$\mathfrak{F}_{09}$	1	1	0	0	0	1	0	0	0	0	1	0	0	1	0	0	0
	2	0	1	0	0	0	1	0	0	-1	0	0	0	0	-1	0	0
	3	1	1	1	0	-1	1	1	0	0	0	1	0	0	0	-1	0
	4	0	0	0	1	-1	1	0	1	0	0	1	1	0	0	1	1
$\mathfrak{F}_{10}$	1	1	0	0	0	1	0	0	0	0	1	0	0	1	0	0	0
	2	0	1	0	0	0	1	0	0	-1	0	0	0	0	-1	0	0
	3	1	0	1	0	0	1	1	0	0	0	1	0	0	0	-1	0
	4	0	0	0	1	0	1	0	1	0	0	1	1	0	0	0	1
$\mathfrak{F}_{11}$	1	1	0	0	0	1	0	0	0	-1	0	0	0	1	0	0	0
	2	0	1	0	0	0	1	0	0	0	-1	0	0	0	-1	0	0
	3	1	0	1	0	0	1	1	0	0	0	1	0	0	0	-1	0
	4	0	0	0	1	0	1	0	1	-1	0	0	1	0	0	0	1
$\mathfrak{F}_{12}$	1	1	0	0	0	1	0	0	0	0	1	0	0	1	0	0	0
	2	0	1	0	0	0	1	0	0	-1	0	0	0	0	-1	0	0
	3	1	0	1	0	0	1	1	0	0	0	1	0	0	0	-1	0
	4	0	0	0	1	0	1	0	1	0	1	1	1	0	-1	0	1
$\mathfrak{F}_{13}$	1	1	0	0	0	1	0	0	0	1	1	0	0	1	0	0	0
	2	0	1	0	0	0	1	0	0	-1	0	0	0	-1	-1	0	0
	3	1	-1	1	0	1	2	1	0	0	0	1	0	0	0	-1	0
	4	0	0	0	1	1	2	0	1	0	0	1	1	0	0	0	1
$\mathfrak{F}_{14}$	1	1	0	0	0	1	0	0	0	1	1	0	0	1	0	0	0
	2	0	1	0	0	0	1	0	0	-1	0	0	0	-1	-1	0	0
	3	1	0	1	0	1	1	1	0	0	0	1	0	0	0	-1	0
	4	1	0	0	1	0	0	0	1	0	0	1	1	0	0	0	1
$\mathfrak{F}_{15}$	1	1	0	0	0	1	0	0	0	0	1	0	0	1	0	0	0
	2	0	1	0	0	0	1	0	0	-1	0	0	0	0	-1	0	0
	3	1	0	1	0	0	1	1	0	0	0	1	0	0	0	-1	0
	4	0	1	0	1	0	1	0	1	0	0	2	1	0	0	0	1
$\mathfrak{F}_{16}$	1	1	0	0	0	1	0	0	0	0	1	0	0	1	0	0	0
	2	0	1	0	0	0	1	0	0	-1	0	0	0	0	-1	0	0
	3	1	0	1	0	0	1	1	0	0	0	1	0	0	0	-1	0
	4	0	1	0	1	0	1	0	1	0	1	2	1	0	-1	0	1
$\mathfrak{F}_{17}$	1	1	0	0	0	1	0	0	0	1	1	0	0	1	0	0	0
	2	0	1	0	0	0	1	0	0	-1	0	0	0	-1	-1	0	0
	3	0	-1	1	0	1	1	1	0	0	0	1	0	0	0	-1	0
	4	0	0	0	1	1	2	0	1	0	0	2	1	0	0	0	1

Table 6

3.5 THE GENERATOR METHOD

1° Now let us develop specific algorithms, by which the problem of classification of chromatic ornamental classes can be solved. In the present section, we shall describe one such algorithm, using the characterization of ornamental groups in terms of generators and relations. When conjoined with the data stored in Tables 4 and 6, this algorithm will in principle enumerate all chromatic ornamental classes and, for each such class, record a representative member.

In the following section, we shall develop a second algorithm, exploiting the identification of chromatic ornamental classes with certain classes of subgroups of ornamental groups. When supplied with the information stored in Tables 3 and 5, this algorithm will also solve the problem of classification.

We shall refer to these algorithms as the _generator method_ and the _subgroup method_, respectively, denoting them by GENMT and SUBMT. The first is relatively easy to implement, and provides highly reliable results. However, it is inefficient, in that it determines the chromatic ornamental classes literally by calculating all ornamental homomorphisms having one of the 17 given ornamental groups as domain, then assembling them in classes by equivalence. In practice, this method yields results of limited range. We have chosen to implement the generator method primarily to produce dependable data against which to check the validity of implementation of the second, more comprehensive method.

The subgroup method is relatively more difficult to implement, but it is efficient. It determines the chromatic ornamental classes by selective calculation of various subgroups of the 17 given ornamental groups. By this method, we have obtained results for a range of colors from 1 through 60.

2° We plan to develop the algorithms GENMT and SUBMT in conventional mathematical terms, without recourse to specific computer languages. However, with the present discussion in mind, one may proceed independently to implement these algorithms in a particular language, for computation by a particular machine.

For GENMT, we shall make use of a suitable diagram to organize the computations. For SUBMT, we shall rely upon verbal analysis to reduce the computations, layer by layer, to the simplest steps.

3° Let us set the framework of computation by introducing an upper bound n^* for the number of colors to be considered. Moreover, let us assume that the number n of colors $(2 \leq n \leq n^*)$ and the domain group $\mathscr{F}_m$ $(1 \leq m \leq 17)$ have been selected. In effect, then, we shall treat the algorithms GENMT and SUBMT themselves as subalgorithms, lying within an ambient algorithm which generates n and m in suitable order.

In Figure 88, one will find a simple diagram illustrating the ambient algorithm. Bold rectangles mark the beginning and the end of the algorithm. Arrowheads direct the flow of computation. Circles stand for reading and writing instructions, rhombii, for tests, and rectangles, for specific operations. The bordered rectangle signifies a subalgorithm; in the present instance, it would be either GENMT or SUBMT.

For example, the rectangle

$$\boxed{n \leftarrow 2}$$

causes the value of n to be set at 2; the rectangle

$$\boxed{n \leftarrow n+1}$$

causes the value of n to be increased by 1. The rhombus

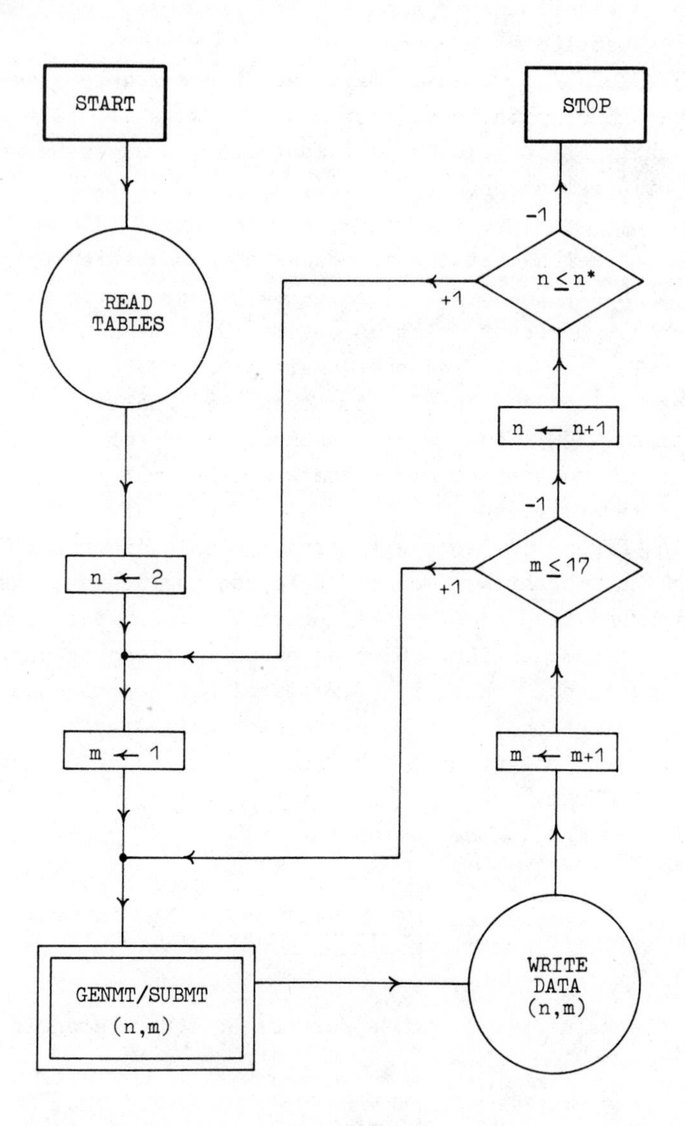
START
STOP
-1
READ
TABLES
n ≤ n*
+1
n ← n+1
-1
n ← 2
m ≤ 17
+1
m ← 1
m ← m+1
GENMT/SUBMT
(n,m)
WRITE
DATA
(n,m)

Figure 88

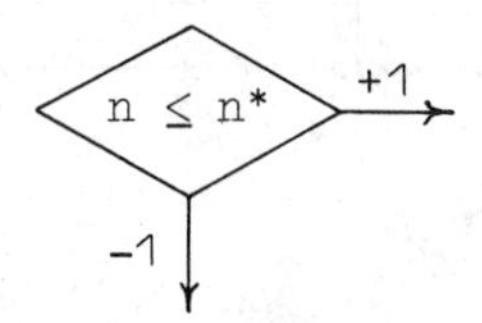

signifies the test of the condition: $n \leq n^*$. When the test proves positive, then the branch marked +1 should be followed, when negative, -1. The circle

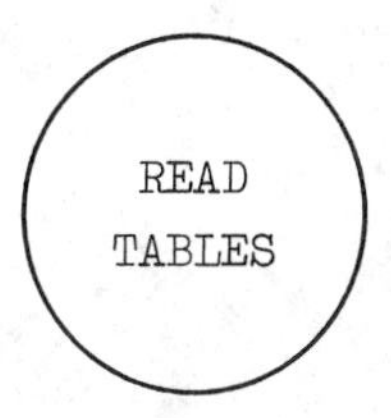

causes certain prepared data files to be read, and rendered available for subsequent operations. For the (sub)algorithm GENMT, Tables 4 and 6 would be appropriate; for the (sub)algorithm SUBMT, Tables 3 and 5, augmented by certain auxiliary data to be described in the next section. The circle

causes a count of the n-chromatic ornamental classes (based upon the domain group $\mathscr{F}_m$) to be placed on file, together with representative members of each of those classes. Of course, responsibility for obtaining such data lies with the (sub)algorithm

GENMT/SUBMT (n,m)

.

4° Let us prepare to design the algorithm GENMT.

Let n and m stand for positive integers satisfying the conditions $2 \leq n \leq n^*$ and $1 \leq m \leq 17$. By $2.5.8^{\circ}$ and 4.4°, we may introduce members

$$\underline{U} = [u, I],$$
$$\underline{V} = [v, I],$$
$$\underline{W} = [0, W],$$
$$\underline{X} = [x, X],$$

of $\mathscr{F}_m$, and members

$$\underline{U}^{\triangle} = [u^{\triangle}, I],$$
$$\underline{V}^{\triangle} = [v^{\triangle}, I],$$
$$\underline{W}^{\triangle} = [0, W^{\triangle}],$$
$$\underline{X}^{\triangle} = [0, X^{\triangle}],$$

of $\mathscr{F}_m^{\triangle}$. [See Tables 3 and 5.] The former generate $\mathscr{F}_m$, and the latter, through the inner automorphisms which they implement, the automorphism group of $\mathscr{F}_m$.

Let f be any mapping carrying $\mathscr{F}_m$ to S_N. Let the images in S_N under f of the indicated generators for $\mathscr{F}_m$ be denoted as follows:

$$\underline{u} = f(\underline{U}),$$
$$\underline{v} = f(\underline{V}),$$
$$\underline{w} = f(\underline{W}),$$
$$\underline{x} = f(\underline{X}).$$

When $1 \leq m \leq 5$, then one should interpret $\underline{x}$ to be ε, the identity permutation on N. Clearly, in order that f be a homomorphism, it is necessary and sufficient that $\underline{u}$, $\underline{v}$, $\underline{w}$, and $\underline{x}$ satisfy the relations displayed in Table 4, the same relations which govern $\underline{U}$, $\underline{V}$, $\underline{W}$, and $\underline{X}$. In order that f be an n-chromatic homomorphism, it is necessary and sufficient that, in addition, $\underline{u}$, $\underline{v}$, $\underline{w}$, and $\underline{x}$ generate a transitive subgroup of S_N.

With reference to 3.4^o, we may now identify the set $\mathbb{F}^n_m$, consisting of all n-chromatic homomorphisms having domain $\mathscr{F}_m$, with the set of all ordered quadruples of permutations on N, just described. Moreover, we may characterize the action of $\mathscr{F}_m^{\Delta} \times S_N$ on $\mathbb{F}^n_m$ in the following manner.

Let f' and f" be any two members of $\mathbb{F}^n_m$, and let $(\underline{u}',\underline{v}',\underline{w}',\underline{x}')$ and $(\underline{u}'',\underline{v}'',\underline{w}'',\underline{x}'')$ be the ordered quadruples of permutations on N with which they are identified. Let (A,ρ) be any member of $\mathscr{F}_m^{\Delta} \times S_N$. From relation (2) in 3.4^o, we obtain $(A,\rho).f' = f''$ iff:

$$\begin{aligned} \rho\underline{u}'\rho^{-1} &= f''(J_A(\underline{U})), \\ \rho\underline{v}'\rho^{-1} &= f''(J_A(\underline{V})), \\ \rho\underline{w}'\rho^{-1} &= f''(J_A(\underline{W})), \\ \rho\underline{x}'\rho^{-1} &= f''(J_A(X)). \end{aligned} \qquad (1)'$$

Of course, J_A may be described as follows:

$$\begin{aligned} A\,\underline{U}\,A^{-1} &= \underline{U}^{eA11}\underline{V}^{eA12}\underline{W}^{eA13}\underline{X}^{eA14}, \\ A\,\underline{V}\,A^{-1} &= \underline{U}^{eA21}\underline{V}^{eA22}\underline{W}^{eA23}\underline{X}^{eA24}, \\ A\,\underline{W}\,A^{-1} &= \underline{U}^{eA31}\underline{V}^{eA32}\underline{W}^{eA33}\underline{X}^{eA34}, \\ A\,\underline{X}\,A^{-1} &= U^{eA41}V^{eA42}W^{eA43}X^{eA44}, \end{aligned}$$

where the eAjk $(1 \leq j,k \leq 4)$ are suitable integers. Relations (1)' now become:

$$\rho \underline{u}'\rho^{-1} = \underline{u}''^{eA11}\underline{v}''^{eA12}\underline{w}''^{eA13}\underline{x}''^{eA14},$$
$$\rho \underline{v}'\rho^{-1} = \underline{u}''^{eA21}\underline{v}''^{eA22}\underline{w}''^{eA23}\underline{x}''^{eA24},$$
$$\rho \underline{w}'\rho^{-1} = \underline{u}''^{eA31}\underline{v}''^{eA32}\underline{w}''^{eA33}\underline{x}''^{eA34}, \qquad (1)''$$
$$\rho \underline{x}'\rho^{-1} = \underline{u}''^{eA41}\underline{v}''^{eA42}\underline{w}''^{eA43}\underline{x}''^{eA44}.$$

When A is any one among $\underline{U}^{\triangle}$, $\underline{V}^{\triangle}$, $\underline{W}^{\triangle}$, and $\underline{X}^{\triangle}$, then the integers $eAjk$ $(1 \leq j,k \leq 4)$ can be found in Table 6.

At this point, let us make use of the fact that the permutations

$$\mu: \quad \begin{pmatrix} 1 & 2 & 3 & \dots & n \\ 2 & 3 & 4 & \dots & 1 \end{pmatrix}$$

and $$\nu: \quad \begin{pmatrix} 1 & 2 & 3 & \dots & n \\ 2 & 1 & 3 & \dots & n \end{pmatrix}$$

generate S_N. [See problem 6 in 9.] Consequently, the members

$$(\underline{U}^{\triangle}, \mu),$$
$$(\underline{V}^{\triangle}, \mu),$$
$$(\underline{W}^{\triangle}, \mu),$$
$$(\underline{X}^{\triangle}, \mu),$$
$$(\underline{U}^{\triangle}, \nu), \qquad (2)$$
$$(\underline{V}^{\triangle}, \nu),$$
$$(\underline{W}^{\triangle}, \nu),$$
$$(\underline{X}^{\triangle}, \nu),$$

of $\mathscr{F}_m{}^{\triangle} \times S_N$ determine the action of $\mathscr{F}_m{}^{\triangle} \times S_N$ on $\mathbb{F}_m^n$.

Now one may expect to calculate the orbits in $\mathbb{F}_m^n$ under $\mathscr{F}_m{}^{\triangle} \times S_N$ by making iterative application of the foregoing eight members of $\mathscr{F}_m{}^{\triangle} \times S_N$, subject to relations (1)".

Clearly, to determine $\mathbb{F}_m^n$, one must be able to produce for inspection, all ordered quadruples of permutations on N. For this purpose, we require an algorithm which will operate upon the set S_N to generate, from a given member of S_N, all members by iteration. Such an algorithm may be characterized as a bijective mapping M carrying S_N to itself, for which the sequence

$$M^i(\varepsilon) \qquad (0 \leq i < n!)$$

displays each member of S_N exactly once. In this context, let us introduce ω to stand for $M^j(\varepsilon)$, where $j = n! - 1$. Clearly, $M(\omega) = \varepsilon$.

Let NEXPM designate an algorithm of the sort just described. Moreover, let us equip the algorithm with a switch, to be set at +1 when the permutation received by the algorithm is ω, and at -1 otherwise. The algorithm will be represented as follows:

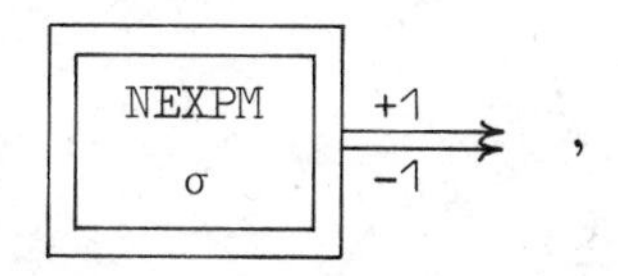

where σ stands for the permutation received. After an application of NEXPM, the flow of computation would proceed along one of two branches, according to the value of the switch.

Now let H be any finite set, and let $\mathscr{P}$ be any finite family of permutations on H. One must be able to calculate the orbits in H under the action of (the group of permutations on H generated by) $\mathscr{P}$, and, for each such orbit, to display a representative member. This calculation would figure in the algorithm GENMT, in two ways. First, it would help to determine the set $\mathbb{F}^n_m$, by indicating whether a given ordered quadruple $(\underline{u},\underline{v},\underline{w},\underline{x})$ of permutations on N generates a transitive subgroup of S_N. For this case, H would be N and $\mathscr{P}$ would be $\{\underline{u},\underline{v},\underline{w},\underline{x}\}$. Second, it would yield the orbits in $\mathbb{F}^n_m$ under the action of

$\mathscr{F}_m^{\triangle} \times S_N$. For this case, H would be $\mathbb{F}_m^n$ and $\mathscr{P}$ would be (in effect) the set of eight members of $\mathscr{F}_m^{\triangle} \times S_N$, displayed in item (2).

Let ORBKT stand for an algorithm which, upon receiving H and $\mathscr{P}$, yields the orbits and representatives of the orbits in H under $\mathscr{P}$. In addition, let us equip ORBKT with a switch, to be set at +1 if the number of orbits proves to be one, and at -1 otherwise. The algorithm will be represented as follows:

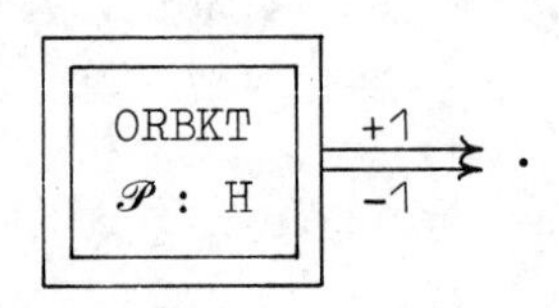

5° For the present, let us assume that the algorithms NEXPM and ORBKT are available, and let us proceed to design the algorithm GENMT. In the subsequent articles 6° and 7°, we will introduce particular realizations of NEXPM and ORBKT.

With reference to Table 4, let us designate the relations governing $(\underline{u},\underline{v},\underline{w},\underline{x})$, in the following manner:

REL1 : $\underline{u}\,\underline{v}\,\underline{u}^{-1} = \underline{v}$

REL2 : $\underline{w}^k = \varepsilon$

REL3 : $\underline{w}\,\underline{u}\,\underline{w}^{-1} =$

REL4 : $\underline{w}\,\underline{v}\,\underline{w}^{-1} =$

REL5 : $\underline{x}^2 =$

REL6 : $\underline{x}\,\underline{u}\,\underline{x}^{-1} =$

REL7 : $\underline{x}\,\underline{v}\,\underline{x}^{-1} =$

REL8 : $\underline{x}\,\underline{w}\,\underline{x}^{-1} =$.

Of course, for all but the first case, the specific content

of each relation depends upon the particular domain group $\mathscr{F}_m$ under consideration. From Table 4, one can obtain the information necessary to complete the description of these relations.

Now let us consider the algorithm GENMT itself. In Figure 89, we have constructed a diagram suitable to explain the routine. The bulk of the diagram serves to determine the set $\mathbb{F}_m^n$. Thus, the rectangle

$$\begin{array}{|l|}\hline \underline{u} \leftarrow \varepsilon \\ \underline{v} \leftarrow \varepsilon \\ \underline{w} \leftarrow \varepsilon \\ \underline{x} \leftarrow \varepsilon \\ \hline \end{array}$$

sets the initial value of the ordered quadruple $(\underline{u},\underline{v},\underline{w},\underline{x})$ at $(\varepsilon,\varepsilon,\varepsilon,\varepsilon)$. The rhombii

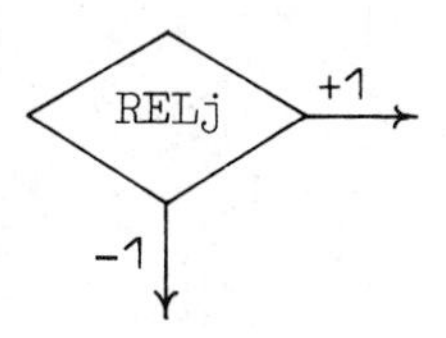

$(1 \leq j \leq 8)$ and the subalgorithm ORBKT test whether a given ordered quadruple $(\underline{u},\underline{v},\underline{w},\underline{x})$ is a member of $\mathbb{F}_m^n$. When all the tests prove positive, the circle

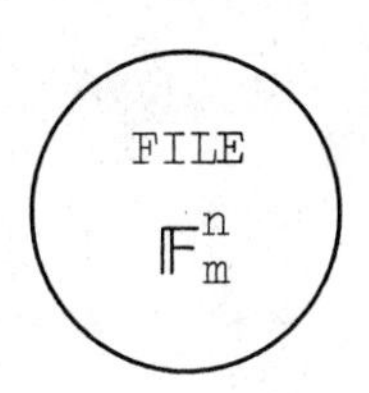

places $(\underline{u},\underline{v},\underline{w},\underline{x})$ on file. The several applications of the subalgorithm NEXPM cause any given ordered quadruple $(\underline{u},\underline{v},\underline{w},\underline{x})$ to be advanced to another, at which point the

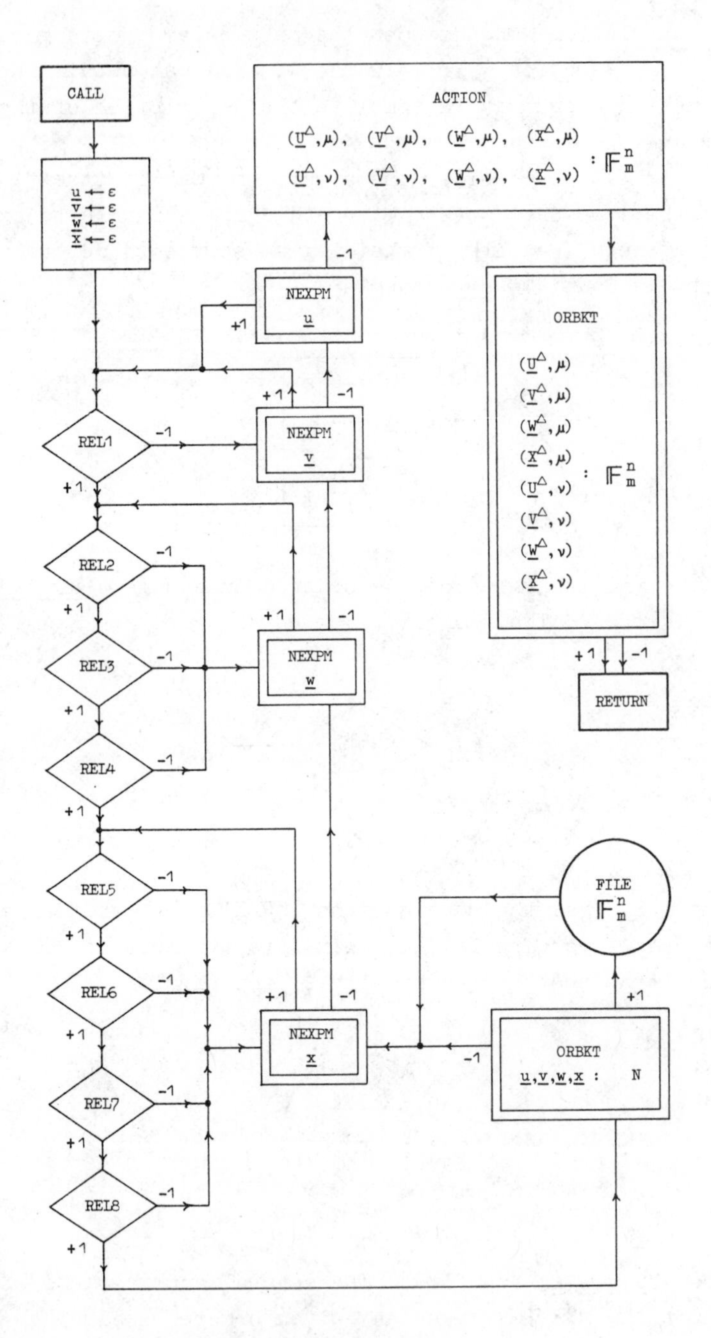
CALL
u ← ε
v ← ε
w ← ε
x ← ε
ACTION
(U△, μ), (V△, μ), (W△, μ), (X△, μ)
(U△, ν), (V△, ν), (W△, ν), (X△, ν) : F n m
NEXPM u
NEXPM v
NEXPM w
NEXPM x
ORBKT
(U△, μ)
(V△, μ)
(W△, μ)
(X△, μ)
(U△, ν)
(V△, ν)
(W△, ν)
(X△, ν)
: F n m
RETURN
REL1
REL2
REL3
REL4
REL5
REL6
REL7
REL8
+1
-1
FILE
F n m
ORBKT
u,v,w,x : N

Figure 89

conducting of tests continues. Upon inspection, one will see that GENMT has been designed to survey the conditions for membership in $\mathbb{F}_m^n$ as efficiently as possible.

When $\mathbb{F}_m^n$ has been determined, then GENMT proceeds to calculate the action of the eight distinguished members of $\mathscr{F}_m^{\triangle} \times S_N$ on $\mathbb{F}_m^n$. [See relations (1)" and item (2), in the preceding article.] In practice, the form of this calculation will depend upon the coding method by which the members of $\mathbb{F}_m^n$ were filed. For example, the members of $\mathbb{F}_m^n$ may be identified in serial order, by an index i ($1 \leq i \leq h$), where h is the number of members of $\mathbb{F}_m^n$. Moreover, the eight distinguished members of $\mathscr{F}_m^{\triangle} \times S_N$ may be identified by an index q ($1 \leq q \leq 8$). The action calculation would determine, for any integers i' and q ($1 \leq i' \leq h$, $1 \leq q \leq 8$), the integer i" ($1 \leq i" \leq h$) such that the action of q upon i' yields i". In the diagram on the opposite page, the action calculation is represented by the rectangle

ACTION

$$\begin{matrix} (\underline{U},\mu), & (\underline{V},\mu), & (\underline{W},\mu), & (\underline{X},\mu) \\ (U,\nu), & (V,\nu), & (W,\nu), & (X,\nu) \end{matrix} : \mathbb{F}_m^n$$

Thereafter, the way is clear for application of the subalgorithm ORBKT, to produce the orbits (and representatives of the orbits) in $\mathbb{F}_m^n$ under the action of $\mathscr{F}_m^{\triangle} \times S_N$.

6° Now let us describe a particular version of the algorithm NEXPM, using the numerical analogue of alphabetization of permutations. We are indebted to D. Alvis for suggesting this approach to the algorithm. For a study of several alternative versions of NEXPM, one might consult the text by E. Reingold.

Given two permutations σ' and $\sigma"$ on N, let us say that σ' <u>precedes</u> σ'' iff there exists a member j of N such that:

> for each member i of N, if $i < j$ then $\sigma'(i) = \sigma''(i)$;
>
> and $\sigma'(j) < \sigma''(j)$.

Clearly, the relation of precedence on S_N is a linear order relation. The initial permutation in S_N would be the identity:

$$\varepsilon : \quad \begin{pmatrix} 1 & 2 & 3 & \dots & n \\ 1 & 2 & 3 & \dots & n \end{pmatrix},$$

and the terminal, the permutation:

$$\omega : \quad \begin{pmatrix} 1 & 2 & 3 & \dots & n \\ n & \dots & 3 & 2 & 1 \end{pmatrix}.$$

Let M stand for the mapping carrying S_N to itself, which assigns to ω the image ε, and which assigns to each permutation σ in $S_N \setminus \{\omega\}$, the immediate successor of σ under the relation of precedence. Beginning with ε, one may generate all permutations on N, by iterative application of M:

$$M^j(\varepsilon) \qquad (0 \leq j < n!).$$

Let us describe in operational terms how to calculate the immediate successor of a given permutation in $S_N \setminus \{\omega\}$. Let σ be any permutation on N, and let j be any member of N. We shall say that j is a <u>transition integer</u> for σ iff:

> $1 \leq j < n$;
>
> for each member i of N, if $j < i < n$ then $\sigma(i+1) < \sigma(i)$;
>
> and $\sigma(j) < \sigma(j+1)$.

Obviously, the terminal permutation ω admits no transition integer. However, for each permutation σ in $S_N \setminus \{\omega\}$, there exists a transition integer, indeed, exactly one. For example, the permutation

$$\kappa : \quad \begin{pmatrix} 1 & 2 & 3 & 4 & 5 & 6 & 7 & 8 \\ 6 & 1 & 4 & 8 & 7 & 5 & 3 & 2 \end{pmatrix}$$

admits 3 as its transition integer.

Now let σ be any permutation in $S_N \setminus \{\omega\}$, and let j be its transition integer. Clearly, there exists precisely one integer k in N such that:

$j < k \leq n$;
$\sigma(j) < \sigma(k)$;
and either $k = n$ or $\sigma(k+1) < \sigma(j)$.

[For the illustrative permutation κ just cited, the value of k would be 6.] Let ρ be the permutation on N determined by the following conditions:

for each member i of N, if $1 \leq i < j$ then $\rho(i) = \sigma(i)$;
$\rho(j) = \sigma(k)$;
and for each member i of N, if $j < i < n$ then $\rho(i) < \rho(i+1)$.

The third condition directs that, after the values of ρ have been assigned to the integers $1, 2, \ldots, j$, in accord with the first two conditions, the remaining values be assigned in ascending order to the integers $j+1, j+2, \ldots, n$. By design, ρ is the immediate successor of σ: $\rho = M(\sigma)$.

For the permutation κ cited earlier, the immediate successor would be the permutation

$$\lambda : \quad \begin{pmatrix} 1 & 2 & 3 & 4 & 5 & 6 & 7 & 8 \\ 6 & 1 & 5 & 2 & 3 & 4 & 7 & 8 \end{pmatrix}.$$

With reference to the foregoing discussion, one should have little difficulty implementing the algorithm NEXPM. As noted in 4°, the algorithm should be supplied with a switch.

7° Finally, let us develop an efficient version of the algorithm ORBKT. We are indebted to R. Mayer for the basic idea of this algorithm. For an alternative version, one may consult the paper by H. Brown. [See also problem 7 in 9.]

Let H be a finite set, and let $\mathscr{P}$ be a finite family of permutations on H. The objectives of ORBKT are to enumerate the orbits in H under (the group of permutations on H generated by) $\mathscr{P}$, and, for each such orbit, to record a representative member.

In practice, H and $\mathscr{P}$ will be identified (by suitable indices) with subsets of Z. Thus, let h stand for the number of members of H, and p, for the number of members of $\mathscr{P}$; and let us simply assume that:

$$H = \{1, 2, 3, \ldots, h\},$$

$$\mathscr{P} = \{1, 2, 3, \ldots, p\}.$$

The action of $\mathscr{P}$ on H must now be characterized as a mapping a carrying $\mathscr{P} \times H$ to H, where, for each integer q $(1 \leq q \leq p)$, the mapping

$$a_q(i \longmapsto a(q,i))$$

carrying H to itself, is a permutation.

The orbits in H under $\mathscr{P}$ can be determined by calculating, in succession, the orbits in H under the families $\{a_1\}$, $\{a_1,a_2\}$, ..., and $\{a_1,a_2,\ldots,a_p\}$. For each integer q $(1 \leq q \leq p)$, the permutation a_q causes various of the orbits in H under $\{a_1,a_2,\ldots,a_{q-1}\}$ to coalesce, producing the orbits in H under $\{a_1,a_2,\ldots,a_{q-1},a_q\}$. We plan to develop the algorithm ORBKT, by analyzing the process of "coalescence of orbits" under a_q $(1 \leq q \leq p)$.

Incidentally, when $q = 1$, then one should interpret $\{a_1,a_2,\ldots,a_{q-1}\}$ to be the empty set. The group of permutations on H generated by that set would be the trivial group, and the corresponding orbits in H would all be singletons.

Now let $\mathcal{J}$ be any partition of H, and let β be any permutation on H. By definition, the members of $\mathcal{J}$ are (nonempty) subsets of H. Moreover, the intersection of any two distinct members of $\mathcal{J}$ is empty, and the union of all the members of $\mathcal{J}$ is H. With reference to $\mathcal{J}$ and β, let us say that a (nonempty) subset S of H is <u>saturated</u> provided that:

$\beta(S) = S$;

and for each set J in $\mathcal{J}$, if $J \cap S$ is not empty then $J \subseteq S$.

The latter condition simply means that S is the union of certain sets in $\mathcal{J}$. Let us say that a (nonempty) subset S of H is <u>minimally saturated</u> provided that:

S is saturated;

and for each saturated subset S' of H, if $S' \subseteq S$ then $S' = S$.

One may describe the minimally saturated subsets of H as those which arise by coalescing certain of the sets in $\mathcal{J}$, as demanded by the action of β. One can readily verify that the minimally saturated subsets of H compose a partition of H. Let that partition be denoted by $\mathcal{I}$.

Now let us consider how to determine $\mathcal{I}$ from $\mathcal{J}$ and β, by efficient calculation. In this way, we shall realize our original objective: to analyze the process of coalescence of orbits under a_q $(1 \le q \le p)$. Indeed, when $\mathcal{J}$ is the orbit structure in H under $\{a_1, a_2, \ldots, a_{q-1}\}$ and when β is a_q, then $\mathcal{I}$ would be the orbit structure in H under $\{a_1, a_2, \ldots, a_{q-1}, a_q\}$.

To discuss the method of calculation, we require a special scheme for representing partitions of H. Of course, the intent of this scheme is to characterize a given partition of H in terms convenient to the description of coalescence of sets. Thus, let $\mathcal{J}$ stand in general for a partition of H. Let n, s, and ℓ be mappings carrying H to

Z, satisfying the following conditions.

$\mathscr{n}$: For each member i of H, the members of the set J in $\mathscr{J}$ to which i belongs, are the following:

$$\mathscr{n}^c(i) \qquad (1 \leq c \leq k),$$

where k is the number of members of J.

$\mathscr{s}$: For each member i of H, $\mathscr{s}(i)$ is the smallest member of the set J in $\mathscr{J}$ to which i belongs.

ℓ: For each member i of H, if $\mathscr{n}(i) = \mathscr{s}(i)$ then $\ell(i) = 1$, while if $\mathscr{n}(i) \neq \mathscr{s}(i)$ then $\ell(i) = 0$.

In effect, then, $\mathscr{n}$ is a permutation on H, for which the cycle structure coincides with $\mathscr{J}$. For each member i of H, one should interpret $\mathscr{n}(i)$ as the "next" member of the set J in $\mathscr{J}$ to which i belongs.

Clearly, for each set J in $\mathscr{J}$, there is exactly one member j of J for which $\ell(j) = 1$.

For each member i of H, one should interpret $\mathscr{s}(i)$ as the "first" member of the set J in $\mathscr{J}$ to which i belongs. Under the cycle structure determined by $\mathscr{n}$, one should then take the unique member j of J for which $\ell(j) = 1$, to be the "last" member of J. The members of J may be displayed as follows:

$$\begin{aligned}
&\mathscr{n}^0(\mathscr{s}(i)) = \mathscr{s}(i), \\
&\mathscr{n}^1(\mathscr{s}(i)), \\
&\vdots \\
&\mathscr{n}^{k-2}(\mathscr{s}(i)), \\
&\mathscr{n}^{k-1}(\mathscr{s}(i)) = j,
\end{aligned}$$

where k is the number of members of J. In this context, $\ell(\mathscr{n}^{k-1}(\mathscr{s}(i))) = 1$.

Now let $\mathscr{J}$ be any partition of H, and let $(\mathscr{n}', \mathscr{s}', \mathscr{l}')$ be an ordered triple of mappings, representing $\mathscr{J}$ in the sense of the preceding discussion. Let β be any permutation on H. We intend to calculate the partition $\mathscr{I}$ of H, determined from $\mathscr{J}$ and β, by considering how $(\mathscr{n}', \mathscr{s}', \mathscr{l}')$ transforms under the action of β. The resulting ordered triple $(\mathscr{n}'', \mathscr{s}'', \mathscr{l}'')$ of mappings will represent the partition $\mathscr{I}$. To describe the transformation of $(\mathscr{n}', \mathscr{s}', \mathscr{l}')$ to $(\mathscr{n}'', \mathscr{s}'', \mathscr{l}'')$, we shall introduce a subordinate algorithm COALQ, which will serve to determine, for a given member i of H, whether (and, if so, in what manner) the sets containing i and $\beta(i)$ should coalesce. Iterative application of COALQ will yield the desired result.

Thus, let i° be any member of H, and let $(\mathscr{n}^{\circ}, \mathscr{s}^{\circ}, \mathscr{l}^{\circ})$ be an ordered triple of mappings of the sort which represents a partition of H. The effect of COALQ is to transform the ordered quadruple $(i^{\circ}, \mathscr{n}^{\circ}, \mathscr{s}^{\circ}, \mathscr{l}^{\circ})$ to another such ordered quadruple, according to the following (complementary) rules.

[1] When $\mathscr{s}^{\circ}(i^{\circ}) < \mathscr{s}^{\circ}(\beta(i^{\circ}))$, then the set J° containing i° and the set K° containing $\beta(i^{\circ})$ are distinct, and hence must coalesce. To achieve this effect, one would modify $\mathscr{n}^{\circ}$ so that it carries the last member of J° to the first member of K°, and the last member of K° to the first member of J°. One would also modify $\mathscr{s}^{\circ}$, by changing the common value of $\mathscr{s}^{\circ}$ on K° to $\mathscr{s}^{\circ}(i^{\circ})$ (so that it coincides with the common value of $\mathscr{s}^{\circ}$ on J°). Moreover, one would change the value of $\mathscr{l}^{\circ}$ at the last member of J°, from 1 to 0.
Finally, one would advance i° to $\mathscr{n}^{\circ}(i^{\circ})$.

[2] When $\mathscr{s}^{\circ}(i^{\circ}) = \mathscr{s}^{\circ}(\beta(i^{\circ}))$, then the sets J° and K° containing i° and $\beta(i^{\circ})$, respectively, are the same, and no modification of $\mathscr{n}^{\circ}$, $\mathscr{s}^{\circ}$, or $\mathscr{l}^{\circ}$ would be necessary.

[2] [x] Under the condition: $\ell^{o}(i^{o}) = 1$, one would recognize that i^{o} is the last member of J^{o}, so that, for J^{o}, the process of coalescence is finished. One would then search through H for the smallest member i, among all members i' having the property that $s^{o}(i^{o}) < s^{o}(i')$.

[2] [x] [+] Having found i, one would change i^{o} to i.

[2] [x] [-] Failing to find i (for the reason that no such members i' exist), one would realize that J^{o} was the last of the sets to be considered, and hence that nothing remained to be done. One would leave i^{o} unchanged.

[2] [y] Under the condition: $\ell^{o}(i^{o}) \neq 1$, one would infer that i^{o} is not the last member of J^{o}. Hence, one would advance i^{o} to $n^{o}(i^{o})$.

Now one can effect the calculation of $\mathcal{I}$ from $\mathcal{J}$ and β, by taking $(i^{o}, n^{o}, s^{o}, \ell^{o})$ initially to be $(1, n', s', \ell')$, and by applying COALQ iteratively until the terminal condition [2][x][-] arises. [In fact, one would apply COALQ h times.] At that point, $(i^{o}, n^{o}, s^{o}, \ell^{o})$ would equal (i'', n'', s'', ℓ''), where the ordered triple (n'', s'', ℓ'') represents the partition $\mathcal{I}$ and where i" happens to be the last member of the last set in $\mathcal{I}$. The calculation would then be finished.

Since the initial value of i^{o} is 1, the various ordered quadruples $(i^{o}, n^{o}, s^{o}, \ell^{o})$ which result from successive applications of COALQ, must all have the property that: $s^{o}(i^{o}) \leq s^{o}(\beta(i^{o}))$. Hence, the conditions which discriminate parts [1] and [2] of COALQ, are sufficient to govern the flow of computation.

Let us illustrate how the algorithm COALQ works. Let h be 8, let $\mathcal{J}$ be the partition of H consisting of the sets

{1,3,8}, {2,7}, {4,6}, {5},

and let β be the permutation

$$\beta: \qquad \begin{pmatrix} 1 & 2 & 3 & 4 & 5 & 6 & 7 & 8 \\ 3 & 2 & 1 & 5 & 4 & 8 & 7 & 6 \end{pmatrix}.$$

Let $\mathcal{J}$ be represented in the compressed form:

$$\mathcal{J}: \qquad (138)(27)(46)(5),$$

which serves at once to define (in terms of cycles) a suitable permutation n, and to display the first (that is, the smallest) and the last members of each cycle. Now we may describe the successive applications of COALQ, as follows:

i°	$(n^{\circ}, s^{\circ}, \ell^{\circ})$	COALQ
1	(138)(27)(46)(5)	[2y]
3	(138)(27)(46)(5)	[2y]
8	(138)(27)(46)(5)	[1]
4	(13846)(27)(5)	[1]
6	(138465)(27)	[2y]
5	(138465)(27)	[2x+]
2	(138465)(27)	[2y]
7	(138465)(27)	[2x-]
7	(138465)(27)	stop

In the column at the left, we have kept a record of the values of i°. In the column at the right, we have noted the particular computational branch of COALQ by which the given line yields the next. From this computation, one can see that $\mathcal{I}$ must be the following:

$\{1,3,8,4,6,5\}, \quad \{2,7\}.$

That is:

$$\mathscr{I}: \qquad (138465)(27).$$

The design of the algorithm ORBKT should now be self-evident. Given the set H and the family $\mathscr{P}$:

$$a_q \qquad (1 \leq q \leq p)$$

of permutations on H, one would proceed as follows to calculate the orbit structure in H under the action of $\mathscr{P}$. One would first introduce the trivial partition $\mathscr{I}_0$ of H:

$$\{1\}, \quad \{2\}, \quad \ldots, \quad \{h\},$$

then calculate the partition $\mathscr{I}_1$ of H determined by $\mathscr{I}_0$ and a_1. In turn, one would calculate the partition $\mathscr{I}_2$ of H determined by $\mathscr{I}_1$ and a_2. After p steps, one would obtain the partition $\mathscr{I}_p$ of H consisting of the orbits in H under $\mathscr{P}$:

$$\begin{array}{lcl} \mathscr{I}_0, \ a_1 & \longrightarrow & \mathscr{I}_1 \ , \\ \mathscr{I}_1, \ a_2 & \longrightarrow & \mathscr{I}_2 \ , \\ \quad \vdots & & \ \vdots \\ \mathscr{I}_{p-1}, a_p & \longrightarrow & \mathscr{I}_p \ . \end{array}$$

At that point, one would have in hand an ordered triple (n, s, ℓ) representing $\mathscr{I}_p$. The number of orbits in H under $\mathscr{P}$ may be determined by counting the members j of H for which $\ell(j) = 1$. Of course, those same members would serve as representatives of the corresponding orbits.

As noted in 4°, the algorithm ORBKT should be supplied with a switch.

3.6 THE SUBGROUP METHOD

1° Let us now concentrate upon the design of the algorithm SUBMT.

Let n be any positive integer for which $2 \leq n \leq n^*$, as proposed in 5.3°. By the following proposition, we shall establish that the problem of determining the n-chromatic ornamental classes coincides with the problem of determining certain classes of subgroups of the groups $\mathscr{F}_m$ $(1 \leq m \leq 17)$.

Let m be any positive integer for which $1 \leq m \leq 17$. Let k be any particular member of N: $1 \leq k \leq n$. As usual, let $\mathbb{F}_m^n$ stand for the set of all n-chromatic homomorphisms having domain $\mathscr{F}_m$.

<u>Proposition</u> <u>26</u>. For each member f of $\mathbb{F}_m^n$, the subset $\mathscr{Q}$ of $\mathscr{F}_m$ consisting of all members C for which $f(C)(k) = k$, is a subgroup having index n in $\mathscr{F}_m$. For each subgroup $\mathscr{Q}$ of $\mathscr{F}_m$ having index n in $\mathscr{F}_m$, there exists a member f of $\mathbb{F}_m^n$ which determines $\mathscr{Q}$, in the sense of the preceding assertion. For any members f' and f" of $\mathbb{F}_m^n$, f' and f" are equivalent iff the subgroups $\mathscr{Q}'$ and $\mathscr{Q}''$ of $\mathscr{F}_m$ which they determine, are conjugate in $\mathscr{F}_m^\Delta$.

Proof. Let f be any member of $\mathbb{F}_m^n$. Clearly, the set $\mathscr{Q}$ is the stabilizer of k under the action

$$((C,j) \longmapsto f(C)(j))$$

of $\mathscr{F}_m$ on N. Hence, it is a subgroup of $\mathscr{F}_m$. Since the range of f acts transitively on N, the action of $\mathscr{F}_m$ on N is transitive. Hence, the index of $\mathscr{Q}$ in $\mathscr{F}_m$ is n.

Now let $\mathscr{Q}$ be any subgroup of $\mathscr{F}_m$, having index n in $\mathscr{F}_m$. Let us introduce a bijective mapping b carrying the set $\mathscr{F}_m/\mathscr{Q}$ of left cosets of $\mathscr{Q}$ in $\mathscr{F}_m$, to N. Let us arrange that $b(\mathscr{Q}) = k$. Applying b, we may transfer the natural action of $\mathscr{F}_m$ on $\mathscr{F}_m/\mathscr{Q}$, to N:

$$C.b(D\mathscr{Q}) = b(CD\mathscr{Q}),$$

where C and D are any members of $\mathscr{F}_m$. Of course, the new action is transitive on N. From this action, we obtain the homomorphism

$$f(C \longmapsto f(C)(j \longmapsto C.j))$$

carrying $\mathscr{F}_m$ to S_N. Since the action of $\mathscr{F}_m$ on N is transitive, the range of f acts transitively on N. It follows that f is a member of $\mathbb{F}_m^n$. For each member C of $\mathscr{F}_m$, we have:

$$\begin{aligned} f(C)(k) = k \quad &\text{iff} \quad b(C\mathscr{Q}) = b(\mathscr{Q}) \\ &\text{iff} \quad C \in \mathscr{Q}. \end{aligned}$$

Hence, the subgroup of $\mathscr{F}_m$ determined by f coincides with $\mathscr{Q}$.

Finally, let f' and f" be members of $\mathbb{F}_m^n$, and let $\mathscr{Q}'$ and $\mathscr{Q}''$ be the subgroups of $\mathscr{F}_m$ which they determine. Let us assume first that f' and f" are equivalent. Thus, let A and ρ be members of $\mathscr{F}_m^{\triangle}$ and S_N, respectively, such that $f''J_A = J_\rho f'$. [See 3.4°.] Let j stand for $\rho(k)$. Since the range of f" acts transitively on N, we may introduce a member D of $\mathscr{F}_m$ such that $f''(D)(j) = k$. For each member C of $\mathscr{F}_m$, we have:

$$\begin{aligned} f'(C)(k) = k \quad &\text{iff} \quad (J_\rho f')(C)(j) = j \\ &\text{iff} \quad (f''J_A)(C)(j) = j \\ &\text{iff} \quad (f''J_D J_A)(C)(k) = k. \end{aligned}$$

Hence, $\mathscr{Q}'' = J_{DA}(\mathscr{Q}')$, so that $\mathscr{Q}'$ and $\mathscr{Q}''$ are conjugate in $\mathscr{F}_m^{\triangle}$.

Conversely, let us assume that $\mathscr{Q}'$ and $\mathscr{Q}''$ are conjugate in $\mathscr{F}_m^{\triangle}$. Thus, let A be a member of $\mathscr{F}_m^{\triangle}$ such that $\mathscr{Q}'' = J_A(\mathscr{Q}')$. For any members C' and C" of $\mathscr{F}_m$:

$$f'(C')(k) = f'(C'')(k)$$
$$\text{iff} \quad C'^{-1}C'' \in \mathscr{Q}'$$
$$\text{iff} \quad (AC'A^{-1})^{-1}(AC''A^{-1}) \in \mathscr{Q}''$$
$$\text{iff} \quad f''(J_A(C'))(k) = f''(J_A(C''))(k).$$

Hence, we may introduce a mapping ρ carrying N to itself, such that:

$$\rho(f'(C)(k)) = f''(J_A(C))(k),$$

where C is any member of $\mathscr{F}_m$. Clearly, ρ is injective, so it must be a permutation on N. Let j be any member of N, and let C be any member of $\mathscr{F}_m$. Let D be a member of $\mathscr{F}_m$ for which $f'(D)(k) = j$. We have:

$$\begin{aligned}
&(\rho^{-1}f''(ACA^{-1})\rho)(j) \\
&\quad = (\rho^{-1}f''(ACA^{-1}))(\rho(f'(D)(k))) \\
&\quad = (\rho^{-1}f''(ACA^{-1}))(f''(ADA^{-1})(k)) \\
&\quad = (\rho^{-1}f''(ACDA^{-1}))(k) \\
&\quad = f'(CD)(k) \\
&\quad = f'(C)(j).
\end{aligned}$$

Hence, $f''J_A = J_\rho f'$, so that f' and f'' are equivalent.
///

2° We have now established that there is a bijective correspondence between the equivalence classes of n-chromatic homomorphisms having domain $\mathscr{F}_m$, and the conjugacy classes in $\mathscr{F}_m^{\triangle}$ of subgroups of $\mathscr{F}_m$ having index n in $\mathscr{F}_m$. In particular, one can construct the homomorphisms from the subgroups, by means of the natural actions of $\mathscr{F}_m$ on the corresponding left coset spaces. Hereafter, we shall direct our attention to the problem of determining sufficiently many subgroups of $\mathscr{F}_m$, to represent the desired conjugacy classes in $\mathcal{F}_m^\triangle$. However, we shall report the results ultimately in terms not of subgroups but of homomorphisms.

3° Let T_m and $\mathcal{G}_m$ denote the translational and linear parts of $\mathcal{F}_m$, respectively. Let $\mathcal{F}_m^\circ$ be any subgroup of $\mathcal{F}_m$, having index n in $\mathcal{F}_m$. Let T_m° and $\mathcal{G}_m^\circ$ denote the translational and linear parts of $\mathcal{F}_m^\circ$, respectively. Of course, T_m° is a subgroup of T_m, and $\mathcal{G}_m^\circ$ is a subgroup of $\mathcal{G}_m$. Moreover, T_m° is invariant under $\mathcal{G}_m^\circ$.

Let n' be the index of T_m° in T_m, and n'', the index of $\mathcal{G}_m^\circ$ in $\mathcal{G}_m$.

<u>Proposition</u> <u>27</u>. Under the preceding conditions, $n = n'n''$.

Proof. Let us introduce the sequences of homomorphisms relating $\mathcal{F}_m$ to T_m and $\mathcal{G}_m$, and $\mathcal{F}_m^\circ$ to T_m° and $\mathcal{G}_m^\circ$, together with the appropriate inclusion mappings:

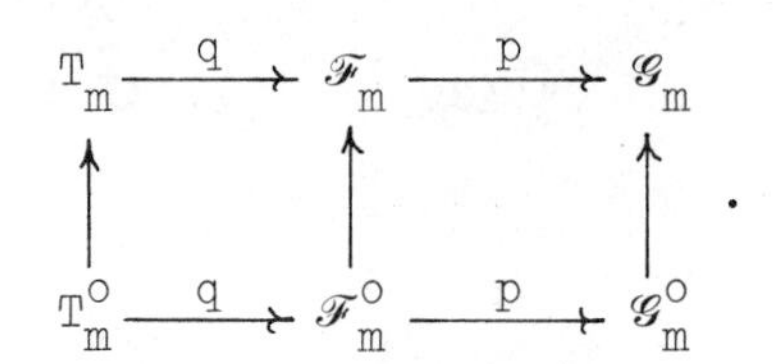

Clearly, the order k° of $\mathcal{G}_m^\circ$ equals the index of $q(T_m^\circ)$ in $\mathcal{F}_m^\circ$. Similarly, the order k of $\mathcal{G}_m$ equals the index of $q(T_m)$ in $\mathcal{F}_m$. Now one may express the index of $q(T_m^\circ)$ in $\mathcal{F}_m$, in two ways: as $k^\circ n$, and as $n'k$. Hence, $k^\circ n = n'k$. Since $n'' = k/k^\circ$, we may conclude that: $n = n'n''$.
///

4° Proposition 27 suggests the general outline of the algorithm SUBMT. Thus, for each factorization of n: $n = n'n''$, one must calculate the subgroups $\mathcal{G}_m^\circ$ of $\mathcal{G}_m$ having index n'' in $\mathcal{G}_m$; for each such subgroup $\mathcal{G}_m^\circ$, one must calculate the subgroups T_m° of T_m which have index n' in T_m and which are invariant under $\mathcal{G}_m^\circ$; and, for any such subgroups $\mathscr{G}_m^o$ and T_m^o, one must find the subgroups $\mathscr{F}_m^o$ of $\mathscr{F}_m$ which are compatible with $(T_m^o, \mathscr{G}_m^o)$, that is, for which the translational part equals T_m^o and the linear part equals $\mathscr{G}_m^o$. Of course, these calculations may, and should be selective. Ultimately, just one subgroup $\mathscr{F}_m^o$ of $\mathscr{F}_m$ from each conjugacy class in $\mathscr{F}_m^\triangle$ will be retained.

Let us refine the foregoing outline of SUBMT.

Let $T_m^\triangle$ and $\mathscr{G}_m^\triangle$ denote the translational and linear parts of $\mathscr{F}_m^\triangle$, respectively. One should recall that:

$$\mathscr{F}_m^\triangle = T_m^\triangle \rtimes \mathscr{G}_m^\triangle .$$

[See 4.]

Let $\mathscr{F}_m^{o\prime}$ and $\mathscr{F}_m^{o\prime\prime}$ be subgroups of $\mathscr{F}_m$, having index n in $\mathscr{F}_m$. Let us assume that $\mathscr{F}_m^{o\prime}$ and $\mathscr{F}_m^{o\prime\prime}$ are conjugate in $\mathscr{F}_m^\triangle$. Hence, we may introduce the following diagram:

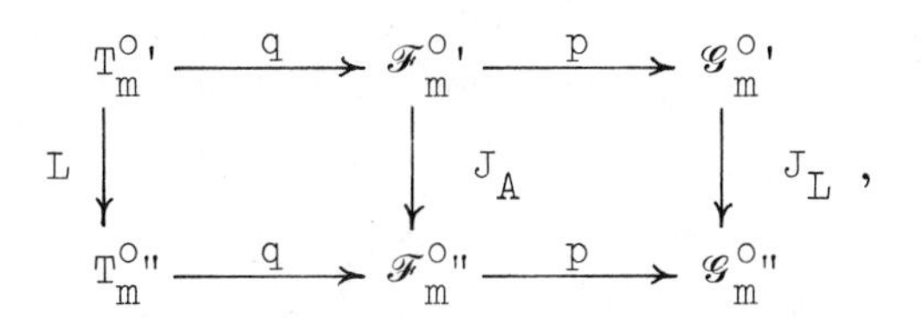

$$\begin{array}{ccccc} T_m^{o\prime} & \xrightarrow{q} & \mathscr{F}_m^{o\prime} & \xrightarrow{p} & \mathscr{G}_m^{o\prime} \\ L\downarrow & & \downarrow J_A & & \downarrow J_L \\ T_m^{o\prime\prime} & \xrightarrow{q} & \mathscr{F}_m^{o\prime\prime} & \xrightarrow{p} & \mathscr{G}_m^{o\prime\prime} \end{array},$$

where A is a suitable member of $\mathscr{F}_m^\triangle$ and where L is the linear part of A. Clearly, $\mathscr{G}_m^{o\prime}$ and $\mathscr{G}_m^{o\prime\prime}$ must be conjugate in $\mathscr{G}_m^\triangle$.

Conversely, let us assume that $\mathscr{G}_m^{o\prime}$ and $\mathscr{G}_m^{o\prime\prime}$ are conjugate in $\mathscr{G}_m^\triangle$. Hence, we may introduce a member L of $\mathscr{G}_m^\triangle$ such that $J_L(\mathscr{G}_m^{o\prime}) = \mathscr{G}_m^{o\prime\prime}$. Taking A to be L, we observe that the groups $\mathscr{F}_m^{o\prime}$ and $J_A(\mathscr{F}_m^{o\prime})$ are conjugate in $\mathscr{F}_m^\triangle$ and that the groups $J_A(\mathscr{F}_m^{o\prime})$ and $\mathscr{F}_m^{o\prime\prime}$ have the same linear part.

The foregoing facts yield the first step in SUBMT.

(a) Given any positive integer n″ which divides n, find one subgroup $\mathscr{G}_m^o$ from each conjugacy class in $\mathscr{G}_m^\triangle$ of subgroups of $\mathscr{G}_m$ having index n″ in $\mathscr{G}_m$.

Now let $\mathcal{G}_m^o$ be any one of the subgroups of $\mathcal{G}_m$ produced by step (**a**). Let $\mathcal{G}_m^{o\triangle}$ denote the normalizer of $\mathcal{G}_m^o$ in $\mathcal{G}_m^{\triangle}$. Let $\mathcal{F}_m^{o\prime}$ and $\mathcal{F}_m^{o\prime\prime}$ be subgroups of $\mathcal{F}_m$ having index n in $\mathcal{F}_m$, and let us assume that the linear parts of both $\mathcal{F}_m^{o\prime}$ and $\mathcal{F}_m^{o\prime\prime}$ equal $\mathcal{G}_m^o$. When $\mathcal{F}_m^{o\prime}$ and $\mathcal{F}_m^{o\prime\prime}$ are conjugate in $\mathcal{F}_m^{\triangle}$, then, by the foregoing diagram:

(*) there exists a member L of $\mathcal{G}_m^{o\triangle}$ such that $T_m^{o\prime\prime} = L(T_m^{o\prime})$.

Conversely, when condition (*) is satisfied, then $\mathcal{F}_m^{o\prime}$ is conjugate in $\mathcal{F}_m^{\triangle}$ to a subgroup of $\mathcal{F}_m$ which has linear part $\mathcal{G}_m^o$ and which has the same translational part as $\mathcal{F}_m^{o\prime\prime}$.

Let us express condition (*) by saying that $T_m^{o\prime}$ and $T_m^{o\prime\prime}$ are <u>conjugate</u> under $\mathcal{G}_m^{o\triangle}$. In terms of this relation, we may formulate the second step in SUBMT.

(**b**) Given $\mathcal{G}_m^o$, find one subgroup T_m^o from each conjugacy class under $\mathcal{G}_m^{o\triangle}$ of those subgroups of T_m which have index n/n" in T_m and which are invariant under $\mathcal{G}_m^o$.

Finally, let $\mathcal{G}_m^o$ and T_m^o be subgroups of $\mathcal{G}_m$ and T_m, respectively, stemming from steps (**a**) and (**b**). Let $\mathcal{G}_m^{oo}$ stand for the subgroup of $\mathcal{G}_m^{o\triangle}$, consisting of all members L for which $L(T_m^o) = T_m^o$. Let $\mathcal{F}_m^{o\prime}$ and $\mathcal{F}_m^{o\prime\prime}$ be subgroups of $\mathcal{F}_m$, having index n in $\mathcal{F}_m$. Let us assume that the translational parts of both $\mathcal{F}_m^{o\prime}$ and $\mathcal{F}_m^{o\prime\prime}$ equal T_m^o, and that the linear parts equal $\mathcal{G}_m^o$. Clearly, $\mathcal{F}_m^{o\prime}$ and $\mathcal{F}_m^{o\prime\prime}$ are conjugate in $\mathcal{F}_m^{\triangle}$ iff they are conjugate in $T_m^{\triangle} \rtimes \mathcal{G}_m^{oo}$. Now we may complete the description of SUBMT, by formulating the third step.

(**c**) Given T_m^o and $\mathcal{G}_m^o$, find one subgroup $\mathcal{F}_m^o$ from each conjugacy class in $T_m^{\triangle} \rtimes \mathcal{G}_m^{oo}$ of those subgroups of $\mathcal{F}_m$ for which the translational part is T_m^o and the linear part is $\mathcal{G}_m^o$.

In some cases, steps (**a**) and (**b**) will be abortive. That is, for a given integer n'' (dividing n), there may be no subgroups of $\mathscr{G}_m$ having index n'' in $\mathscr{G}_m$; and, for a given group $\mathscr{G}_m^o$, there may be no subgroups of T_m which have index n/n'' in T_m and which are invariant under $\mathscr{G}_m^o$.

Moreover, for the nonsymmorphic cases, step (**c**) may be abortive. However, for the symmorphic cases, step (**c**) proves always to be productive. In particular, for given groups T_m^o and $\mathscr{G}_m^o$, the group $T_m^o \rtimes \mathscr{G}_m^o$ will satisfy the conditions of that step.

5° Now let us discuss the means for implementing the foregoing steps (**a**), (**b**), and (**c**). We shall consider first the cases in which $1 \leq m \leq 2$, then the cases in which $3 \leq m \leq 17$. This division is a matter of convenience: the former cases are amenable to certain shortcuts, which substantially reduce the computation involved.

Throughout the discussion, we shall make use of the generators

$$\underline{U} = [u, I],$$
$$\underline{V} = [v, I],$$
$$\underline{W} = [0, W],$$
$$\underline{X} = [x, X],$$

for $\mathscr{F}_m$, from Table 3, and the members

$$\underline{U}^\triangle = [u^\triangle, I],$$
$$\underline{V}^\triangle = [v^\triangle, I],$$
$$\underline{W}^\triangle = [0, W^\triangle],$$
$$\underline{X}^\triangle = [0, X^\triangle],$$

of $\mathscr{F}_m^\triangle$, from Table 5, serving to generate the automorphism group of $\mathscr{F}_m$.

Let us assume that $m = 1$. For simplicity, we shall drop the subscript m from the foregoing symbols. In this case, $\mathscr{G}$ is $\mathscr{Z}_1$, $T^\triangle$ is R^2, and $\mathscr{G}^\triangle$ is $\mathscr{L}_T$. Clearly, the subgroups $\mathscr{F}^o$ of $\mathscr{F}$ having index n in $\mathscr{F}$, must be of the form: $T^o \rtimes \mathscr{Z}_1$, where T^o is a subgroup of T having index n in T. Moreover, two such subgroups $\mathscr{F}^{o\prime}$ and $\mathscr{F}^{o\prime\prime}$ of $\mathscr{F}$ are conjugate in $R^2 \rtimes \mathscr{L}_T$ iff $T^{o\prime}$ and $T^{o\prime\prime}$ are conjugate under $\mathscr{L}_T$. However, the conjugacy classes under $\mathscr{L}_T$ of subgroups of T having index n in T, may be identified with the square factors of n, that is, with the positive integers μ for which μ^2 divides n. [See problem 8 in 9.] For each such factor μ, the subgroup

$$T^\mu = Z.u^\mu + Z.v^\mu$$

of T is a representative of the corresponding conjugacy class, where

$$u^\mu = \mu.u, \qquad v^\mu = \frac{n}{\mu}.v .$$

As a result, the groups

$$\mathscr{F}^\mu = T^\mu \rtimes \mathscr{Z}_1 \qquad (\mu^2 | n)$$

represent the various conjugacy classes in $\mathscr{F}^\triangle$ of subgroups of $\mathscr{F}$ having index n in $\mathscr{F}$. To obtain the n-chromatic homomorphism f determined by $\mathscr{F}^\mu$, one need only calculate the actions of $\underline{U}$ and $\underline{V}$ on the left coset space $\mathscr{F}/\mathscr{F}^\mu$. Thus, the members

$$[i.u + j.v, I] \qquad (0 \leq i < \mu, \quad 0 \leq j < \frac{n}{\mu})$$

of $\mathscr{F}$ represent the various left cosets of $\mathscr{F}^\mu$ in $\mathscr{F}$, and the actions of $\underline{U}$ and $\underline{V}$ on $\mathscr{F}/\mathscr{F}^\mu$ can be described as follows:

$$\begin{aligned} \underline{U}(i,j) &= (i+1,j) \\ \underline{V}(i,j) &= (i,j+1) \end{aligned} \qquad (\mathrm{mod}(\mu, \tfrac{n}{\mu})) , \qquad (1)$$

where $0 \leq i < p$ and $0 \leq j < \frac{n}{p}$.

Now let us assume that $m = 2$. Let us again drop the subscript m. For this case, $\mathscr{G}$ is $\mathcal{Z}_2$, $T^\triangle$ is $\frac{1}{2}.T$, and $\mathscr{G}^\triangle$ is $\mathscr{L}_T$. When $n'' = 1$, then step (**a**) in SUBMT yields just one group $\mathscr{G}^o$, namely, $\mathcal{Z}_2$. When $n'' = 2$ (which would entail that n be even), then step (**a**) yields just the group $\mathcal{Z}_1$. In both cases, $\mathscr{G}^{o\triangle}$ would be $\mathscr{L}_T$. For all other values of n'', step (**a**) would be abortive.

Let us consider the case in which $n'' = 1$. By problem 8 in 9, step (**b**) will yield one subgroup T^p of T for each square factor p of n; let it be

$$T^p = Z.u^p + Z.v^p,$$

where

$$u^p = p.u, \qquad v^p = \frac{n}{p}.v .$$

For each subgroup $\mathscr{F}^o$ of $\mathscr{F}$ having linear part $\mathcal{Z}_2$ and translational part T, one may introduce a member t of T such that $[t,-I]$ is contained in $\mathscr{F}^o$. Let s stand for $-\frac{1}{2}.t$. Clearly, s is contained in $T^\triangle$, and:

$$[s,I]\mathscr{F}^o[s,I]^{-1} = T \rtimes \mathcal{Z}_2,$$

so that $\mathscr{F}^o$ and $T \rtimes \mathcal{Z}_2$ are conjugate in $\frac{1}{2}.T \rtimes \mathscr{G}^{oo}$. We infer that step (**c**) in SUBMT will yield just one subgroup of $\mathscr{F}$; let it be

$$\mathscr{F}^p = T^p \rtimes \mathcal{Z}_2.$$

To obtain the n-chromatic homomorphism f determined by $\mathscr{F}^p$, one should repeat the observations made earlier, under the assumption that $m = 1$. The action of $\underline{U}$ and $\underline{V}$ on $\mathscr{F}/\mathscr{F}^p$ can be described as before. [See relations (1).] The action of $\underline{W}$ can be described as follows:

$$\underline{W}(i,j) = (-i,-j) \qquad (\mathrm{mod}(p, \tfrac{n}{p})) , \tag{2}$$

where $0 \leq i < \rho$ and $0 \leq j < \frac{n}{\rho}$.

Let us consider the case in which $n'' = 2$. Step (**b**) will produce one subgroup T^{ρ} for each square factor ρ of $n/2$; let it be

$$T^{\rho} = Z.u^{\rho} + Z.v^{\rho},$$

where

$$u^{\rho} = \rho.u, \qquad v = \frac{n}{2\rho}.v .$$

For each such subgroup T^{ρ} of T, step (**c**) will yield one subgroup $\mathscr{F}^{\rho}$ of $\mathscr{F}$, namely, $T^{\rho} \rtimes \mathscr{Z}_1$. The members

$$[i.u + j.v, (-1)^k I] \qquad (0 \leq i < \rho, \ 0 \leq j < \frac{n}{2\rho}, \ 0 \leq k < 2)$$

of $\mathscr{F}$ represent the various left cosets of $\mathscr{F}^{\rho}$ in $\mathscr{F}$. The actions of $\underline{U}$, $\underline{V}$, and $\underline{W}$ on $\mathscr{F}/\mathscr{F}^{\rho}$ can be described as follows:

$$\begin{aligned} \underline{U}(i,j,k) &= (i+1, j, k) \\ \underline{V}(i,j,k) &= (i, j+1, k) \qquad (\mathrm{mod}(\rho, \tfrac{n}{2\rho}, 2)), \qquad (3) \\ \underline{W}(i,j,k) &= (-i,-j,k+1) \end{aligned}$$

where $0 \leq i < \rho$, $0 \leq j < \frac{n}{2\rho}$, and $0 \leq k < 2$.

The foregoing discussion shows that, for the group $\mathscr{F}_1$, there are $\sigma(n)$ n-chromatic ornamental classes, where $\sigma(n)$ stands for the number of square factors of n. For the group $\mathscr{F}_2$, there are $\sigma(n)$ n-chromatic ornamental classes when n is odd; there are $\sigma(n) + \sigma(\frac{n}{2})$ such classes when n is even. Representatives of these classes can be calculated from relations (1), (2), and (3). Hence, for the groups $\mathscr{F}_1$ and $\mathscr{F}_2$, the algorithm SUBMT reduces to a survey of the square factors of given positive integers, and to modular arithmetic. In these cases, then, the problems of implementation are trivial.

6° For the remainder of this section, we shall assume that $3 \le m \le 17$. Under this assumption, the various groups $\mathcal{G}_m$ and $\mathcal{G}_m^\triangle$ must all be subgroups of one or the other of the groups

$$\mathcal{D}_4(X'), \quad \mathcal{D}_6(X''),$$

where X' and X" are suitable reflections. [See Tables 3 and 5.] Obviously, the groups produced by step (**a**) in the algorithm SUBMT, must also be included in one or the other of those groups. Hence, we may expect to implement step (**a**) by enumerating and then comparing the subgroups of $\mathcal{D}_4(X')$ and $\mathcal{D}_6(X'')$.

There are in fact 10 subgroups of $\mathcal{D}_4(X')$ and 16 subgroups of $\mathcal{D}_6(X'')$. Let them be denoted as follows:

$$\mathcal{H}'_\ell, \qquad 1 \le \ell \le 10,$$
$$\mathcal{H}''_\ell, \qquad 1 \le \ell \le 16.$$

In Figure 90, we have presented a schematic display of these groups, numbering them from 1 through 26:

$$\mathcal{H}_\ell = \mathcal{H}'_\ell, \qquad 1 \le \ell \le 10,$$
$$\mathcal{H}_\ell = \mathcal{H}''_{\ell-10}, \qquad 11 \le \ell \le 26.$$

For each case, the representative diagram suggests a generating rotation W (by an arc-arrow) and a generating reflection X (by a bold line, standing for the axis of X). Of course, in some cases, W reduces to a null arc-arrow (signifying that W = I), and, in some cases, X is missing. We have inserted the conventional bases {u',v'} and {u",v"} for the underlying quadratic and hexagonal lattices, in order to distinguish cases which would otherwise be identified. The fact is that certain groups are

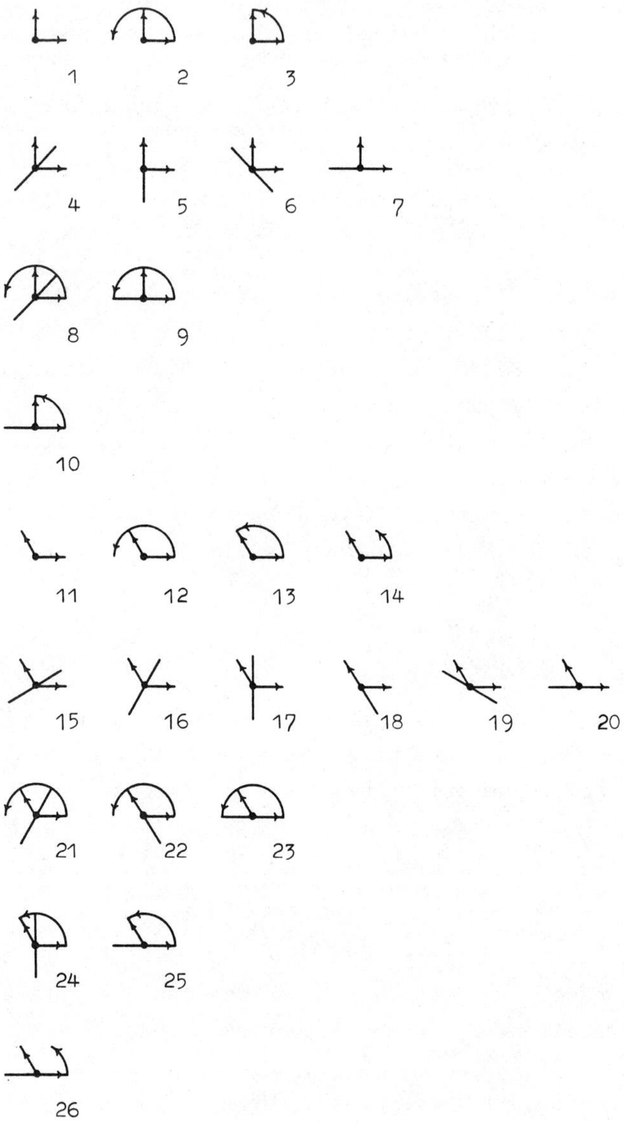

Figure 90

subgroups of both $\mathscr{D}_4(X')$ and $\mathscr{D}_6(X'')$, but, for our purposes, it is better to regard them as distinct.

In Table 7, we have listed the members of $\mathscr{D}_4(X')$ and $\mathscr{D}_6(X'')$. The first 8 are the members of $\mathscr{D}_4(X')$, the last 12, the members of $\mathscr{D}_6(X'')$. They stand in matrix form, relative to the bases $\{u',v'\}$ and $\{u'',v''\}$, respectively. In Table 8, we have listed the members of the groups $\mathscr{H}_\ell$ $(1 \leq \ell \leq 26)$, using the code established in Table 7. For each case, the rotations in $\mathscr{H}_\ell$ appear first, then the reflections (if there be any):

$$W, \quad W^2, \quad \ldots, \quad W^k = I; \quad WX, \quad \ldots, \quad W^{k-1}X, \quad W^kX = X.$$

The thirteenth entry is exceptional, in that it is the order of $\mathscr{H}_\ell$.

In Table 9, we have identified the linear parts $\mathscr{G}_m$ of the groups $\mathscr{F}_m$ $(3 \leq m \leq 17)$, among the groups $\mathscr{H}_\ell$ $(1 \leq \ell \leq 26)$:

$$\mathscr{G}_m = \mathscr{H}_{\ell(m)}.$$

Finally, in Table 10, we have assembled the data necessary to implement step (**a**) in SUBMT. This table consists of 26 rows, indexed by the integer ℓ $(1 \leq \ell \leq 26)$, and of 15 columns, indexed by the integer m $(3 \leq m \leq 17)$. The entries in the table are integers, bounded by -26 and +26. The entry $\varepsilon(\ell,m)$ in the ℓ-th row and m-th column serves to discriminate the case in which $\mathscr{H}_\ell$ is a subgroup of $\mathscr{G}_m$ from the case in which it is not. Moreover, in the former case, $\varepsilon(\ell,m)$ points to the normalizer of $\mathscr{H}_\ell$ in $\mathscr{G}_m^\triangle$, anticipating the next step (step (**b**)) in SUBMT. Specifically:

> when $\mathscr{H}_\ell$ is not a subgroup of $\mathscr{G}_m$, then $\varepsilon(\ell,m) = 0$;
> when $\mathscr{H}_\ell$ is a subgroup of $\mathscr{G}_m$, then $1 \leq |\varepsilon(\ell,m)| \leq 26$ and $\mathscr{H}_{|\varepsilon(\ell,m)|}$ is the normalizer of $\mathscr{H}_\ell$ in $\mathscr{G}_m^\triangle$.

$\mathscr{D}_4(X')$		$\mathscr{D}_6(X'')$	
1	$\begin{bmatrix} 0 & -1 \\ 1 & 0 \end{bmatrix}$	9	$\begin{bmatrix} 1 & -1 \\ 1 & 0 \end{bmatrix}$
2	$\begin{bmatrix} -1 & 0 \\ 0 & -1 \end{bmatrix}$	10	$\begin{bmatrix} 0 & -1 \\ 1 & -1 \end{bmatrix}$
3	$\begin{bmatrix} 0 & 1 \\ -1 & 0 \end{bmatrix}$	11	$\begin{bmatrix} -1 & 0 \\ 0 & -1 \end{bmatrix}$
4	$\begin{bmatrix} 1 & 0 \\ 0 & 1 \end{bmatrix}$	12	$\begin{bmatrix} -1 & 1 \\ -1 & 0 \end{bmatrix}$
5	$\begin{bmatrix} 0 & 1 \\ 1 & 0 \end{bmatrix}$	13	$\begin{bmatrix} 0 & 1 \\ -1 & 1 \end{bmatrix}$
6	$\begin{bmatrix} -1 & 0 \\ 0 & 1 \end{bmatrix}$	14	$\begin{bmatrix} 1 & 0 \\ 0 & 1 \end{bmatrix}$
7	$\begin{bmatrix} 0 & -1 \\ -1 & 0 \end{bmatrix}$	15	$\begin{bmatrix} 1 & 0 \\ 1 & -1 \end{bmatrix}$
8	$\begin{bmatrix} 1 & 0 \\ 0 & -1 \end{bmatrix}$	16	$\begin{bmatrix} 0 & 1 \\ 1 & 0 \end{bmatrix}$
		17	$\begin{bmatrix} -1 & 1 \\ 0 & 1 \end{bmatrix}$
		18	$\begin{bmatrix} -1 & 0 \\ -1 & 1 \end{bmatrix}$
		19	$\begin{bmatrix} 0 & -1 \\ -1 & 0 \end{bmatrix}$
		20	$\begin{bmatrix} 1 & -1 \\ 0 & -1 \end{bmatrix}$

Table 7

$\mathcal{H}_l$	1	2	3	4	5	6	7	8	9	10	11	12	order
l = 1	4												1
2	2	4											2
3	1	2	3	4									4
4	4	5											2
5	4	6											2
6	4	7											2
7	4	8											2
8	2	4	7	5									4
9	2	4	6	8									4
10	1	2	3	4	5	6	7	8					8
11	14												1
12	11	14											2
13	10	12	14										3
14	9	10	11	12	13	14							6
15	14	15											2
16	14	16											2
17	14	17											2
18	14	18											2
19	14	19											2
20	14	20											2
21	11	14	19	16									4
22	11	14	15	18									4
23	11	14	17	20									4
24	10	12	14	19	15	17							6
25	10	12	14	16	18	20							6
26	9	10	11	12	13	14	15	16	17	18	19	20	12

Table 8

m	3	4	5	6	7	8	9	10	11	12	13	14	15	16	17
l(m)	13	3	14	4	7	7	8	9	9	9	25	24	10	10	26

Table 9

ℓ \ m	3	4	5	6	7	8	9	10	11	12	13	14	15	16	17
1	0	10	0	8	9	9	10	10	9	10	0	0	10	10	0
2	0	10	0	0	0	0	10	10	9	10	0	0	10	10	0
3	0	10	0	0	0	0	0	0	0	0	0	0	10	10	0
4	0	0	0	8	0	0	8	0	0	0	0	0	8	8	0
5	0	0	0	0	0	0	0	-9	9	-9	0	0	-9	-9	0
6	0	0	0	0	0	0	-8	0	0	0	0	0	-8	-8	0
7	0	0	0	0	9	9	0	9	9	9	0	0	9	9	0
8	0	0	0	0	0	0	10	0	0	0	0	0	10	10	0
9	0	0	0	0	0	0	0	10	9	10	0	0	10	10	0
10	0	0	0	0	0	0	0	0	0	0	0	0	10	10	0
11	26	0	26	0	0	0	0	0	0	0	26	26	0	0	26
12	0	0	26	0	0	0	0	0	0	0	0	0	0	0	26
13	26	0	26	0	0	0	0	0	0	0	26	26	0	0	26
14	0	0	26	0	0	0	0	0	0	0	0	0	0	0	26
15	0	0	0	0	0	0	0	0	0	0	0	-22	0	0	-22
16	0	0	0	0	0	0	0	0	0	0	-21	0	0	0	-21
17	0	0	0	0	0	0	0	0	0	0	0	23	0	0	23
18	0	0	0	0	0	0	0	0	0	0	-22	0	0	0	-22
19	0	0	0	0	0	0	0	0	0	0	0	-21	0	0	-21
20	0	0	0	0	0	0	0	0	0	0	23	0	0	0	23
21	0	0	0	0	0	0	0	0	0	0	0	0	0	0	-21
22	0	0	0	0	0	0	0	0	0	0	0	0	0	0	-22
23	0	0	0	0	0	0	0	0	0	0	0	0	0	0	23
24	0	0	0	0	0	0	0	0	0	0	0	26	0	0	26
25	0	0	0	0	0	0	0	0	0	0	26	0	0	0	26
26	0	0	0	0	0	0	0	0	0	0	0	0	0	0	26

Table 10

From the family of groups $\mathscr{H}_\ell$ $(1 \leq \ell \leq 26)$ which are subgroups of $\mathscr{G}_m$, we have (more or less arbitrarily) selected one group to represent each conjugacy class in $\mathscr{G}_m^{\triangle}$. For these groups, $\varepsilon(\ell, m)$ is positive, for the remainder, negative.

With Tables 7, 8, 9, and 10 on record, one may proceed to implement step (**a**), in the following manner. As initial data, one would have values of n $(2 \leq n \leq n^*)$ and m $(3 \leq m \leq 17)$. From Table 9, one would obtain $\mathcal{G}_m$. With reference to Table 10, one would then survey the integers ℓ $(1 \leq \ell \leq 26)$, to find the ones for which $0 < \varepsilon(\ell,m)$. For these values of ℓ, one would calculate the index n" of $\mathcal{H}_\ell$ in $\mathcal{G}_m$, using the thirteenth column of Table 8, and one would determine whether or not n" divides n. In these terms, one can describe the groups produced by step (**a**):

$$\mathcal{H}_\ell: \qquad 0 < \varepsilon(\ell,m), \qquad n''|n.$$

For each of them, the value of $\varepsilon(\ell,m)$ would yield the normalizer of $\mathcal{H}_\ell$ in $\mathcal{G}_m^\triangle$:

$$\mathcal{G}_m^o = \mathcal{H}_\ell, \qquad \mathcal{G}_m^{o\triangle} = \mathcal{H}_{\varepsilon(\ell,m)}.$$

7° Now let $\mathcal{G}_m^o$ be any one of the groups produced by step (**a**). Let n' stand for n/n". To implement step (**b**) in SUBMT, one must be able to describe all subgroups T_m^o of T_m which have index n' in T_m and which are invariant under $\mathcal{G}_m^o$, and one must be able to select from these subgroups, one representative from each conjugacy class under $\mathcal{G}_m^{o\triangle}$.

Let T_m^o be any subgroup of T_m having index n' in T_m, and let us introduce a basis for T_m^o of the form:

$$\begin{aligned} u^o &= p.u \quad , \\ v^o &= r.u + s.v, \end{aligned} \tag{1}$$

where p, r, and s are integers for which $0 < p$, $0 < s$, $0 \leq r < p$, and $ps = n'$. [See problem 8 in 9.] In point of fact, the integers p, r, and s are uniquely determined by T_m^o, relative to the given basis {u,v} for T_m. Consequently, we can describe all subgroups of T_m having index n' in T_m, by surveying the divisors of n'. The number of such subgroups of T_m would be the sum of the divisors of n".

Let L be any member of $\mathcal{G}_m^o$. Let the matrix for L relative to $\{u,v\}$ be the following:

$$L \longleftrightarrow \begin{bmatrix} a & c \\ b & d \end{bmatrix}.$$

In order that T_m^o be invariant under L, it is necessary and sufficient that the matrix

$$\begin{bmatrix} p & r \\ 0 & s \end{bmatrix}^{-1} \begin{bmatrix} a & c \\ b & d \end{bmatrix} \begin{bmatrix} p & r \\ 0 & s \end{bmatrix}$$

have integral entries, that is, that the (necessarily integral) entries in the matrix

$$\begin{bmatrix} s & -r \\ 0 & p \end{bmatrix} \begin{bmatrix} a & c \\ b & d \end{bmatrix} \begin{bmatrix} p & r \\ 0 & s \end{bmatrix}$$

be divisible by n'. [See problem 10 in 9.] Consequently, one can determine whether or not T_m^o is invariant under $\mathcal{G}_m^o$, by applying the foregoing criterion to the various members L of $\mathcal{G}_m^o$. Of course, it would be sufficient to test just the first and last members of $\mathcal{G}_m^o$, as listed in Table 8, since they generate the whole group.

Now let $\mathbb{T}_m^o$ stand for the (finite) family consisting of all subgroups of T_m which have index n' in T_m and which are invariant under $\mathcal{G}_m^o$. Let $T_m^{o\prime}$ and $T_m^{o\prime\prime}$ be any members of $\mathbb{T}_m^o$, and let L be any member of $\mathcal{G}_m^{o\triangle}$. Let

$$\begin{bmatrix} p' & r' \\ 0 & s' \end{bmatrix} \quad \text{and} \quad \begin{bmatrix} p'' & r'' \\ 0 & s'' \end{bmatrix}$$

be the matrices which determine $T_m^{o\prime}$ and $T_m^{o\prime\prime}$, respectively, in the sense of relations (1). Let the matrix for L relative to $\{u,v\}$ be the following:

$$L \longleftrightarrow \begin{bmatrix} a & c \\ b & d \end{bmatrix}.$$

In order that $L(T_m^{o\prime}) = T_m^{o\prime\prime}$, it is necessary and sufficient that the (necessarily integral) entries in the matrix

$$\begin{bmatrix} s'' & -r'' \\ 0 & p'' \end{bmatrix}\begin{bmatrix} a & c \\ b & d \end{bmatrix}\begin{bmatrix} p' & r' \\ 0 & s' \end{bmatrix}$$

be divisible by n'. [See problem 10 in 9.] Using this criterion, one can easily calculate the action of $\mathcal{G}_m^{o\Delta}$ on $\mathbb{T}_m^o$. One may then apply the algorithm ORBKT, to obtain representatives of each of the conjugacy classes in $\mathbb{T}_m^o$ under $\mathcal{G}_m^{o\Delta}$. [See 5.7°.] To this end, it would be sufficient to know the actions of just the first and last members of $\mathcal{G}_m^{o\Delta}$, as listed in Table 8.

8° Finally, let us consider how to implement step **(c)** in SUBMT.

Let $\mathcal{G}_m^o$ and T_m^o be subgroups of $\mathcal{G}_m$ and T_m, respectively, produced by steps **(a)** and **(b)**. Let $\mathcal{F}_m^o$ be a subgroup of $\mathcal{F}_m$ for which the translational part equals T_m^o and the linear part equals $\mathcal{G}_m^o$. Let $\{u^o, v^o\}$ be the basis for T_m^o, described in relations (1) of the preceding article:

$$\begin{aligned} u^o &= p.u \quad , \\ v^o &= r.u + s.v, \end{aligned} \tag{1}$$

where $0 < p$, $0 < s$, $0 \le r < p$, and $ps = n'$. The members of $\mathcal{F}_m$ may be presented in the form:

$$[t, I][c_L, L],$$

where t is any member of T_m, where L is any member of $\mathcal{G}_m$, and where:

$$\begin{aligned} c_L &= 0 \quad \text{when } L \in \mathcal{G}_m^+, \\ c_L &= x \quad \text{when } L \in \mathcal{G}_m^-. \end{aligned}$$

Hence, the members of $\mathscr{F}_m^o$ must be of the form:

$$[t^o, I][b_{L^o} + c_{L^o}, L^o],$$

where t^o is any member of T_m^o, where L^o is any member of $\mathscr{G}_m^o$, and where b_{L^o} is a suitable member of T_m. Obviously, one may confine the values of b_{L^o} to some set of representatives of the cosets of T_m^o in T_m. Let us agree to draw the values of b_{L^o} from the set

$$S = \{i.u + j.v : 0 \leq i < \rho, \ 0 \leq j < \sigma\}.$$

[See problem 8 in 9.]

For convenience, let a_{L^o} stand for the sum of b_{L^o} and c_{L^o}:

$$a_{L^o} = b_{L^o} + c_{L^o}. \tag{2}$$

Now let W^o be the generator for $\mathscr{G}_m^{o+}$, listed as the first member of $\mathscr{G}_m^o$ in Table 8. Of course:

$$(W^o)^k = I, \tag{3}$$

where k is the order of $\mathscr{G}_m^{o+}$. For the cases in which $\mathscr{G}_m^{o-}$ is not empty, let X^o be the reflection listed as the last member of $\mathscr{G}_m^o$ in Table 8. For the contrary cases, one should delete all references to X^o. [Incidentally, one can determine whether or not $\mathscr{G}_m^{o-}$ is empty by testing whether or not the member of $\mathscr{G}_m^o$ listed last in Table 8, equals I. Moreover, one can locate the last member of $\mathscr{G}_m^o$ by reading the order of $\mathscr{G}_m^o$ from the thirteenth column of Table 8.] Of course:

$$(X^o)^2 = I, \tag{4}$$

$$X^o W^o X^o W^o = I. \tag{5}$$

Since:

$$[a_{W^o}, W^o]^k = [a_{W^o} + W^o(a_{W^o}) + \ldots + (W^o)^{k-1}(a_{W^o}),\ (W^o)^k],$$

$$[a_{X^o}, X^o]^2 = [a_{X^o} + X^o(a_{X^o}),\ (X^o)^2],$$

and $[a_{X^o}, X^o][a_{W^o}, W^o][a_{X^o}, X^o][a_{W^o}, W^o]$

$$= [a_{X^o} + X^o(a_{W^o}) + X^oW^o(a_{X^o}) + X^oW^oX^o(a_{W^o}),\ X^oW^oX^oW^o],$$

we may apply relations (3), (4), and (5) to obtain:

(**i**) $a_{W^o} + W^o(a_{W^o}) + \ldots + (W^o)^{k-1}(a_{W^o}) \ \varepsilon\ T^o_m$,

(**j**) $a_{X^o} + X^o(a_{X^o}) \ \varepsilon\ T^o_m$,

and (**k**) $a_{X^o} + X^o(a_{W^o}) + X^oW^o(a_{X^o}) + X^oW^oX^o(a_{W^o}) \ \varepsilon\ T^o_m$.

Actually, when $1 < k$ then condition (**i**) will always be satisfied, because

$$a_{W^o} + W^o(a_{W^o}) + \ldots + (W^o)^{k-1}(a_{W^o})$$

remains fixed under W^o and hence must equal 0. When $k = 1$, then b_{W^o} would have to be 0.

<u>Proposition</u> <u>28</u>. For each subgroup $\mathscr{F}^o_m$ of $\mathscr{F}_m$ having translational part T^o_m and linear part $\mathscr{G}^o_m$, the members b_{W^o} and b_{X^o} of S for which $[a_{W^o}, W^o]$ and $[a_{X^o}, X^o]$ are contained in $\mathscr{F}^o_m$, satisfy conditions (**i**), (**j**), and (**k**). Conversely, for any members b_{W^o} and b_{X^o} of S satisfying conditions (**i**), (**j**), and (**k**), there is precisely one subgroup $\mathscr{F}^o_m$ of $\mathscr{F}_m$ having translational part T^o_m and linear part $\mathscr{G}^o_m$, and containing $[a_{W^o}, W^o]$ and $[a_{X^o}, X^o]$.

When $\mathscr{G}^o_m = \mathscr{G}^{o+}_m$, then one should delete all references to X^o.

One can establish the foregoing proposition by straightforward (though tedious) computation. To develop a sharper argument, one may consult problems 13 and 14 in 9, where the proposition has been reformulated in terms of "crossed-homomorphisms."

Now one can identify the family of all subgroups $\mathscr{F}_m^o$ of $\mathscr{F}_m$ having translational part T_m^o and linear part $\mathscr{G}_m^o$, with the subset S^{oo} of $S \times S$ which consists of all ordered pairs (b_{W^o}, b_{X^o}) satisfying conditions (**i**), (**j**), and (**k**).

When $\mathscr{G}_m^o = \mathscr{G}_m^{o+}$, then one should interpret b_{X^o} to be 0.

Of course, for application of conditions (**i**), (**j**), and (**k**), one must be able to judge whether or not a given member of R^2 is contained in T_m^o. Thus, let z be any member of R^2:

$$z = a.u + b.v,$$

where a and b are any real numbers. In order that z be contained in T_m^o, it is necessary and sufficient that the entries in the matrix

$$\begin{bmatrix} p & r \\ 0 & s \end{bmatrix}^{-1} \begin{bmatrix} a \\ b \end{bmatrix}$$

be integral. [See problem 10 in 9.] Actually, the various members z of R^2 which figure in the present (and in all subsequent) calculations, are members of $\frac{1}{6}.T_m$:

$$\frac{1}{6}.T_m = \{z \in R^2 : 6.z \in T_m\}.$$

[See Tables 1, 3, and 5.] Hence, the foregoing condition for membership in T_m^o is equivalent to this, that the necessarily integral entries in the matrix

$$\begin{bmatrix} s & -r \\ 0 & p \end{bmatrix} \begin{bmatrix} 6a \\ 6b \end{bmatrix}$$

be multiples of $6n'$.

To complete the implementation of step (c), one must calculate the group $\mathcal{G}_m^{oo}$, consisting of all members L of $\mathcal{G}_m^{o\Delta}$ for which $L(T_m^o) = T_m^o$; one must then determine the action of $T_m^{\Delta} \rtimes \mathcal{G}_m^{oo}$ on S^{oo}; finally, one may enumerate the resulting orbits in S^{oo} and exhibit representatives of them, by application of the algorithm ORBKT.

Thus, let t^{Δ} be any member of T_m^{Δ} and let L^{oo} be any member of $\mathcal{G}_m^{oo}$. We intend to calculate the actions of $[t^{\Delta},I]$ and $[0,L^{oo}]$ on S^{oo}. Let $(b_W o',b_X o')$ and $(b_W o'',b_X o'')$ be any two members of S^{oo}, and let $\mathcal{F}_m^{o}{}'$ and $\mathcal{F}_m^{o}{}''$ be the subgroups of $\mathcal{F}_m$ to which they correspond. Let V^o stand either for W^o or for X^o. We have:

$$[t^{\Delta}, I][a_V o',V^o][t^{\Delta}, I]^{-1}$$

$$= [(t^{\Delta} + a_V o' - V^o(t^{\Delta}) - a_V o'') + a_V o'', V^o],$$

and $$[0,L^{oo}][a_V o',V^o][0,L^{oo}]^{-1}$$

$$= [(L^{oo}(a_V o') - a_V oo'') + a_V oo'', V^{oo}],$$

where V^{oo} stands for $L^{oo}V^oL^{oo-1}$. In this context, $a_V o'$, $a_V o''$, and $a_V oo''$ are determined by the foregoing relation (2), and $b_V oo''$ would be the member of S such that

$$[b_V oo'' + c_V oo, V^{oo}]$$

is contained in $\mathcal{F}_m^{o}{}''$. Hence, $J_{[t^{\Delta},I]}(\mathcal{F}_m^{o}{}') = \mathcal{F}_m^{o}{}''$ iff:

$$\begin{aligned} t^{\Delta} + a_W o' - W^o(t^{\Delta}) - a_W o'' &\in T_m^o, \\ t^{\Delta} + a_X o' - X^o(t^{\Delta}) - a_X o'' &\in T_m^o; \end{aligned} \qquad (6)$$

and $J_{[0,L^{oo}]}(\mathcal{F}_m^{o}{}') = \mathcal{F}_m^{o}{}''$ iff:

$$L^{oo}(a_{W^o}') - a_{L^{oo}W^oL^{oo-1}}'' \;\varepsilon\; T_m^o,$$
$$L^{oo}(a_{X^o}') - a_{L^{oo}X^oL^{oo-1}}'' \;\varepsilon\; T_m^o. \tag{7}$$

Under conditions (6), we would conclude that:

$$[t^{\triangle}, I].(b_{W^o}', b_{X^o}') = (b_{W^o}'', b_{X^o}''); \tag{8}$$

under conditions (7):

$$[0, L^{oo}].(b_{W^o}', b_{X^o}') = (b_{W^o}'', b_{X^o}''). \tag{9}$$

Of course, conditions (7) presume prior computation of $a_{W^{oo}}''$ and $a_{X^{oo}}''$, where:

$$W^{oo} = L^{oo}W^oL^{oo-1},$$
$$X^{oo} = L^{oo}X^oL^{oo-1}.$$

This computation should be based upon the facts that W^{oo} must equal either W^o or W^{o-1}, and that X^{oo} must be of the form: $W^{oj}X^o$ $(1 \leq j \leq k)$. Since:

$$[a_{W^o}'', W^o]^{-1} = [(-W^{o-1}(a_{W^o}'') - a_{W^{o-1}}'') + a_{W^{o-1}}'', W^{o-1}],$$

and $[a_{W^o}'', W^o]^j [a_{X^o}'', X^o]$

$$= [(\textstyle\sum_{i=0}^{j-1} W^{oi}(a_{W^o}'') + W^{oj}(a_{X^o}'') - a_{W^{oj}X^o}'') + a_{W^{oj}X^o}'', W^{oj}X^o],$$

one may infer that:

$$-W^{o-1}(a_{W^o}'') - a_{W^{o-1}}'' \;\varepsilon\; T_m^o,$$

and $(\sum_{i=0}^{j-1} W^{oi}(a_{W^o}'')) + W^{oj}(a_{X^o}'') - a_{W^{oj}X^o}'' \;\varepsilon\; T_m^o.$

Hence, in conditions (7), one may replace $a_W o o''$ either by $a_W o''$ or by

$$-W^{o-1}(a_W o''),$$

and one may replace $a_X o o''$ by

$$a_W o'' + W^o(a_W o'') + \ldots + W^{o(j-1)}(a_W o'') + W^{oj}(a_X o'').$$

By appropriate choice of the members $t^\triangle$ of T_m^o and L^{oo} of $\mathscr{G}_m^{oo}$, we may initiate application of ORBKT. To be specific, let us select $u^\triangle$, $v^\triangle$, W^{oo}, and X^{oo}, where $u^\triangle$ and $v^\triangle$ are the members of $T_m{}^\triangle$ displayed in Table 5, and where W^{oo} and X^{oo} are suitable generators for $\mathscr{G}_m^{oo}$. Such generators for $\mathscr{G}_m^{oo}$ can be obtained by surveying the members L of $\mathscr{G}_m^{o\triangle}$ (as listed in Table 8), retaining the first and last of those for which $L(T_m^o) = T_m^o$. [Incidentally, it might happen that $(\mathscr{G}_m^{oo})^-$ is empty, in which case the foregoing survey would yield: $X^{oo} = I$.] The members

$$[u^\triangle, I],$$
$$[v^\triangle, I],$$
$$[0, W^{oo}],$$
$$[0, X^{oo}],$$

of $T_m{}^\triangle \rtimes \mathscr{G}_m^{oo}$ would serve to generate $T_m{}^\triangle \rtimes \mathscr{G}_m^{oo}$, and the actions of these members on S^{oo} would provide the initial data for application of ORBKT.

9° Now let $\mathscr{F}_m^o$ be any one of the subgroups of $\mathscr{F}_m$ having index n in $\mathscr{F}_m$, produced by steps (**a**), (**b**), and (**c**). With reference to Proposition 26, let us consider how to calculate the n-chromatic homomorphism f carrying $\mathscr{F}_m$ to S_N, which corresponds to $\mathscr{F}_m^o$. Of course, it is sufficient to calculate:

$$\underline{u} = f(\underline{U}),$$
$$\underline{v} = f(\underline{V}),$$
$$\underline{w} = f(\underline{W}),$$
$$\underline{x} = f(\underline{X}).$$

We shall first identify a subset $\mathscr{N}$ of $\mathscr{F}_m$ containing precisely one member of each left coset of $\mathscr{F}_m^o$ in $\mathscr{F}_m$. Thus, let k^o and k stand for the orders of $\mathscr{G}_m^o$ and $\mathscr{G}_m$, respectively. For the cases:

$$\mathscr{G}_m^+ = \mathscr{G}_m,$$

$$\mathscr{G}_m^+ \neq \mathscr{G}_m \quad \text{and} \quad \mathscr{G}_m^{o+} \neq \mathscr{G}_m^o,$$

we shall take $\mathscr{N}$ to be the set:

$$\{[W^{\mathfrak{u}}(i.u + j.v,\ I][c_{W^{\mathfrak{u}}},\ W^{\mathfrak{u}}]: \\ 0 \leq i < \mathfrak{p},\ \ 0 \leq j < \mathfrak{s},\ \ 0 \leq \mathfrak{u} < k/k^o\};$$

for the cases:

$$\mathscr{G}_m^+ \neq \mathscr{G}_m \quad \text{but} \quad \mathscr{G}_m^{o+} = \mathscr{G}_m^o,$$

we shall take $\mathscr{N}$ to be the set:

$$\{[W^{\mathfrak{u}}X^{\mathfrak{v}}(i.u + j.v),\ I][c_{W^{\mathfrak{u}}X^{\mathfrak{v}}},\ W^{\mathfrak{u}}X^{\mathfrak{v}}]: \\ 0 \leq i < \mathfrak{p},\ \ 0 \leq j < \mathfrak{s},\ \ 0 \leq \mathfrak{u} < k/2k^o,\ \ 0 \leq \mathfrak{v} < 2\}.$$

For such a set $\mathscr{N}$, one may introduce a bijective mapping

$$b(C \longmapsto j = b(C))$$

carrying $\mathscr{N}$ to N. In terms of this mapping, the n-chromatic homomorphism f can be characterized as follows. Let B denote any one among $\underline{U}$, $\underline{V}$, $\underline{W}$, and $\underline{X}$, let C' and C" be any members of $\mathscr{N}$, and let j' and j" be the corresponding members of N: $j' = b(C')$, $j'' = b(C'')$. Then:

$$f(B)(j') = j'' \quad \text{iff} \quad C''^{-1}BC' \in \mathscr{F}_m^o.$$

To apply the foregoing condition, one must be able to judge whether or not a given member of $\mathscr{F}_m$ is contained in $\mathscr{F}_m^o$. Thus, let A be any member of $\mathscr{F}_m$: $A = [t,L]$. In order that A be contained in $\mathscr{F}_m^o$, it is necessary and sufficient that:

$$(\lambda) \qquad L \in \mathscr{G}_m^o$$

and $$(\tau) \qquad t - a_L \in T_m^o.$$

One can test condition (λ) simply by surveying the members of $\mathscr{G}_m^o$, as listed in Table 8. When the test proves positive, one would calculate a_L and then test condition (τ), by the methods described in 8°. In particular, L would have to be either of the form: $(W^o)^u$ $(1 \leq u \leq k^o)$, in which case one should replace a_L in condition (τ) by

$$a_{W^o} + W^o(a_{W^o}) + \ldots + (W^o)^{u-1}(a_{W^o}),$$

or else of the form: $(W^o)^u(X^o)^v$ $(1 \leq u \leq k^o/2,\ 0 \leq v < 2)$, in which case one should replace a_L by

$$a_{W^o} + W^o(a_{W^o}) + \ldots + (W^o)^{u-1}(a_{W^o}) + v.(W^o)^u(a_{X^o}).$$

10° In review, we observe that the algorithm SUBMT should be composed of the subordinate algorithms:

LINPT (n,m)

implementing step (**a**),

TRNPT

$(n,m, \mathscr{G}_m^o)$

implementing step (**b**),

EXTEN

$(n,m, \mathscr{G}_m^o, T_m^o)$

implementing step (**c**), and

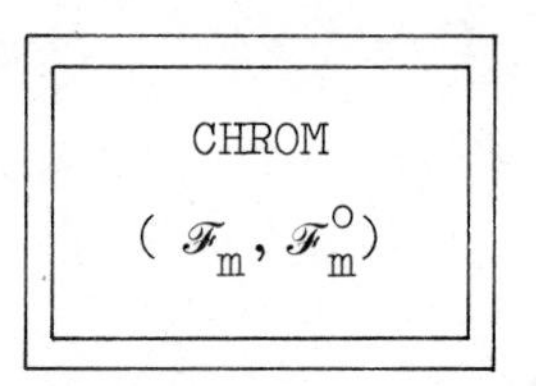

which calculates the n-chromatic homomorphism corresponding to the groups $\mathscr{F}_m$ and $\mathscr{F}_m^o$. Moreover, the commentary in 7°, 8°, and 9° shows that these algorithms themselves should be supported by algorithms which test whether or not two given lattices in R^2 are equal and whether or not a given member of R^2 is contained in a given lattice in R^2. [See problems 9 and 10 in 9.]

3.7 DATA

1° Guided by the discussion in the preceding two sections, we have implemented the algorithms GENMT and SUBMT in FORTRAN 77. The resulting programs were executed by a PDP 11/70, equipped with a UNIX operating system. We obtained data under GENMT, for the range:

$$2 \leq n \leq 4, \qquad 1 \leq m \leq 17,$$

under SUBMT, for the range:

$$2 \leq n \leq 60, \qquad 1 \leq m \leq 17.$$

The former data served to check the validity of implementation of the latter algorithm. Under SUBMT, the chromatic homomorphisms were calculated only for the range:

$$2 \leq n \leq 8, \qquad 1 \leq m \leq 17.$$

In Table 11, we have recorded the raw counts of the chromatic ornamental classes, in Table 12, the specific representative chromatic homomorphisms. The formats are self-evident.

For comparison, one should consult the papers by M. Senechal and by J. D. Jarratt and R. L. E. Schwarzenberger.

ORNAMENTAL CLASS			COLOR NUMBER 1	2	3	4	5	6	7	8	9	10	11	12
1	$\mathbb{Z}_1^a$	p1	1	1	1	2	1	1	1	2	2	1	1	2
2	$\mathbb{Z}_2^a$	p2	1	2	1	3	1	2	1	4	2	2	1	3
3	$\mathbb{Z}_3^h$	p3	1	0	2	1	0	1	1	0	3	0	0	4
4	$\mathbb{Z}_4^q$	p4	1	2	0	5	1	2	0	9	1	4	0	9
5	$\mathbb{Z}_6^h$	p6	1	1	2	1	0	5	1	1	3	0	0	8
6	$\mathbb{D}_1^r$	cm	1	3	2	7	2	7	2	13	3	8	2	17
7	$\mathbb{D}_1^o$	pm	1	5	2	10	2	11	2	16	3	12	2	23
8	$\mathbb{D}_1^o$	pg	1	2	2	4	2	5	2	7	3	6	2	11
9	$\mathbb{D}_2^r$	cmm	1	5	1	11	1	8	1	21	2	9	1	22
10	$\mathbb{D}_2^o$	pmm	1	5	1	13	1	9	1	21	2	10	1	25
11	$\mathbb{D}_2^o$	pmg	1	5	2	11	2	11	2	19	3	12	2	26
12	$\mathbb{D}_2^o$	pgg	1	2	1	4	1	4	1	7	2	5	1	9
13	$\mathbb{D}_3'$	p31m	1	1	2	1	0	5	0	1	3	0	0	7
14	$\mathbb{D}_3''$	p3m1	1	1	2	1	0	4	0	1	3	0	0	7
15	$\mathbb{D}_4^q$	p4m	1	5	0	13	0	2	0	28	1	3	0	16
16	$\mathbb{D}_4^q$	p4g	1	3	0	7	0	2	0	13	1	3	0	10
17	$\mathbb{D}_6^h$	p6m	1	3	2	2	0	11	0	3	3	0	0	20
TOTALS			17	46	23	96	14	90	15	166	40	75	13	219

Table 11

(1)

ORNAMENTAL CLASS			COLOR NUMBER 13	14	15	16	17	18	19	20	21	22	23	24
1	$\mathbb{Z}_1^a\llcorner$	p1	1	1	1	3	1	2	1	2	1	1	1	2
2	$\mathbb{Z}_2^a\llcorner$	p2	1	2	1	5	1	4	1	3	1	2	1	4
3	$\mathbb{Z}_3^h\llcorner$	p3	1	0	2	1	0	3	1	0	4	0	0	5
4	$\mathbb{Z}_4^q\llcorner$	p4	1	3	0	16	1	6	0	14	0	4	0	21
5	$\mathbb{Z}_6^h\llcorner$	p6	1	1	2	1	0	10	1	0	4	0	0	16
6	$\mathbb{D}_1^r\llcorner$	cm	2	9	4	21	2	14	2	20	4	11	2	33
7	$\mathbb{D}_1^o\llcorner$	pm	2	13	4	24	2	20	2	26	4	15	2	39
8	$\mathbb{D}_1^o\llcorner$	pg	2	7	4	12	2	11	2	14	4	9	2	21
9	$\mathbb{D}_2^r\llcorner$	cmm	1	9	2	33	1	15	1	26	2	10	1	45
10	$\mathbb{D}_2^o\llcorner$	pmm	1	10	2	32	1	16	1	28	2	11	1	45
11	$\mathbb{D}_2^o\llcorner$	pmg	2	13	4	30	2	20	2	30	4	15	2	48
12	$\mathbb{D}_2^o\llcorner$	pgg	1	5	2	12	1	8	1	12	2	6	1	18
13	$\mathbb{D}_3^{'}\llcorner$	p31m	0	1	2	1	0	8	0	0	2	0	0	13
14	$\mathbb{D}_3^{''}\llcorner$	p3m1	0	1	2	1	0	7	0	0	2	0	0	12
15	$\mathbb{D}_4^q\llcorner$	p4m	0	2	0	45	0	7	0	18	0	2	0	46
16	$\mathbb{D}_4^q\llcorner$	p4g	0	2	0	23	0	5	0	12	0	2	0	24
17	$\mathbb{D}_6^h\llcorner$	p6m	0	1	2	2	0	18	0	0	2	0	0	41
TOTALS			16	80	34	262	14	174	15	205	38	88	13	433

Table 11

(2)

ORNAMENTAL CLASS			COLOR NUMBER											
			25	26	27	28	29	30	31	32	33	34	35	36
1	$\mathbb{Z}_1^a$	p1	2	1	2	2	1	1	1	3	1	1	1	4
2	$\mathbb{Z}_2^a$	p2	2	2	2	3	1	2	1	6	1	2	1	6
3	$\mathbb{Z}_3^h$	p3	1	0	5	1	0	4	1	0	3	0	0	9
4	$\mathbb{Z}_4^q$	p4	2	6	0	14	1	8	0	27	0	7	0	24
5	$\mathbb{Z}_6^h$	p6	1	1	5	1	0	10	1	1	3	0	0	21
6	$\mathbb{D}_1^r$	cm	3	12	4	23	2	22	2	33	4	14	4	36
7	$\mathbb{D}_1^o$	pm	3	16	4	29	2	30	2	36	4	18	4	45
8	$\mathbb{D}_1^o$	pg	3	10	4	17	2	18	2	21	4	12	4	27
9	$\mathbb{D}_2^r$	cmm	2	11	2	27	1	20	1	50	2	12	2	43
10	$\mathbb{D}_2^o$	pmm	2	12	2	29	1	22	1	45	2	13	2	46
11	$\mathbb{D}_2^o$	pmg	3	16	4	34	2	30	2	47	4	18	4	53
12	$\mathbb{D}_2^o$	pgg	2	7	2	13	1	12	1	20	2	8	2	21
13	$\mathbb{D}_3^{\prime}$	p31m	1	1	4	0	0	6	0	1	2	0	0	14
14	$\mathbb{D}_3^{\prime\prime}$	p3m1	1	1	4	0	0	6	0	1	2	0	0	14
15	$\mathbb{D}_4^q$	p4m	1	3	0	17	0	4	0	66	0	3	0	31
16	$\mathbb{D}_4^q$	p4g	1	3	0	11	0	4	0	35	0	3	0	19
17	$\mathbb{D}_6^h$	p6m	1	1	4	1	0	14	0	3	2	0	0	39
TOTALS			31	103	48	222	14	213	15	395	36	111	24	452

Table 11

(3)

ORNAMENTAL CLASS			COLOR NUMBER 37	38	39	40	41	42	43	44	45	46	47	48
1	$\mathbb{Z}_1^a\llcorner$	p1	1	1	1	2	1	1	1	2	2	1	1	3
2	$\mathbb{Z}_2^a\llcorner$	p2	1	2	1	4	1	2	1	3	2	2	1	5
3	$\mathbb{Z}_3^h\llcorner$	p3	1	0	5	0	0	5	1	0	6	0	0	10
4	$\mathbb{Z}_4^q\llcorner$	p4	1	6	0	30	1	10	0	19	1	7	0	43
5	$\mathbb{Z}_6^h\llcorner$	p6	1	1	5	0	0	15	1	0	6	0	0	26
6	$\mathbb{D}_1^r\llcorner$	cm	2	15	4	40	2	26	2	29	6	17	2	57
7	$\mathbb{D}_1^o\llcorner$	pm	2	19	4	46	2	34	2	35	6	21	2	63
8	$\mathbb{D}_1^o\llcorner$	pg	2	13	4	28	2	22	2	23	6	15	2	39
9	$\mathbb{D}_2^r\llcorner$	cmm	1	12	2	53	1	22	1	32	3	13	1	78
10	$\mathbb{D}_2^o\llcorner$	pmm	1	13	2	51	1	24	1	33	3	14	1	72
11	$\mathbb{D}_2^o\llcorner$	pmg	2	19	4	58	2	34	2	42	6	21	2	82
12	$\mathbb{D}_2^o\llcorner$	pgg	1	8	2	24	1	14	1	17	3	9	1	34
13	$\mathbb{D}_3'\llcorner$	p31m	0	1	2	0	0	9	0	0	4	0	0	17
14	$\mathbb{D}_3''\llcorner$	p3m1	0	1	2	0	0	8	0	0	4	0	0	17
15	$\mathbb{D}_4^q\llcorner$	p4m	0	2	0	51	0	4	0	18	0	2	0	85
16	$\mathbb{D}_4^q\llcorner$	p4g	0	2	0	29	0	4	0	12	0	2	0	47
17	$\mathbb{D}_6^h\llcorner$	p6m	0	1	2	0	0	16	0	0	4	0	0	63
TOTALS			16	116	40	416	14	250	15	265	62	124	13	741

Table 11

(4)

ORNAMENTAL CLASS			COLOR NUMBER											
			49	50	51	52	53	54	55	56	57	58	59	60
1	$\mathbb{Z}_1^a \llcorner$	p1	2	2	1	2	1	2	1	2	1	1	1	2
2	$\mathbb{Z}_2^a \llcorner$	p2	2	4	1	3	1	4	1	4	1	2	1	3
3	$\mathbb{Z}_3^h \llcorner$	p3	2	0	4	1	0	8	0	0	6	0	0	10
4	$\mathbb{Z}_4^q \llcorner$	p4	1	12	0	24	1	12	0	34	0	10	0	38
5	$\mathbb{Z}_6^h \llcorner$	p6	2	1	4	1	0	22	0	1	6	0	0	26
6	$\mathbb{D}_1^r \llcorner$	cm	3	23	4	32	2	30	4	47	4	20	2	58
7	$\mathbb{D}_1^o \llcorner$	pm	3	29	4	38	2	38	4	53	4	24	2	70
8	$\mathbb{D}_1^o \llcorner$	pg	3	20	4	26	2	26	4	35	4	18	2	46
9	$\mathbb{D}_2^r \llcorner$	cmm	2	20	2	36	1	24	2	58	2	15	1	64
10	$\mathbb{D}_2^o \llcorner$	pmm	2	21	2	36	1	26	2	55	2	16	1	66
11	$\mathbb{D}_2^o \llcorner$	pmg	3	29	4	46	2	38	4	68	4	24	2	84
12	$\mathbb{D}_2^o \llcorner$	pgg	2	13	2	20	1	16	2	28	2	11	1	34
13	$\mathbb{D}_3^{'} \llcorner$	p31m	1	1	2	0	0	12	0	1	2	0	0	14
14	$\mathbb{D}_3^{''} \llcorner$	p3m1	1	1	2	0	0	11	0	1	2	0	0	14
15	$\mathbb{D}_4^q \llcorner$	p4m	1	8	0	20	0	4	0	51	0	3	0	36
16	$\mathbb{D}_4^q \llcorner$	p4g	1	6	0	14	0	4	0	29	0	3	0	24
17	$\mathbb{D}_6^h \llcorner$	p6m	1	3	2	1	0	26	0	1	2	0	0	38
TOTALS			32	193	38	300	14	303	24	468	42	147	13	627

Table 11

(5)

n	m	k		1	2	3	4	5	6	7	8
2	1	1	$\underline{u}$	1	2						
			$\underline{v}$	2	1						
			$\underline{w}$	1	2						
			$\underline{x}$	-	-						
	2	1	$\underline{u}$	1	2						
			$\underline{v}$	1	2						
			$\underline{w}$	2	1						
			$\underline{x}$	-	-						
		2	$\underline{u}$	1	2						
			$\underline{v}$	2	1						
			$\underline{w}$	1	2						
			$\underline{x}$	-	-						
	4	1	$\underline{u}$	1	2						
			$\underline{v}$	1	2						
			$\underline{w}$	2	1						
			$\underline{x}$	-	-						
		2	$\underline{u}$	2	1						
			$\underline{v}$	2	1						
			$\underline{w}$	1	2						
			$\underline{x}$	-	-						
	5	1	$\underline{u}$	1	2						
			$\underline{v}$	1	2						
			$\underline{w}$	2	1						
			$\underline{x}$	-	-						
	6	1	$\underline{u}$	1	2						
			$\underline{v}$	1	2						
			$\underline{w}$	1	2						
			$\underline{x}$	2	1						
		2	$\underline{u}$	2	1						
			$\underline{v}$	2	1						
			$\underline{w}$	1	2						
			$\underline{x}$	1	2						

n	m	k		1	2	3	4	5	6	7	8
2	6	3	$\underline{u}$	2	1						
			$\underline{v}$	2	1						
			$\underline{w}$	1	2						
			$\underline{x}$	2	1						
	7	1	$\underline{u}$	1	2						
			$\underline{v}$	1	2						
			$\underline{w}$	1	2						
			$\underline{x}$	2	1						
		2	$\underline{u}$	1	2						
			$\underline{v}$	2	1						
			$\underline{w}$	1	2						
			$\underline{x}$	1	2						
		3	$\underline{u}$	2	1						
			$\underline{v}$	1	2						
			$\underline{w}$	1	2						
			$\underline{x}$	1	2						
		4	$\underline{u}$	2	1						
			$\underline{v}$	1	2						
			$\underline{w}$	1	2						
			$\underline{x}$	2	1						
		5	$\underline{u}$	2	1						
			$\underline{v}$	2	1						
			$\underline{w}$	1	2						
			$\underline{x}$	1	2						
	8	1	$\underline{u}$	1	2						
			$\underline{v}$	1	2						
			$\underline{w}$	1	2						
			$\underline{x}$	2	1						
		2	$\underline{u}$	1	2						
			$\underline{v}$	2	1						
			$\underline{w}$	1	2						
			$\underline{x}$	1	2						

Table 12

(1)

n	m	k		1 2 3 4 5 6 7 8
2	9	1	u	1 2
			v	1 2
			w	1 2
			x	2 1
		2	u	1 2
			v	1 2
			w	2 1
			x	1 2
		3	u	2 1
			v	2 1
			w	1 2
			x	1 2
		4	u	2 1
			v	2 1
			w	1 2
			x	2 1
		5	u	2 1
			v	2 1
			w	2 1
			x	1 2
	10	1	u	1 2
			v	1 2
			w	1 2
			x	2 1
		2	u	1 2
			v	1 2
			w	2 1
			x	1 2
		3	u	1 2
			v	2 1
			w	1 2
			x	2 1

n	m	k		1 2 3 4 5 6 7 8
2	10	4	u	1 2
			v	2 1
			w	1 2
			x	1 2
		5	u	2 1
			v	2 1
			w	1 2
			x	1 2
	11	1	u	1 2
			v	1 2
			w	1 2
			x	2 1
		2	u	1 2
			v	1 2
			w	2 1
			x	2 1
		3	u	1 2
			v	1 2
			w	2 1
			x	1 2
		4	u	1 2
			v	2 1
			w	1 2
			x	2 1
		5	u	1 2
			v	2 1
			w	1 2
			x	1 2
	12	1	u	1 2
			v	1 2
			w	1 2
			x	2 1

Table 12

(2)

n	m	k		1	2	3	4	5	6	7	8
2	12	2	u	1	2						
			v	1	2						
			w	2	1						
			x	1	2						
	13	1	u	1	2						
			v	1	2						
			w	1	2						
			x	2	1						
	14	1	u	1	2						
			v	1	2						
			w	1	2						
			x	2	1						
	15	1	u	1	2						
			v	1	2						
			w	1	2						
			x	2	1						
		2	u	1	2						
			v	1	2						
			w	2	1						
			x	2	1						
		3	u	1	2						
			v	1	2						
			w	2	1						
			x	1	2						
		4	u	2	1						
			v	2	1						
			w	1	2						
			x	2	1						
		5	u	2	1						
			v	2	1						
			w	1	2						
			x	1	2						

n	m	k		1	2	3	4	5	6	7	8
2	16	1	u	1	2						
			v	1	2						
			w	1	2						
			x	2	1						
		2	u	1	2						
			v	1	2						
			w	2	1						
			x	2	1						
		3	u	1	2						
			v	1	2						
			w	2	1						
			x	1	2						
	17	1	u	1	2						
			v	1	2						
			w	1	2						
			x	2	1						
		2	u	1	2						
			v	1	2						
			w	2	1						
			x	2	1						
		3	u	1	2						
			v	1	2						
			w	2	1						
			x	1	2						
3	1	1	u	1	2	3					
			v	2	3	1					
			w	1	2	3					
			x	-	-	-					
	2	1	u	1	2	3					
			v	2	3	1					
			w	1	3	2					
			x	-	-	-					

Table 12

(3)

n	m	k		1	2	3	4	5	6	7	8
3	3	1	u	1	2	3					
			v	1	2	3					
			w	2	3	1					
			x	-	-	-					
		2	u	2	3	1					
			v	2	3	1					
			w	1	2	3					
			x	-	-	-					
	5	1	u	1	2	3					
			v	1	2	3					
			w	2	3	1					
			x	-	-	-					
		2	u	2	3	1					
			v	2	3	1					
			w	1	3	2					
			x	-	-	-					
	6	1	u	2	3	1					
			v	3	1	2					
			w	1	2	3					
			x	1	3	2					
		2	u	2	3	1					
			v	2	3	1					
			w	1	2	3					
			x	1	2	3					
	7	1	u	1	2	3					
			v	2	3	1					
			w	1	2	3					
			x	1	3	2					
		2	u	2	3	1					
			v	1	2	3					
			w	1	2	3					
			x	1	2	3					

n	m	k		1	2	3	4	5	6	7	8
3	8	1	u	1	2	3					
			v	2	3	1					
			w	1	2	3					
			x	1	3	2					
		2	u	2	3	1					
			v	1	2	3					
			w	1	2	3					
			x	3	1	2					
	9	1	u	2	3	1					
			v	3	1	2					
			w	1	3	2					
			x	1	3	2					
	10	1	u	1	2	3					
			v	2	3	1					
			w	1	3	2					
			x	1	3	2					
	11	1	u	1	2	3					
			v	2	3	1					
			w	1	3	2					
			x	1	3	2					
		2	u	2	3	1					
			v	1	2	3					
			w	1	3	2					
			x	3	1	2					
	12	1	u	1	2	3					
			v	2	3	1					
			w	1	3	2					
			x	3	2	1					
	13	1	u	1	2	3					
			v	1	2	3					
			w	2	3	1					
			x	1	3	2					

Table 12

(4)

n	m	k		1 2 3 4 5 6 7 8
3	13	2	u	2 3 1
			v	2 3 1
			w	1 2 3
			x	1 2 3
	14	1	u	1 2 3
			v	1 2 3
			w	2 3 1
			x	1 3 2
		2	u	2 3 1
			v	2 3 1
			w	1 2 3
			x	1 3 2
	17	1	u	1 2 3
			v	1 2 3
			w	2 3 1
			x	1 3 2
		2	u	2 3 1
			v	2 3 1
			w	1 3 2
			x	1 2 3
4	1	1	u	1 2 3 4
			v	2 3 4 1
			w	1 2 3 4
			x	- - - -
		2	u	2 1 4 3
			v	3 4 1 2
			w	1 2 3 4
			x	- - - -
	2	1	u	1 2 3 4
			v	2 1 4 3
			w	3 4 1 2
			x	- - - -

n	m	k		1 2 3 4 5 6 7 8
4	2	2	u	1 2 3 4
			v	2 3 4 1
			w	1 4 3 2
			x	- - - -
		3	u	2 1 4 3
			v	3 4 1 2
			w	1 2 3 4
			x	- - - -
	3	1	u	2 1 4 3
			v	3 4 1 2
			w	1 3 4 2
			x	- - - -
	4	1	u	1 2 3 4
			v	1 2 3 4
			w	2 3 4 1
			x	- - - -
		2	u	1 2 4 3
			v	2 1 3 4
			w	3 4 1 2
			x	- - - -
		3	u	2 1 4 3
			v	2 1 4 3
			w	3 4 1 2
			x	- - - -
		4	u	2 1 4 3
			v	2 1 4 3
			w	3 4 2 1
			x	- - - -
		5	u	2 1 4 3
			v	3 4 1 2
			w	1 3 2 4
			x	- - - -

Table 12

(5)

n	m	k		1 2 3 4 5 6 7 8
4	5	1	u	2 1 4 3
			v	3 4 1 2
			w	1 4 2 3
			x	- - - -
	6	1	u	1 2 4 3
			v	2 1 3 4
			w	1 2 3 4
			x	3 4 1 2
		2	u	2 1 4 3
			v	2 1 4 3
			w	1 2 3 4
			x	3 4 1 2
		3	u	2 1 4 3
			v	3 4 1 2
			w	1 2 3 4
			x	1 3 2 4
		4	u	2 3 4 1
			v	4 1 2 3
			w	1 2 3 4
			x	1 4 3 2
		5	u	2 3 4 1
			v	4 1 2 3
			w	1 2 3 4
			x	4 3 2 1
		6	u	2 3 4 1
			v	2 3 4 1
			w	1 2 3 4
			x	1 2 3 4
		7	u	2 3 4 1
			v	2 3 4 1
			w	1 2 3 4
			x	3 4 1 2

n	m	k		1 2 3 4 5 6 7 8
4	7	1	u	1 2 3 4
			v	2 1 4 3
			w	1 2 3 4
			x	3 4 1 2
		2	u	2 1 4 3
			v	1 2 3 4
			w	1 2 3 4
			x	3 4 1 2
		3	u	2 1 4 3
			v	2 1 4 3
			w	1 2 3 4
			x	3 4 1 2
		4	u	1 2 3 4
			v	2 3 4 1
			w	1 2 3 4
			x	1 4 3 2
		5	u	2 1 4 3
			v	3 4 1 2
			w	1 2 3 4
			x	1 2 3 4
		6	u	2 1 4 3
			v	3 4 1 2
			w	1 2 3 4
			x	2 1 4 3
		7	u	2 1 4 3
			v	3 4 2 1
			w	1 2 3 4
			x	1 2 4 3
		8	u	2 3 4 1
			v	1 2 3 4
			w	1 2 3 4
			x	1 2 3 4

Table 12

(6)

n	m	k		1	2	3	4	5	6	7	8
4	7	9	u	2	3	4	1				
			v	1	2	3	4				
			w	1	2	3	4				
			x	3	4	1	2				
		10	u	2	3	4	1				
			v	3	4	1	2				
			w	1	2	3	4				
			x	1	2	3	4				
	8	1	u	1	2	3	4				
			v	2	1	4	3				
			w	1	2	3	4				
			x	3	4	1	2				
		2	u	2	1	4	3				
			v	1	2	3	4				
			w	1	2	3	4				
			x	3	4	2	1				
		3	u	2	1	4	3				
			v	2	1	4	3				
			w	1	2	3	4				
			x	3	4	2	1				
		4	u	1	2	3	4				
			v	2	3	4	1				
			w	1	2	3	4				
			x	1	4	3	2				
	9	1	u	1	2	3	4				
			v	1	2	3	4				
			w	2	1	4	3				
			x	3	4	1	2				
		2	u	1	2	4	3				
			v	2	1	3	4				
			w	1	2	3	4				
			x	3	4	1	2				

n	m	k		1	2	3	4	5	6	7	8
4	9	3	u	2	1	4	3				
			v	2	1	4	3				
			w	1	2	3	4				
			x	3	4	1	2				
		4	u	2	1	4	3				
			v	2	1	4	3				
			w	2	1	4	3				
			x	3	4	1	2				
		5	u	2	1	4	3				
			v	2	1	4	3				
			w	3	4	1	2				
			x	1	2	3	4				
		6	u	2	1	4	3				
			v	2	1	4	3				
			w	3	4	1	2				
			x	2	1	4	3				
		7	u	2	1	4	3				
			v	3	4	1	2				
			w	1	2	3	4				
			x	1	3	2	4				
		8	u	2	3	4	1				
			v	4	1	2	3				
			w	1	4	3	2				
			x	3	2	1	4				
		9	u	2	3	4	1				
			v	4	1	2	3				
			w	1	4	3	2				
			x	1	4	3	2				
		10	u	2	3	4	1				
			v	4	1	2	3				
			w	4	3	2	1				
			x	2	1	4	3				

Table 12

(7)

n	m	k	1 2 3 4 5 6 7 8
4	9	11	u 2 3 4 1
			v 4 1 2 3
			w 4 3 2 1
			x 4 3 2 1
	10	1	u 1 2 3 4
			v 1 2 3 4
			w 2 1 4 3
			x 3 4 1 2
		2	u 1 2 3 4
			v 2 1 4 3
			w 1 2 3 4
			x 3 4 1 2
		3	u 2 1 4 3
			v 2 1 4 3
			w 1 2 3 4
			x 3 4 1 2
		4	u 1 2 3 4
			v 2 1 4 3
			w 3 4 1 2
			x 1 2 3 4
		5	u 2 1 4 3
			v 1 2 3 4
			w 3 4 1 2
			x 1 2 3 4
		6	u 2 1 4 3
			v 1 2 3 4
			w 3 4 1 2
			x 2 1 4 3
		7	u 2 1 4 3
			v 2 1 4 3
			w 3 4 1 2
			x 1 2 3 4
4	10	8	u 2 1 4 3
			v 3 4 1 2
			w 1 2 3 4
			x 1 2 3 4
		9	u 2 1 4 3
			v 3 4 1 2
			w 1 2 3 4
			x 4 3 2 1
		10	u 2 1 4 3
			v 3 4 1 2
			w 1 2 3 4
			x 3 4 1 2
		11	u 1 2 3 4
			v 2 3 4 1
			w 1 4 3 2
			x 3 2 1 4
		12	u 1 2 3 4
			v 2 3 4 1
			w 1 4 3 2
			x 1 4 3 2
		13	u 2 1 4 3
			v 3 4 2 1
			w 1 2 4 3
			x 1 2 4 3
	11	1	u 1 2 3 4
			v 1 2 3 4
			w 2 1 4 3
			x 3 4 1 2
		2	u 1 2 3 4
			v 2 1 4 3
			w 1 2 3 4
			x 3 4 1 2

Table 12

(8)

n	m	k		1 2 3 4 5 6 7 8
4	11	3	u	2 1 4 3
			v	1 2 3 4
			w	1 2 4 3
			x	3 4 2 1
		4	u	2 1 4 3
			v	2 1 4 3
			w	1 2 4 3
			x	3 4 2 1
		5	u	1 2 3 4
			v	2 1 4 3
			w	3 4 1 2
			x	3 4 1 2
		6	u	1 2 3 4
			v	2 1 4 3
			w	3 4 1 2
			x	4 3 2 1
		7	u	2 1 4 3
			v	1 2 3 4
			w	3 4 1 2
			x	4 3 1 2
		8	u	2 1 4 3
			v	2 1 4 3
			w	3 4 1 2
			x	4 3 1 2
		9	u	1 2 3 4
			v	2 1 4 3
			w	3 4 1 2
			x	1 2 3 4
		10	u	1 2 3 4
			v	2 3 4 1
			w	1 4 3 2
			x	3 2 1 4

n	m	k		1 2 3 4 5 6 7 8
4	11	11	u	1 2 3 4
			v	2 3 4 1
			w	1 4 3 2
			x	1 4 3 2
	12	1	u	1 2 3 4
			v	1 2 3 4
			w	2 1 4 3
			x	3 4 1 2
		2	u	1 2 3 4
			v	2 1 4 3
			w	1 2 4 3
			x	3 4 1 2
		3	u	2 1 4 3
			v	2 1 4 3
			w	1 2 3 4
			x	3 4 2 1
		4	u	1 2 3 4
			v	2 1 4 3
			w	3 4 1 2
			x	1 2 4 3
	13	1	u	2 1 4 3
			v	3 4 1 2
			w	1 3 4 2
			x	1 2 4 3
	14	1	u	2 1 4 3
			v	3 4 1 2
			w	1 3 4 2
			x	1 2 4 3
	15	1	u	1 2 3 4
			v	1 2 3 4
			w	2 1 4 3
			x	3 4 1 2

Table 12

(9)

n	m	k		1 2 3 4 5 6 7 8
4	15	2	u	2 1 4 3
			v	2 1 4 3
			w	1 2 3 4
			x	3 4 1 2
		3	u	1 2 3 4
			v	1 2 3 4
			w	2 3 4 1
			x	4 3 2 1
		4	u	1 2 3 4
			v	1 2 3 4
			w	2 3 4 1
			x	1 4 3 2
		5	u	2 1 4 3
			v	2 1 4 3
			w	3 4 1 2
			x	3 4 1 2
		6	u	2 1 4 3
			v	2 1 4 3
			w	3 4 1 2
			x	4 3 2 1
		7	u	2 1 4 3
			v	2 1 4 3
			w	3 4 2 1
			x	4 3 2 1
		8	u	1 2 4 3
			v	2 1 3 4
			w	3 4 1 2
			x	2 1 4 3
		9	u	1 2 4 3
			v	2 1 3 4
			w	3 4 1 2
			x	1 2 3 4

n	m	k		1 2 3 4 5 6 7 8
4	15	10	u	2 1 4 3
			v	2 1 4 3
			w	3 4 1 2
			x	1 2 3 4
		11	u	2 1 4 3
			v	2 1 4 3
			w	3 4 2 1
			x	1 2 4 3
		12	u	2 1 4 3
			v	3 4 1 2
			w	1 3 2 4
			x	1 2 3 4
		13	u	2 1 4 3
			v	3 4 1 2
			w	1 3 2 4
			x	4 3 2 1
	16	1	u	1 2 3 4
			v	1 2 3 4
			w	2 1 4 3
			x	3 4 1 2
		2	u	2 1 4 3
			v	2 1 4 3
			w	1 2 4 3
			x	3 4 2 1
		3	u	1 2 3 4
			v	1 2 3 4
			w	2 3 4 1
			x	4 3 2 1
		4	u	1 2 3 4
			v	1 2 3 4
			w	2 3 4 1
			x	1 4 3 2

Table 12

(10)

n	m	k		1	2	3	4	5	6	7	8
4	16	5	u	2	1	4	3				
			v	2	1	4	3				
			w	3	4	1	2				
			x	4	3	1	2				
		6	u	2	1	4	3				
			v	2	1	4	3				
			w	3	4	2	1				
			x	3	4	2	1				
		7	u	2	1	4	3				
			v	2	1	4	3				
			w	3	4	2	1				
			x	4	3	1	2				
	17	1	u	1	2	3	4				
			v	1	2	3	4				
			w	2	1	4	3				
			x	3	4	1	2				
		2	u	2	1	4	3				
			v	3	4	1	2				
			w	1	4	2	3				
			x	1	2	4	3				
5	1	1	u	1	2	3	4	5			
			v	2	3	4	5	1			
			w	1	2	3	4	5			
			x	-	-	-	-	-			
	2	1	u	1	2	3	4	5			
			v	2	3	4	5	1			
			w	1	5	4	3	2			
			x	-	-	-	-	-			
	4	1	u	2	3	4	5	1			
			v	4	5	1	2	3			
			w	1	4	2	5	3			
			x	-	-	-	-	-			

n	m	k		1	2	3	4	5	6	7	8
5	6	1	u	2	3	4	5	1			
			v	5	1	2	3	4			
			w	1	2	3	4	5			
			x	1	5	4	3	2			
		2	u	2	3	4	5	1			
			v	2	3	4	5	1			
			w	1	2	3	4	5			
			x	1	2	3	4	5			
	7	1	u	1	2	3	4	5			
			v	2	3	4	5	1			
			w	1	2	3	4	5			
			x	1	5	4	3	2			
		2	u	2	3	4	5	1			
			v	1	2	3	4	5			
			w	1	2	3	4	5			
			x	1	2	3	4	5			
	8	1	u	1	2	3	4	5			
			v	2	3	4	5	1			
			w	1	2	3	4	5			
			x	1	5	4	3	2			
		2	u	2	3	4	5	1			
			v	1	2	3	4	5			
			w	1	2	3	4	5			
			x	4	5	1	2	3			
	9	1	u	2	3	4	5	1			
			v	5	1	2	3	4			
			w	1	5	4	3	2			
			x	1	5	4	3	2			
	10	1	u	1	2	3	4	5			
			v	2	3	4	5	1			
			w	1	5	4	3	2			
			x	1	5	4	3	2			

Table 12

(11)

n	m	k		1 2 3 4 5 6 7 8
5	11	1	u	1 2 3 4 5
			v	2 3 4 5 1
			w	1 5 4 3 2
			x	1 5 4 3 2
		2	u	2 3 4 5 1
			v	1 2 3 4 5
			w	1 5 4 3 2
			x	4 5 1 2 3
	12	1	u	1 2 3 4 5
			v	2 3 4 5 1
			w	1 5 4 3 2
			x	4 3 2 1 5
6	1	1	u	1 2 3 4 5 6
			v	2 3 4 5 6 1
			w	1 2 3 4 5 6
			x	- - - - - - -
	2	1	u	1 2 3 4 5 6
			v	2 3 1 5 6 4
			w	4 6 5 1 3 2
			x	- - - - - - -
		2	u	1 2 3 4 5 6
			v	2 3 4 5 6 1
			w	1 6 5 4 3 2
			x	- - - - - - -
	3	1	u	1 2 4 3 6 5
			v	2 1 3 4 6 5
			w	3 4 5 6 1 2
			x	- - - - - - -
	4	1	u	1 2 3 6 4 5
			v	2 3 1 4 5 6
			w	4 5 6 1 3 2
			x	- - - - - - -

n	m	k		1 2 3 4 5 6 7 8
6	4	2	u	2 3 1 5 6 4
			v	3 1 2 5 6 4
			w	4 5 6 1 3 2
			x	- - - - - - -
	5	1	u	1 2 3 4 5 6
			v	1 2 3 4 5 6
			w	2 3 4 5 6 1
			x	- - - - - - -
		2	u	1 2 4 3 6 5
			v	2 1 4 3 5 6
			w	3 4 5 6 1 2
			x	- - - - - - -
		3	u	1 2 4 3 6 5
			v	2 1 4 3 5 6
			w	3 4 5 6 2 1
			x	- - - - - - -
		4	u	2 3 1 6 4 5
			v	2 3 1 6 4 5
			w	4 5 6 1 2 3
			x	- - - - - - -
		5	u	2 3 1 6 4 5
			v	2 3 1 6 4 5
			w	4 5 6 3 1 2
			x	- - - - - - -
	6	1	u	1 2 3 5 6 4
			v	2 3 1 4 5 6
			w	1 2 3 4 5 6
			x	4 5 6 1 2 3
		2	u	2 3 1 6 4 5
			v	3 1 2 5 6 4
			w	1 2 3 4 5 6
			x	4 5 6 1 2 3

Table 12

(12)

n	m	k		1 2 3 4 5 6 7 8
6	6	3	u	2 3 1 5 6 4
			v	2 3 1 5 6 4
			w	1 2 3 4 5 6
			x	4 5 6 1 2 3
		4	u	2 3 4 5 6 1
			v	6 1 2 3 4 5
			w	1 2 3 4 5 6
			x	1 6 5 4 3 2
		5	u	2 3 4 5 6 1
			v	6 1 2 3 4 5
			w	1 2 3 4 5 6
			x	6 5 4 3 2 1
		6	u	2 3 4 5 6 1
			v	2 3 4 5 6 1
			w	1 2 3 4 5 6
			x	1 2 3 4 5 6
		7	u	2 3 4 5 6 1
			v	2 3 4 5 6 1
			w	1 2 3 4 5 6
			x	4 5 6 1 2 3
	7	1	u	1 2 3 4 5 6
			v	2 3 1 6 4 5
			w	1 2 3 4 5 6
			x	4 5 6 1 2 3
		2	u	2 3 1 5 6 4
			v	1 2 3 4 5 6
			w	1 2 3 4 5 6
			x	4 5 6 1 2 3
		3	u	2 3 1 5 6 4
			v	3 1 2 5 6 4
			w	1 2 3 4 5 6
			x	4 5 6 1 2 3

n	m	k		1 2 3 4 5 6 7 8
6	7	4	u	1 2 3 4 5 6
			v	2 3 4 5 6 1
			w	1 2 3 4 5 6
			x	1 6 5 4 3 2
		5	u	2 1 4 3 6 5
			v	3 4 5 6 1 2
			w	1 2 3 4 5 6
			x	1 2 5 6 3 4
		6	u	2 1 4 3 6 5
			v	3 4 5 6 1 2
			w	1 2 3 4 5 6
			x	2 1 6 5 4 3
		7	u	2 1 4 3 6 5
			v	3 4 5 6 2 1
			w	1 2 3 4 5 6
			x	1 2 6 5 4 3
		8	u	2 3 1 5 6 4
			v	4 5 6 1 2 3
			w	1 2 3 4 5 6
			x	1 2 3 4 5 6
		9	u	2 3 4 5 6 1
			v	1 2 3 4 5 6
			w	1 2 3 4 5 6
			x	1 2 3 4 5 6
		10	u	2 3 4 5 6 1
			v	1 2 3 4 5 6
			w	1 2 3 4 5 6
			x	4 5 6 1 2 3
		11	u	2 3 4 5 6 1
			v	4 5 6 1 2 3
			w	1 2 3 4 5 6
			x	1 2 3 4 5 6

Table 12

(13)

n	m	k		1 2 3 4 5 6 7 8
6	8	1	u	1 2 3 4 5 6
			v	2 3 1 6 4 5
			w	1 2 3 4 5 6
			x	4 5 6 1 2 3
		2	u	2 3 1 5 6 4
			v	1 2 3 4 5 6
			w	1 2 3 4 5 6
			x	4 5 6 2 3 1
		3	u	2 3 1 5 6 4
			v	3 1 2 5 6 4
			w	1 2 3 4 5 6
			x	4 5 6 2 3 1
		4	u	1 2 3 4 5 6
			v	2 3 4 5 6 1
			w	1 2 3 4 5 6
			x	1 6 5 4 3 2
		5	u	2 3 1 5 6 4
			v	4 5 6 1 2 3
			w	1 2 3 4 5 6
			x	3 1 2 6 4 5
	9	1	u	1 2 3 5 6 4
			v	2 3 1 4 5 6
			w	1 3 2 4 6 5
			x	4 5 6 1 2 3
		2	u	2 3 1 6 4 5
			v	3 1 2 5 6 4
			w	1 3 2 4 6 5
			x	4 5 6 1 2 3
		3	u	2 3 1 6 4 5
			v	3 1 2 5 6 4
			w	4 5 6 1 2 3
			x	1 3 2 4 6 5

n	m	k		1 2 3 4 5 6 7 8
6	9	4	u	2 3 1 6 4 5
			v	2 3 1 6 4 5
			w	4 5 6 1 2 3
			x	1 2 3 4 5 6
		5	u	2 3 4 5 6 1
			v	6 1 2 3 4 5
			w	1 6 5 4 3 2
			x	1 6 5 4 3 2
		6	u	2 3 4 5 6 1
			v	6 1 2 3 4 5
			w	1 6 5 4 3 2
			x	4 3 2 1 6 5
		7	u	2 3 4 5 6 1
			v	6 1 2 3 4 5
			w	6 5 4 3 2 1
			x	3 2 1 6 5 4
		8	u	2 3 4 5 6 1
			v	6 1 2 3 4 5
			w	6 5 4 3 2 1
			x	6 5 4 3 2 1
	10	1	u	1 2 3 4 5 6
			v	2 3 1 6 4 5
			w	1 3 2 4 6 5
			x	4 5 6 1 2 3
		2	u	2 3 1 5 6 4
			v	3 1 2 5 6 4
			w	1 3 2 4 6 5
			x	4 5 6 1 2 3
		3	u	1 2 3 4 5 6
			v	2 3 1 6 4 5
			w	4 5 6 1 2 3
			x	1 3 2 4 6 5

Table 12

(14)

n	m	k		1 2 3 4 5 6 7 8
6	10	4	u	2 3 1 6 4 5
			v	1 2 3 4 5 6
			w	4 5 6 1 2 3
			x	1 2 3 4 5 6
		5	u	2 1 4 3 6 5
			v	3 4 5 6 1 2
			w	1 2 5 6 3 4
			x	1 2 5 6 3 4
		6	u	2 1 4 3 6 5
			v	3 4 5 6 1 2
			w	1 2 5 6 3 4
			x	2 1 6 5 4 3
		7	u	1 2 3 4 5 6
			v	2 3 4 5 6 1
			w	1 6 5 4 3 2
			x	4 3 2 1 6 5
		8	u	1 2 3 4 5 6
			v	2 3 4 5 6 1
			w	1 6 5 4 3 2
			x	1 6 5 4 3 2
		9	u	2 1 4 3 6 5
			v	3 4 5 6 2 1
			w	1 2 6 5 4 3
			x	1 2 6 5 4 3
	11	1	u	1 2 3 4 5 6
			v	2 3 1 6 4 5
			w	1 3 2 4 6 5
			x	4 5 6 1 2 3
		2	u	2 3 1 5 6 4
			v	1 2 3 4 5 6
			w	1 3 2 6 5 4
			x	4 5 6 2 3 1
6	11	3	u	2 3 1 5 6 4
			v	3 1 2 5 6 4
			w	1 3 2 6 5 4
			x	4 5 6 2 3 1
		4	u	1 2 3 4 5 6
			v	2 3 1 6 4 5
			w	4 5 6 1 2 3
			x	4 5 6 1 2 3
		5	u	2 3 1 6 4 5
			v	1 2 3 4 5 6
			w	4 5 6 1 2 3
			x	6 5 4 1 3 2
		6	u	1 2 3 4 5 6
			v	2 3 1 6 4 5
			w	4 5 6 1 2 3
			x	1 3 2 4 6 5
		7	u	2 3 1 6 4 5
			v	1 2 3 4 5 6
			w	4 5 6 1 2 3
			x	3 1 2 5 6 4
		8	u	1 2 3 4 5 6
			v	2 3 4 5 6 1
			w	1 6 5 4 3 2
			x	4 3 2 1 6 5
		9	u	1 2 3 4 5 6
			v	2 3 4 5 6 1
			w	1 6 5 4 3 2
			x	1 6 5 4 3 2
		10	u	2 3 1 5 6 4
			v	4 5 6 1 2 3
			w	1 3 2 4 6 5
			x	6 4 5 3 1 2

Table 12

(15)

n	m	k		1	2	3	4	5	6	7	8
6	11	11	u	2	3	1	5	6	4		
			v	4	5	6	1	2	3		
			w	1	3	2	4	6	5		
			x	3	1	2	6	4	5		
	12	1	u	1	2	3	4	5	6		
			v	2	3	1	6	4	5		
			w	1	3	2	5	4	6		
			x	4	5	6	1	2	3		
		2	u	2	3	1	5	6	4		
			v	3	1	2	5	6	4		
			w	1	3	2	5	4	6		
			x	4	5	6	2	3	1		
		3	u	1	2	3	4	5	6		
			v	2	3	1	6	4	5		
			w	4	5	6	1	2	3		
			x	1	3	2	6	5	4		
		4	u	2	3	1	6	4	5		
			v	1	2	3	4	5	6		
			w	4	5	6	1	2	3		
			x	3	1	2	5	6	4		
	13	1	u	1	2	3	4	5	6		
			v	1	2	3	4	5	6		
			w	2	3	1	5	6	4		
			x	4	6	5	1	3	2		
		2	u	2	3	1	5	6	4		
			v	2	3	1	5	6	4		
			w	1	2	3	4	5	6		
			x	4	5	6	1	2	3		
		3	u	2	3	1	5	6	4		
			v	2	3	1	5	6	4		
			w	3	1	2	5	6	4		
			x	4	5	6	1	2	3		

n	m	k		1	2	3	4	5	6	7	8
6	13	4	u	1	2	4	3	6	5		
			v	2	1	3	4	6	5		
			w	3	4	5	6	1	2		
			x	1	2	5	6	3	4		
		5	u	1	2	4	3	6	5		
			v	2	1	3	4	6	5		
			w	3	4	5	6	1	2		
			x	2	1	6	5	4	3		
	14	1	u	1	2	3	4	5	6		
			v	1	2	3	4	5	6		
			w	2	3	1	5	6	4		
			x	4	6	5	1	3	2		
		2	u	2	3	1	6	4	5		
			v	2	3	1	6	4	5		
			w	1	2	3	4	5	6		
			x	4	5	6	1	2	3		
		3	u	1	2	4	3	6	5		
			v	2	1	3	4	6	5		
			w	3	4	5	6	1	2		
			x	1	2	5	6	3	4		
		4	u	1	2	4	3	6	5		
			v	2	1	3	4	6	5		
			w	3	4	5	6	1	2		
			x	2	1	6	5	4	3		
	15	1	u	2	3	1	5	6	4		
			v	3	1	2	5	6	4		
			w	4	5	6	1	3	2		
			x	4	5	6	1	2	3		
		2	u	1	2	3	6	4	5		
			v	2	3	1	4	5	6		
			w	4	5	6	1	3	2		
			x	1	3	2	4	5	6		

Table 12

(16)

n	m	k		1	2	3	4	5	6	7	8
6	16	1	u	2	3	1	5	6	4		
			v	3	1	2	5	6	4		
			w	4	5	6	1	3	2		
			x	5	6	4	1	2	3		
		2	u	1	2	3	6	4	5		
			v	2	3	1	4	5	6		
			w	4	5	6	1	3	2		
			x	3	2	1	5	6	4		
	17	1	u	1	2	3	4	5	6		
			v	1	2	3	4	5	6		
			w	2	3	1	5	6	4		
			x	4	6	5	1	3	2		
		2	u	2	3	1	5	6	4		
			v	2	3	1	5	6	4		
			w	1	3	2	4	6	5		
			x	4	5	6	1	2	3		
		3	u	1	2	3	4	5	6		
			v	1	2	3	4	5	6		
			w	2	3	4	5	6	1		
			x	4	3	2	1	6	5		
		4	u	1	2	3	4	5	6		
			v	1	2	3	4	5	6		
			w	2	3	4	5	6	1		
			x	1	6	5	4	3	2		
		5	u	1	2	4	3	6	5		
			v	2	1	4	3	5	6		
			w	3	4	5	6	1	2		
			x	1	2	5	6	3	4		
		6	u	1	2	4	3	6	5		
			v	2	1	4	3	5	6		
			w	3	4	5	6	1	2		
			x	2	1	6	5	4	3		

n	m	k		1	2	3	4	5	6	7	8
6	17	7	u	1	2	4	3	6	5		
			v	2	1	4	3	5	6		
			w	3	4	5	6	2	1		
			x	1	2	6	5	4	3		
		8	u	1	2	4	3	6	5		
			v	2	1	4	3	5	6		
			w	3	4	5	6	2	1		
			x	2	1	5	6	3	4		
		9	u	2	3	1	6	4	5		
			v	2	3	1	6	4	5		
			w	4	5	6	1	2	3		
			x	4	6	5	1	3	2		
		10	u	2	3	1	6	4	5		
			v	2	3	1	6	4	5		
			w	4	5	6	3	1	2		
			x	6	5	4	3	2	1		
		11	u	2	3	1	6	4	5		
			v	2	3	1	6	4	5		
			w	4	5	6	1	2	3		
			x	1	2	3	4	5	6		
7	1	1	u	1	2	3	4	5	6	7	
			v	2	3	4	5	6	7	1	
			w	1	2	3	4	5	6	7	
			x	-	-	-	-	-	-	-	
	2	1	u	1	2	3	4	5	6	7	
			v	2	3	4	5	6	7	1	
			w	1	7	6	5	4	3	2	
			x	-	-	-	-	-	-	-	
	3	1	u	2	3	4	5	6	7	1	
			v	5	6	7	1	2	3	4	
			w	1	5	2	6	3	7	4	
			x	-	-	-	-	-	-	-	

Table 12

(17)

n	m	k		1	2	3	4	5	6	7	8
7	5	1	u	2	3	4	5	6	7	1	
			v	5	6	7	1	2	3	4	
			w	1	6	4	2	7	5	3	
			x	-	-	-	-	-	-	-	
	6	1	u	2	3	4	5	6	7	1	
			v	7	1	2	3	4	5	6	
			w	1	2	3	4	5	6	7	
			x	1	7	6	5	4	3	2	
		2	u	2	3	4	5	6	7	1	
			v	2	3	4	5	6	7	1	
			w	1	2	3	4	5	6	7	
			x	1	2	3	4	5	6	7	
	7	1	u	1	2	3	4	5	6	7	
			v	2	3	4	5	6	7	1	
			w	1	2	3	4	5	6	7	
			x	1	7	6	5	4	3	2	
		2	u	2	3	4	5	6	7	1	
			v	1	2	3	4	5	6	7	
			w	1	2	3	4	5	6	7	
			x	1	2	3	4	5	6	7	
	8	1	u	1	2	3	4	5	6	7	
			v	2	3	4	5	6	7	1	
			w	1	2	3	4	5	6	7	
			x	1	7	6	5	4	3	2	
		2	u	2	3	4	5	6	7	1	
			v	1	2	3	4	5	6	7	
			w	1	2	3	4	5	6	7	
			x	5	6	7	1	2	3	4	
	9	1	u	2	3	4	5	6	7	1	
			v	7	1	2	3	4	5	6	
			w	1	7	6	5	4	3	2	
			x	1	7	6	5	4	3	2	

n	m	k		1	2	3	4	5	6	7	8
7	10	1	u	1	2	3	4	5	6	7	
			v	2	3	4	5	6	7	1	
			w	1	7	6	5	4	3	2	
			x	1	7	6	5	4	3	2	
	11	1	u	1	2	3	4	5	6	7	
			v	2	3	4	5	6	7	1	
			w	1	7	6	5	4	3	2	
			x	1	7	6	5	4	3	2	
		2	u	2	3	4	5	6	7	1	
			v	1	2	3	4	5	6	7	
			w	1	7	6	5	4	3	2	
			x	5	6	7	1	2	3	4	
	12	1	u	1	2	3	4	5	6	7	
			v	2	3	4	5	6	7	1	
			w	1	7	6	5	4	3	2	
			x	5	4	3	2	1	7	6	
8	1	1	u	1	2	3	4	5	6	7	8
			v	2	3	4	5	6	7	8	1
			w	1	2	3	4	5	6	7	8
			x	-	-	-	-	-	-	-	-
		2	u	2	1	4	3	6	5	8	7
			v	3	4	5	6	7	8	1	2
			w	1	2	3	4	5	6	7	8
			x	-	-	-	-	-	-	-	-
	2	1	u	1	2	3	4	5	6	7	8
			v	2	3	4	1	6	7	8	5
			w	5	8	7	6	1	4	3	2
			x	-	-	-	-	-	-	-	-
		2	u	2	1	4	3	6	5	8	7
			v	3	4	1	2	7	8	5	6
			w	5	6	7	8	1	2	3	4
			x	-	-	-	-	-	-	-	-

Table 12

(18)

n	m	k		1	2	3	4	5	6	7	8
8	2	3	u	1	2	3	4	5	6	7	8
			v	2	3	4	5	6	7	8	1
			w	1	8	7	6	5	4	3	2
			x	-	-	-	-	-	-	-	-
		4	u	2	1	4	3	6	5	8	7
			v	3	4	5	6	7	8	1	2
			w	1	2	7	8	5	6	3	4
			x	-	-	-	-	-	-	-	-
	4	1	u	1	2	4	3	5	6	8	7
			v	2	1	3	4	6	5	7	8
			w	3	4	5	6	7	8	1	2
			x	-	-	-	-	-	-	-	-
		2	u	2	1	4	3	6	5	8	7
			v	2	1	4	3	6	5	8	7
			w	3	4	5	6	7	8	1	2
			x	-	-	-	-	-	-	-	-
		3	u	2	1	4	3	7	8	5	6
			v	3	4	1	2	6	5	8	7
			w	5	6	7	8	3	4	1	2
			x	-	-	-	-	-	-	-	-
		4	u	2	1	4	3	7	8	5	6
			v	3	4	1	2	6	5	8	7
			w	5	6	7	8	1	2	3	4
			x	-	-	-	-	-	-	-	-
		5	u	1	2	3	4	8	5	6	7
			v	2	3	4	1	5	6	7	8
			w	5	6	7	8	1	4	3	2
			x	-	-	-	-	-	-	-	-
		6	u	2	1	4	3	8	7	5	6
			v	3	4	2	1	6	5	8	7
			w	5	6	7	8	1	2	4	3
			x	-	-	-	-	-	-	-	-

n	m	k		1	2	3	4	5	6	7	8
8	4	7	u	2	3	4	1	6	7	8	5
			v	4	1	2	3	6	7	8	5
			w	5	6	7	8	1	4	3	2
			x	-	-	-	-	-	-	-	-
		8	u	2	3	4	1	6	7	8	5
			v	4	1	2	3	6	7	8	5
			w	5	6	7	8	4	3	2	1
			x	-	-	-	-	-	-	-	-
		9	u	2	3	4	1	6	7	8	5
			v	5	6	7	8	3	4	1	2
			w	1	5	3	7	4	8	2	6
			x	-	-	-	-	-	-	-	-
	5	1	u	2	1	4	3	7	8	5	6
			v	3	4	1	2	8	7	6	5
			w	5	6	7	8	1	3	4	2
			x	-	-	-	-	-	-	-	-
	6	1	u	2	1	4	3	7	8	5	6
			v	3	4	1	2	6	5	8	7
			w	1	2	3	4	5	6	7	8
			x	5	6	7	8	1	2	3	4
		2	u	1	2	3	4	6	7	8	5
			v	2	3	4	1	5	6	7	8
			w	1	2	3	4	5	6	7	8
			x	5	6	7	8	1	2	3	4
		3	u	2	3	4	1	8	5	6	7
			v	4	1	2	3	6	7	8	5
			w	1	2	3	4	5	6	7	8
			x	5	6	7	8	1	2	3	4
		4	u	2	1	4	3	7	8	6	5
			v	3	4	2	1	6	5	8	7
			w	1	2	3	4	5	6	7	8
			x	5	6	7	8	1	2	3	4

Table 12

(19)

n	m	k		1	2	3	4	5	6	7	8
8	6	5	u	2	3	4	1	6	7	8	5
			v	2	3	4	1	6	7	8	5
			w	1	2	3	4	5	6	7	8
			x	5	6	7	8	1	2	3	4
		6	u	2	3	4	1	6	7	8	5
			v	5	6	7	8	3	4	1	2
			w	1	2	3	4	5	6	7	8
			x	6	4	8	2	7	1	5	3
		7	u	2	3	4	1	6	7	8	5
			v	5	6	7	8	3	4	1	2
			w	1	2	3	4	5	6	7	8
			x	1	5	3	7	2	6	4	8
		8	u	2	3	4	5	6	7	8	1
			v	8	1	2	3	4	5	6	7
			w	1	2	3	4	5	6	7	8
			x	1	8	7	6	5	4	3	2
		9	u	2	3	4	5	6	7	8	1
			v	8	1	2	3	4	5	6	7
			w	1	2	3	4	5	6	7	8
			x	8	7	6	5	4	3	2	1
		10	u	2	3	4	5	6	7	8	1
			v	6	7	8	1	2	3	4	5
			w	1	2	3	4	5	6	7	8
			x	1	6	3	8	5	2	7	4
		11	u	2	3	4	5	6	7	8	1
			v	4	5	6	7	8	1	2	3
			w	1	2	3	4	5	6	7	8
			x	1	4	7	2	5	8	3	6
		12	u	2	3	4	5	6	7	8	1
			v	2	3	4	5	6	7	8	1
			w	1	2	3	4	5	6	7	8
			x	1	2	3	4	5	6	7	8

n	m	k		1	2	3	4	5	6	7	8
8	6	13	u	2	3	4	5	6	7	8	1
			v	2	3	4	5	6	7	8	1
			w	1	2	3	4	5	6	7	8
			x	5	6	7	8	1	2	3	4
	7	1	u	1	2	3	4	5	6	7	8
			v	2	3	4	1	8	5	6	7
			w	1	2	3	4	5	6	7	8
			x	5	6	7	8	1	2	3	4
		2	u	2	1	4	3	6	5	8	7
			v	3	4	1	2	7	8	5	6
			w	1	2	3	4	5	6	7	8
			x	5	6	7	8	1	2	3	4
		3	u	2	1	4	3	6	5	8	7
			v	3	4	2	1	8	7	5	6
			w	1	2	3	4	5	6	7	8
			x	5	6	7	8	1	2	3	4
		4	u	2	3	4	1	6	7	8	5
			v	1	2	3	4	5	6	7	8
			w	1	2	3	4	5	6	7	8
			x	5	6	7	8	1	2	3	4
		5	u	2	3	4	1	6	7	8	5
			v	3	4	1	2	7	8	5	6
			w	1	2	3	4	5	6	7	8
			x	5	6	7	8	1	2	3	4
		6	u	2	3	4	1	6	7	8	5
			v	4	1	2	3	6	7	8	5
			w	1	2	3	4	5	6	7	8
			x	5	6	7	8	1	2	3	4
		7	u	1	2	3	4	5	6	7	8
			v	2	3	4	5	6	7	8	1
			w	1	2	3	4	5	6	7	8
			x	1	8	7	6	5	4	3	2

Table 12

(20)

n	m	k		1	2	3	4	5	6	7	8
8	7	8	u	2	1	4	3	6	5	8	7
			v	3	4	5	6	7	8	1	2
			w	1	2	3	4	5	6	7	8
			x	1	2	7	8	5	6	3	4
		9	u	2	1	4	3	6	5	8	7
			v	3	4	5	6	7	8	1	2
			w	1	2	3	4	5	6	7	8
			x	2	1	8	7	6	5	4	3
		10	u	2	1	4	3	6	5	8	7
			v	3	4	5	6	7	8	2	1
			w	1	2	3	4	5	6	7	8
			x	1	2	8	7	6	5	4	3
		11	u	2	3	4	1	6	7	8	5
			v	5	6	7	8	1	2	3	4
			w	1	2	3	4	5	6	7	8
			x	1	2	3	4	5	6	7	8
		12	u	2	3	4	1	6	7	8	5
			v	5	6	7	8	1	2	3	4
			w	1	2	3	4	5	6	7	8
			x	3	4	1	2	7	8	5	6
		13	u	2	3	4	1	6	7	8	5
			v	5	6	7	8	3	4	1	2
			w	1	2	3	4	5	6	7	8
			x	1	2	3	4	7	8	5	6
		14	u	2	3	4	5	6	7	8	1
			v	1	2	3	4	5	6	7	8
			w	1	2	3	4	5	6	7	8
			x	1	2	3	4	5	6	7	8
		15	u	2	3	4	5	6	7	8	1
			v	1	2	3	4	5	6	7	8
			w	1	2	3	4	5	6	7	8
			x	5	6	7	8	1	2	3	4

n	m	k		1	2	3	4	5	6	7	8
8	7	16	u	2	3	4	5	6	7	8	1
			v	5	6	7	8	1	2	3	4
			w	1	2	3	4	5	6	7	8
			x	1	2	3	4	5	6	7	8
	8	1	u	1	2	3	4	5	6	7	8
			v	2	3	4	1	8	5	6	7
			w	1	2	3	4	5	6	7	8
			x	5	6	7	8	1	2	3	4
		2	u	2	1	4	3	6	5	8	7
			v	3	4	1	2	7	8	5	6
			w	1	2	3	4	5	6	7	8
			x	5	6	7	8	2	1	4	3
		3	u	2	1	4	3	6	5	8	7
			v	3	4	2	1	8	7	5	6
			w	1	2	3	4	5	6	7	8
			x	5	6	7	8	2	1	4	3
		4	u	2	3	4	1	6	7	8	5
			v	1	2	3	4	5	6	7	8
			w	1	2	3	4	5	6	7	8
			x	5	6	7	8	2	3	4	1
		5	u	2	3	4	1	6	7	8	5
			v	3	4	1	2	7	8	5	6
			w	1	2	3	4	5	6	7	8
			x	5	6	7	8	2	3	4	1
		6	u	2	3	4	1	6	7	8	5
			v	4	1	2	3	6	7	8	5
			w	1	2	3	4	5	6	7	8
			x	5	6	7	8	2	3	4	1
		7	u	1	2	3	4	5	6	7	8
			v	2	3	4	5	6	7	8	1
			w	1	2	3	4	5	6	7	8
			x	1	8	7	6	5	4	3	2

Table 12

(21)

n	m	k		1	2	3	4	5	6	7	8
8	9	1	u	1	2	3	4	6	5	8	7
			v	2	1	4	3	5	6	7	8
			w	3	4	1	2	7	8	5	6
			x	5	6	7	8	1	2	3	4
		2	u	2	1	4	3	6	5	8	7
			v	2	1	4	3	6	5	8	7
			w	3	4	1	2	7	8	5	6
			x	5	6	7	8	1	2	3	4
		3	u	2	1	4	3	7	8	5	6
			v	3	4	1	2	6	5	8	7
			w	3	4	1	2	7	8	5	6
			x	5	6	7	8	1	2	3	4
		4	u	2	1	4	3	7	8	5	6
			v	3	4	1	2	6	5	8	7
			w	1	2	3	4	5	6	7	8
			x	5	6	7	8	1	2	3	4
		5	u	1	2	3	4	6	7	8	5
			v	2	3	4	1	5	6	7	8
			w	1	4	3	2	5	8	7	6
			x	5	6	7	8	1	2	3	4
		6	u	2	1	4	3	7	8	6	5
			v	3	4	2	1	6	5	8	7
			w	1	2	4	3	5	6	8	7
			x	5	6	7	8	1	2	3	4
		7	u	2	3	4	1	8	5	6	7
			v	4	1	2	3	6	7	8	5
			w	1	4	3	2	5	8	7	6
			x	5	6	7	8	1	2	3	4
		8	u	2	3	4	1	8	5	6	7
			v	4	1	2	3	6	7	8	5
			w	4	3	2	1	8	7	6	5
			x	5	6	7	8	1	2	3	4

n	m	k		1	2	3	4	5	6	7	8
8	9	9	u	2	1	4	3	6	5	8	7
			v	3	4	1	2	7	8	5	6
			w	5	6	7	8	1	2	3	4
			x	1	3	2	4	5	7	6	8
		10	u	2	3	4	1	8	5	6	7
			v	4	1	2	3	6	7	8	5
			w	5	6	7	8	1	2	3	4
			x	1	4	3	2	5	8	7	6
		11	u	2	3	4	1	8	5	6	7
			v	4	1	2	3	6	7	8	5
			w	5	6	7	8	1	2	3	4
			x	4	3	2	1	8	7	6	5
		12	u	2	3	4	1	8	5	6	7
			v	2	3	4	1	8	5	6	7
			w	5	6	7	8	1	2	3	4
			x	1	2	3	4	5	6	7	8
		13	u	2	3	4	1	8	5	6	7
			v	2	3	4	1	8	5	6	7
			w	5	6	7	8	1	2	3	4
			x	3	4	1	2	7	8	5	6
		14	u	2	3	4	1	6	7	8	5
			v	5	6	7	8	3	4	1	2
			w	1	4	3	2	7	6	5	8
			x	6	4	8	2	7	1	5	3
		15	u	2	3	4	1	6	7	8	5
			v	5	6	7	8	3	4	1	2
			w	1	4	3	2	7	6	5	8
			x	3	7	1	5	4	8	2	6
		16	u	2	3	4	1	6	7	8	5
			v	5	6	7	8	3	4	1	2
			w	1	4	3	2	7	6	5	8
			x	1	5	3	7	2	6	4	8

Table 12

(22)

n	m	k		1	2	3	4	5	6	7	8
8	9	17	u	2	3	4	5	6	7	8	1
			v	6	7	8	1	2	3	4	5
			w	1	8	7	6	5	4	3	2
			x	1	6	3	8	5	2	7	4
		18	u	2	3	4	5	6	7	8	1
			v	8	1	2	3	4	5	6	7
			w	1	8	7	6	5	4	3	2
			x	1	8	7	6	5	4	3	2
		19	u	2	3	4	5	6	7	8	1
			v	8	1	2	3	4	5	6	7
			w	1	8	7	6	5	4	3	2
			x	5	4	3	2	1	8	7	6
		20	u	2	3	4	5	6	7	8	1
			v	8	1	2	3	4	5	6	7
			w	8	7	6	5	4	3	2	1
			x	8	7	6	5	4	3	2	1
		21	u	2	3	4	5	6	7	8	1
			v	8	1	2	3	4	5	6	7
			w	8	7	6	5	4	3	2	1
			x	4	3	2	1	8	7	6	5
	10	1	u	1	2	3	4	5	6	7	8
			v	2	1	4	3	6	5	8	7
			w	3	4	1	2	7	8	5	6
			x	5	6	7	8	1	2	3	4
		2	u	2	1	4	3	6	5	8	7
			v	2	1	4	3	6	5	8	7
			w	3	4	1	2	7	8	5	6
			x	5	6	7	8	1	2	3	4
		3	u	2	1	4	3	6	5	8	7
			v	3	4	1	2	7	8	5	6
			w	1	2	3	4	5	6	7	8
			x	5	6	7	8	1	2	3	4

n	m	k		1	2	3	4	5	6	7	8
8	10	4	u	1	2	3	4	5	6	7	8
			v	2	3	4	1	8	5	6	7
			w	1	4	3	2	5	8	7	6
			x	5	6	7	8	1	2	3	4
		5	u	2	1	4	3	6	5	8	7
			v	3	4	2	1	8	7	5	6
			w	1	2	4	3	5	6	8	7
			x	5	6	7	8	1	2	3	4
		6	u	2	3	4	1	6	7	8	5
			v	4	1	2	3	6	7	8	5
			w	1	4	3	2	5	8	7	6
			x	5	6	7	8	1	2	3	4
		7	u	1	2	3	4	5	6	7	8
			v	2	3	4	1	8	5	6	7
			w	5	6	7	8	1	2	3	4
			x	1	4	3	2	5	8	7	6
		8	u	2	1	4	3	6	5	8	7
			v	3	4	1	2	7	8	5	6
			w	5	6	7	8	1	2	3	4
			x	1	2	3	4	5	6	7	8
		9	u	2	1	4	3	6	5	8	7
			v	3	4	1	2	7	8	5	6
			w	5	6	7	8	1	2	3	4
			x	2	1	4	3	6	5	8	7
		10	u	2	1	4	3	6	5	8	7
			v	3	4	2	1	8	7	5	6
			w	5	6	7	8	1	2	3	4
			x	1	2	4	3	5	6	8	7
		11	u	2	3	4	1	8	5	6	7
			v	1	2	3	4	5	6	7	8
			w	5	6	7	8	1	2	3	4
			x	1	2	3	4	5	6	7	8

Table 12

(23)

n	m	k		1 2 3 4 5 6 7 8
8	10	12	u	2 3 4 1 8 5 6 7
			v	1 2 3 4 5 6 7 8
			w	5 6 7 8 1 2 3 4
			x	3 4 1 2 7 8 5 6
		13	u	2 3 4 1 8 5 6 7
			v	3 4 1 2 7 8 5 6
			w	5 6 7 8 1 2 3 4
			x	1 2 3 4 5 6 7 8
		14	u	2 1 4 3 6 5 8 7
			v	3 4 5 6 7 8 1 2
			w	1 2 7 8 5 6 3 4
			x	5 6 3 4 1 2 7 8
		15	u	2 1 4 3 6 5 8 7
			v	3 4 5 6 7 8 1 2
			w	1 2 7 8 5 6 3 4
			x	1 2 7 8 5 6 3 4
		16	u	2 1 4 3 6 5 8 7
			v	3 4 5 6 7 8 1 2
			w	1 2 7 8 5 6 3 4
			x	6 5 4 3 2 1 8 7
		17	u	2 1 4 3 6 5 8 7
			v	3 4 5 6 7 8 1 2
			w	1 2 7 8 5 6 3 4
			x	2 1 8 7 6 5 4 3
		18	u	2 3 4 1 6 7 8 5
			v	5 6 7 8 3 4 1 2
			w	1 4 3 2 7 6 5 8
			x	1 2 3 4 7 8 5 6
		19	u	1 2 3 4 5 6 7 8
			v	2 3 4 5 6 7 8 1
			w	1 8 7 6 5 4 3 2
			x	5 4 3 2 1 8 7 6

n	m	k		1 2 3 4 5 6 7 8
8	10	20	u	1 2 3 4 5 6 7 8
			v	2 3 4 5 6 7 8 1
			w	1 8 7 6 5 4 3 2
			x	1 8 7 6 5 4 3 2
		21	u	2 1 4 3 6 5 8 7
			v	3 4 5 6 7 8 2 1
			w	1 2 8 7 6 5 4 3
			x	1 2 8 7 6 5 4 3
	11	1	u	1 2 3 4 5 6 7 8
			v	2 1 4 3 6 5 8 7
			w	3 4 1 2 7 8 5 6
			x	5 6 7 8 1 2 3 4
		2	u	2 1 4 3 6 5 8 7
			v	1 2 3 4 5 6 7 8
			w	3 4 1 2 8 7 6 5
			x	5 6 7 8 2 1 4 3
		3	u	2 1 4 3 6 5 8 7
			v	2 1 4 3 6 5 8 7
			w	3 4 1 2 8 7 6 5
			x	5 6 7 8 2 1 4 3
		4	u	1 2 3 4 5 6 7 8
			v	2 3 4 1 8 5 6 7
			w	1 4 3 2 5 8 7 6
			x	5 6 7 8 1 2 3 4
		5	u	2 1 4 3 6 5 8 7
			v	3 4 1 2 7 8 5 6
			w	1 2 3 4 6 5 8 7
			x	5 6 7 8 2 1 4 3
		6	u	2 1 4 3 6 5 8 7
			v	3 4 2 1 8 7 5 6
			w	1 2 4 3 6 5 7 8
			x	5 6 7 8 2 1 4 3

Table 12

(24)

n	m	k		1 2 3 4 5 6 7 8
8	11	7	u	2 3 4 1 6 7 8 5
			v	1 2 3 4 5 6 7 8
			w	1 4 3 2 8 7 6 5
			x	5 6 7 8 2 3 4 1
		8	u	2 3 4 1 6 7 8 5
			v	3 4 1 2 7 8 5 6
			w	1 4 3 2 8 7 6 5
			x	5 6 7 8 2 3 4 1
		9	u	2 3 4 1 6 7 8 5
			v	4 1 2 3 6 7 8 5
			w	1 4 3 2 8 7 6 5
			x	5 6 7 8 2 3 4 1
		10	u	1 2 3 4 5 6 7 8
			v	2 3 4 1 8 5 6 7
			w	5 6 7 8 1 2 3 4
			x	5 6 7 8 1 2 3 4
		11	u	1 2 3 4 5 6 7 8
			v	2 3 4 1 8 5 6 7
			w	5 6 7 8 1 2 3 4
			x	7 8 5 6 3 4 1 2
		12	u	2 1 4 3 6 5 8 7
			v	3 4 1 2 7 8 5 6
			w	5 6 7 8 1 2 3 4
			x	6 5 8 7 1 2 3 4
		13	u	2 1 4 3 6 5 8 7
			v	3 4 1 2 7 8 5 6
			w	5 6 7 8 1 2 3 4
			x	8 7 6 5 3 4 1 2
		14	u	2 1 4 3 6 5 8 7
			v	3 4 2 1 8 7 5 6
			w	5 6 7 8 1 2 3 4
			x	6 5 8 7 1 2 3 4

n	m	k		1 2 3 4 5 6 7 8
8	11	15	u	2 3 4 1 8 5 6 7
			v	1 2 3 4 5 6 7 8
			w	5 6 7 8 1 2 3 4
			x	8 7 6 5 1 4 3 2
		16	u	2 3 4 1 8 5 6 7
			v	3 4 1 2 7 8 5 6
			w	5 6 7 8 1 2 3 4
			x	8 7 6 5 1 4 3 2
		17	u	1 2 3 4 5 6 7 8
			v	2 3 4 1 8 5 6 7
			w	5 6 7 8 1 2 3 4
			x	1 4 3 2 5 8 7 6
		18	u	1 2 3 4 5 6 7 8
			v	2 3 4 5 6 7 8 1
			w	1 8 7 6 5 4 3 2
			x	5 4 3 2 1 8 7 6
		19	u	1 2 3 4 5 6 7 8
			v	2 3 4 5 6 7 8 1
			w	1 8 7 6 5 4 3 2
			x	1 8 7 6 5 4 3 2
	12	1	u	1 2 3 4 5 6 7 8
			v	2 1 4 3 6 5 8 7
			w	3 4 1 2 8 7 6 5
			x	5 6 7 8 1 2 3 4
		2	u	2 1 4 3 6 5 8 7
			v	2 1 4 3 6 5 8 7
			w	3 4 1 2 7 8 5 6
			x	5 6 7 8 2 1 4 3
		3	u	2 1 4 3 6 5 8 7
			v	3 4 1 2 7 8 5 6
			w	1 2 3 4 8 7 6 5
			x	5 6 7 8 2 1 4 3

Table 12

(25)

n	m	k		1	2	3	4	5	6	7	8
8	12	4	u	1	2	3	4	5	6	7	8
			v	2	3	4	1	8	5	6	7
			w	1	4	3	2	6	5	8	7
			x	5	6	7	8	1	2	3	4
		5	u	2	1	4	3	6	5	8	7
			v	3	4	2	1	8	7	5	6
			w	1	2	4	3	8	7	6	5
			x	5	6	7	8	2	1	4	3
		6	u	2	3	4	1	6	7	8	5
			v	4	1	2	3	6	7	8	5
			w	1	4	3	2	7	6	5	8
			x	5	6	7	8	2	3	4	1
		7	u	1	2	3	4	5	6	7	8
			v	2	3	4	1	8	5	6	7
			w	5	6	7	8	1	2	3	4
			x	1	4	3	2	8	7	6	5
	13	1	u	2	1	4	3	6	5	8	7
			v	3	4	1	2	8	7	6	5
			w	1	3	4	2	5	8	6	7
			x	5	6	7	8	1	2	3	4
	14	1	u	2	1	4	3	6	5	8	7
			v	3	4	1	2	8	7	6	5
			w	1	3	4	2	5	8	6	7
			x	5	6	7	8	1	2	3	4
	15	1	u	1	2	3	4	5	6	7	8
			v	1	2	3	4	5	6	7	8
			w	2	3	4	1	6	7	8	5
			x	5	8	7	6	1	4	3	2
		2	u	1	2	4	3	5	6	8	7
			v	2	1	3	4	6	5	7	8
			w	3	4	1	2	7	8	5	6
			x	5	6	7	8	1	2	3	4

n	m	k		1	2	3	4	5	6	7	8
8	15	3	u	2	1	4	3	6	5	8	7
			v	2	1	4	3	6	5	8	7
			w	3	4	1	2	7	8	5	6
			x	5	6	7	8	1	2	3	4
		4	u	2	1	4	3	6	5	8	7
			v	2	1	4	3	6	5	8	7
			w	3	4	2	1	7	8	6	5
			x	5	6	8	7	1	2	4	3
		5	u	2	1	4	3	6	5	8	7
			v	3	4	1	2	7	8	5	6
			w	1	3	2	4	5	7	6	8
			x	5	6	7	8	1	2	3	4
		6	u	2	1	4	3	6	5	8	7
			v	2	1	4	3	6	5	8	7
			w	3	4	5	6	7	8	1	2
			x	7	8	5	6	3	4	1	2
		7	u	2	1	4	3	6	5	8	7
			v	2	1	4	3	6	5	8	7
			w	3	4	5	6	7	8	1	2
			x	8	7	6	5	4	3	2	1
		8	u	1	2	4	3	5	6	8	7
			v	2	1	3	4	6	5	7	8
			w	3	4	5	6	7	8	1	2
			x	1	2	7	8	5	6	3	4
		9	u	2	1	3	4	6	5	7	8
			v	1	2	4	3	5	6	8	7
			w	3	4	5	6	7	8	1	2
			x	1	2	7	8	5	6	3	4
		10	u	2	1	3	4	6	5	7	8
			v	1	2	4	3	5	6	8	7
			w	3	4	5	6	7	8	1	2
			x	2	1	8	7	6	5	4	3

Table 12

(26)

n	m	k		1	2	3	4	5	6	7	8
8	15	11	u	2	1	4	3	6	5	8	7
			v	2	1	4	3	6	5	8	7
			w	3	4	5	6	7	8	1	2
			x	1	2	7	8	5	6	3	4
		12	u	2	1	4	3	7	8	5	6
			v	3	4	1	2	6	5	8	7
			w	5	6	7	8	1	2	3	4
			x	5	7	6	8	1	3	2	4
		13	u	2	3	4	1	6	7	8	5
			v	4	1	2	3	6	7	8	5
			w	5	6	7	8	1	4	3	2
			x	7	8	5	6	3	4	1	2
		14	u	2	3	4	1	6	7	8	5
			v	4	1	2	3	6	7	8	5
			w	5	6	7	8	1	4	3	2
			x	5	6	7	8	1	2	3	4
		15	u	2	3	4	1	6	7	8	5
			v	4	1	2	3	6	7	8	5
			w	5	6	7	8	4	3	2	1
			x	7	8	5	6	3	4	1	2
		16	u	2	3	4	1	6	7	8	5
			v	4	1	2	3	6	7	8	5
			w	5	6	7	8	4	3	2	1
			x	5	6	7	8	1	2	3	4
		17	u	2	1	4	3	7	8	5	6
			v	3	4	1	2	6	5	8	7
			w	5	6	7	8	3	4	1	2
			x	3	4	1	2	5	6	7	8
		18	u	2	1	4	3	7	8	5	6
			v	3	4	1	2	6	5	8	7
			w	5	6	7	8	3	4	1	2
			x	1	2	3	4	7	8	5	6

n	m	k		1	2	3	4	5	6	7	8
8	15	19	u	2	1	4	3	7	8	5	6
			v	3	4	1	2	6	5	8	7
			w	5	6	7	8	3	4	1	2
			x	2	1	4	3	8	7	6	5
		20	u	2	1	4	3	7	8	5	6
			v	3	4	1	2	6	5	8	7
			w	5	6	7	8	1	2	3	4
			x	1	2	3	4	5	6	7	8
		21	u	2	1	4	3	7	8	5	6
			v	3	4	1	2	6	5	8	7
			w	5	6	7	8	1	2	3	4
			x	4	3	2	1	8	7	6	5
		22	u	2	1	4	3	7	8	5	6
			v	3	4	1	2	6	5	8	7
			w	5	6	7	8	1	2	3	4
			x	3	4	1	2	7	8	5	6
		23	u	1	2	3	4	8	5	6	7
			v	2	3	4	1	5	6	7	8
			w	5	6	7	8	1	4	3	2
			x	3	2	1	4	7	8	5	6
		24	u	1	2	3	4	8	5	6	7
			v	2	3	4	1	5	6	7	8
			w	5	6	7	8	1	4	3	2
			x	1	4	3	2	5	6	7	8
		25	u	2	1	4	3	8	7	5	6
			v	3	4	2	1	6	5	8	7
			w	5	6	7	8	1	2	4	3
			x	2	1	3	4	6	5	8	7
		26	u	2	1	4	3	8	7	5	6
			v	3	4	2	1	6	5	8	7
			w	5	6	7	8	1	2	4	3
			x	1	2	4	3	5	6	7	8

Table 12

(27)

n	m	k		1	2	3	4	5	6	7	8
8	15	27	u	2	3	4	1	6	7	8	5
			v	5	6	7	8	3	4	1	2
			w	1	5	3	7	4	8	2	6
			x	1	2	3	4	7	8	5	6
		28	u	2	3	4	1	6	7	8	5
			v	5	6	7	8	3	4	1	2
			w	1	5	3	7	4	8	2	6
			x	3	4	1	2	5	6	7	8
	16	1	u	1	2	3	4	5	6	7	8
			v	1	2	3	4	5	6	7	8
			w	2	3	4	1	6	7	8	5
			x	5	8	7	6	1	4	3	2
		2	u	1	2	4	3	5	6	8	7
			v	2	1	3	4	6	5	7	8
			w	3	4	1	2	8	7	5	6
			x	5	6	7	8	1	2	4	3
		3	u	2	1	4	3	6	5	8	7
			v	2	1	4	3	6	5	8	7
			w	3	4	1	2	8	7	6	5
			x	5	6	7	8	2	1	4	3
		4	u	2	1	4	3	6	5	8	7
			v	2	1	4	3	6	5	8	7
			w	3	4	2	1	8	7	5	6
			x	5	6	8	7	2	1	3	4
		5	u	2	1	4	3	6	5	8	7
			v	3	4	1	2	7	8	5	6
			w	1	3	2	4	6	8	5	7
			x	5	6	7	8	2	1	4	3
		6	u	2	1	4	3	6	5	8	7
			v	2	1	4	3	6	5	8	7
			w	3	4	5	6	7	8	1	2
			x	8	7	5	6	4	3	1	2
8	16	7	u	2	1	4	3	6	5	8	7
			v	2	1	4	3	6	5	8	7
			w	3	4	5	6	7	8	1	2
			x	7	8	6	5	3	4	2	1
		8	u	1	2	4	3	5	6	8	7
			v	2	1	3	4	6	5	7	8
			w	3	4	5	6	7	8	1	2
			x	1	2	7	8	6	5	4	3
		9	u	2	1	4	3	7	8	5	6
			v	3	4	1	2	6	5	8	7
			w	5	6	7	8	3	4	1	2
			x	6	8	5	7	4	2	3	1
		10	u	2	3	4	1	6	7	8	5
			v	4	1	2	3	6	7	8	5
			w	5	6	7	8	1	4	3	2
			x	8	5	6	7	3	4	1	2
		11	u	2	3	4	1	6	7	8	5
			v	4	1	2	3	6	7	8	5
			w	5	6	7	8	1	4	3	2
			x	6	7	8	5	1	2	3	4
		12	u	2	3	4	1	6	7	8	5
			v	4	1	2	3	6	7	8	5
			w	5	6	7	8	4	3	2	1
			x	8	5	6	7	3	4	1	2
		13	u	2	3	4	1	6	7	8	5
			v	4	1	2	3	6	7	8	5
			w	5	6	7	8	4	3	2	1
			x	6	7	8	5	1	2	3	4
	17	1	u	2	1	4	3	6	5	8	7
			v	3	4	1	2	8	7	6	5
			w	1	4	2	3	5	7	8	6
			x	5	6	7	8	1	2	3	4

Table 12

(28)

n	m	k		1	2	3	4	5	6	7	8
8	17	2	u	2	1	4	3	7	8	5	6
			v	3	4	1	2	8	7	6	5
			w	5	6	7	8	1	3	4	2
			x	5	7	6	8	1	3	2	4
		3	u	2	1	4	3	7	8	5	6
			v	3	4	1	2	8	7	6	5
			w	5	6	7	8	1	3	4	2
			x	1	2	4	3	5	8	7	6
			u								
			v								
			w								
			x								
			u								
			v								
			w								
			x								
			u								
			v								
			w								
			x								
			u								
			v								
			w								
			x								
			u								
			v								
			w								
			x								

n	m	k		1	2	3	4	5	6	7	8
			u								
			v								
			w								
			x								
			u								
			v								
			w								
			x								
			u								
			v								
			w								
			x								
			u								
			v								
			w								
			x								
			u								
			v								
			w								
			x								
			u								
			v								
			w								
			x								
			u								
			v								
			w								
			x								

Table 12

(29)

3.8 FOLIO OF EXAMPLES

1° Using the data in Table 12, we have constructed 4-chromatic plane ornaments illustrating each of the 96 4-chromatic ornamental classes. The "colors" have been rendered as follows:

color	code
1	
2	. . . .
3	o o o o o o o o o o o o
4	- - - - - - - - - - - -

The various figures are distinguished by labels of the form: 4.m.k, where $\mathscr{F}_m$ is the relevant ornamental group and where k is the counter used in Table 12.

4.01.01
4.01.02

4.02.01

4.02.02

4.02.03

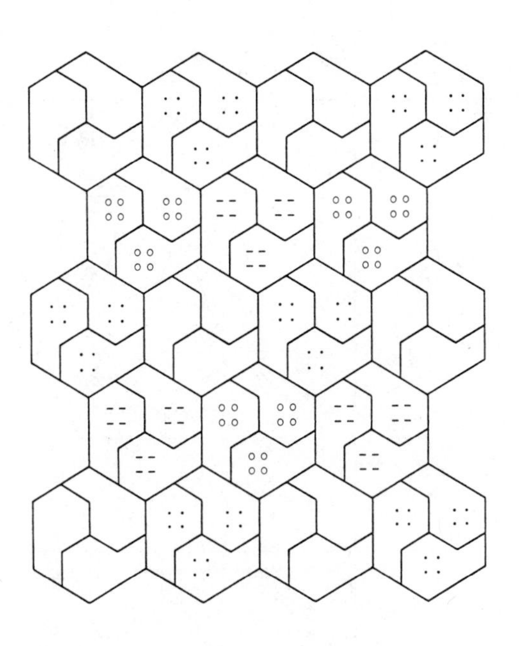

4.03.01

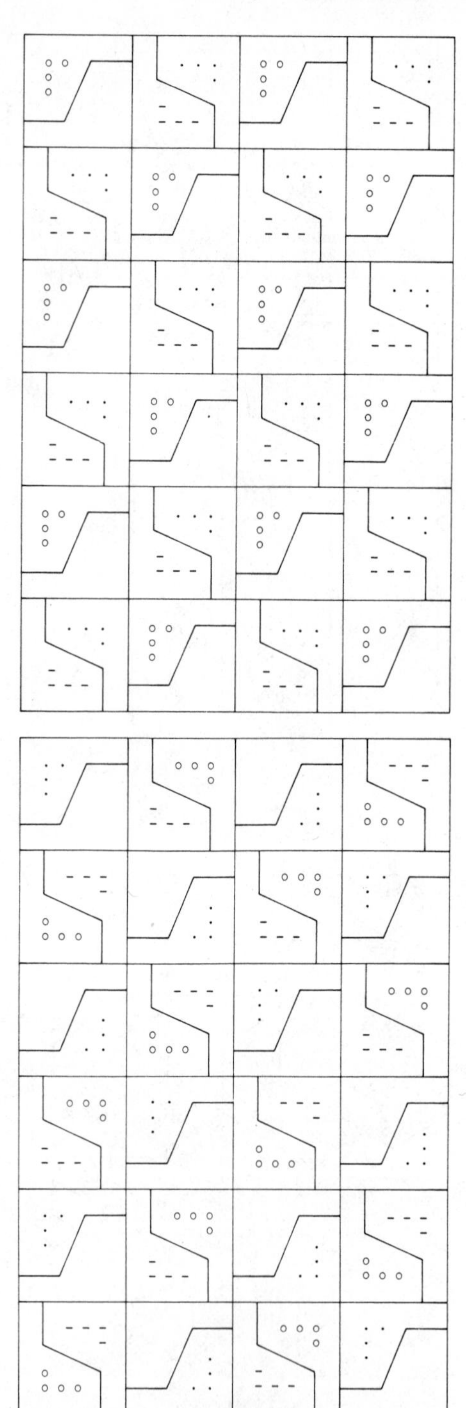

4.04.01

4.04.02

4.04.03

4.04.04

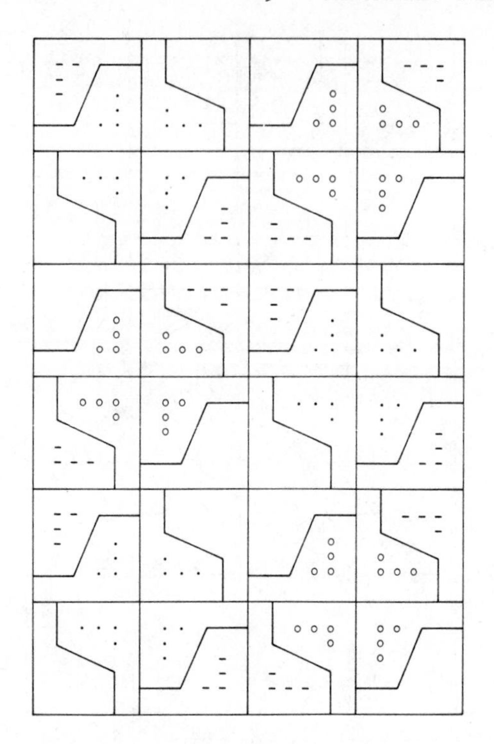

4.04.05

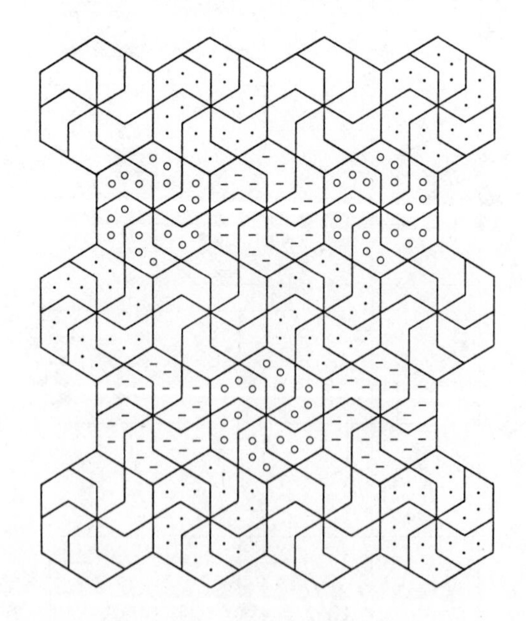

4.05.01

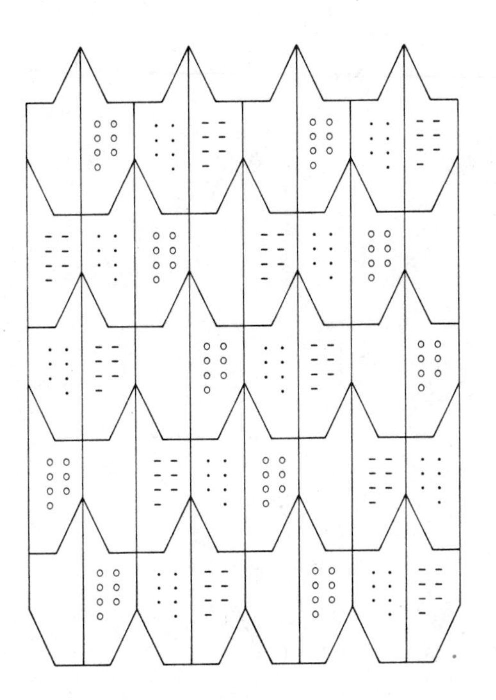

4.06.01

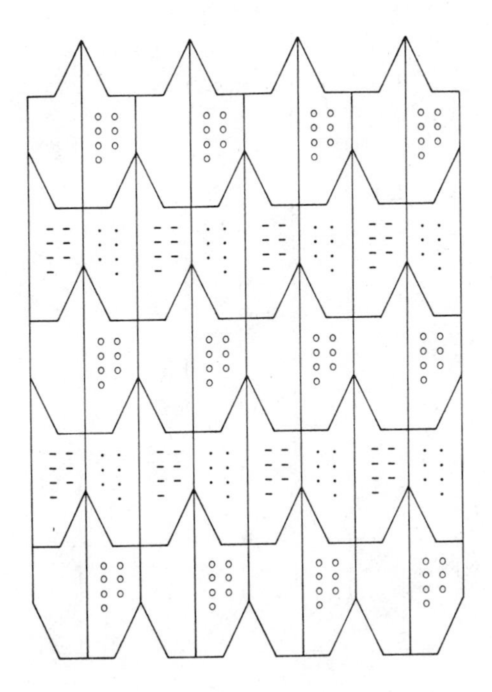

4.06.02

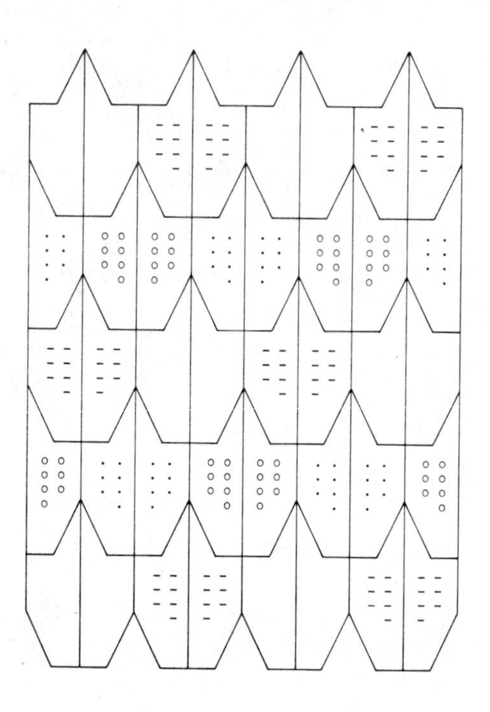

4.06.03

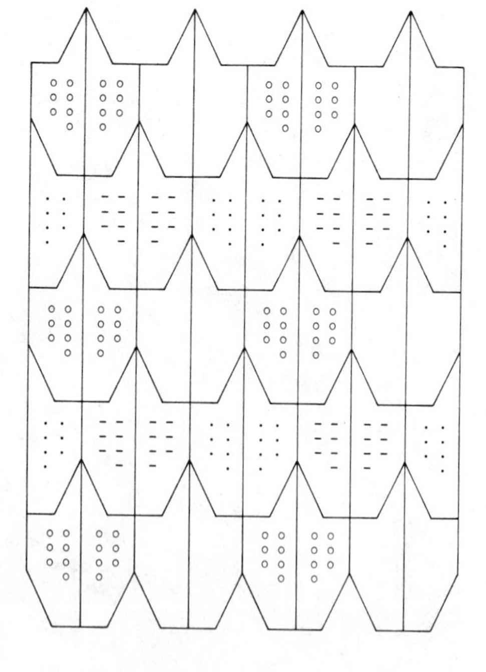

4.06.04

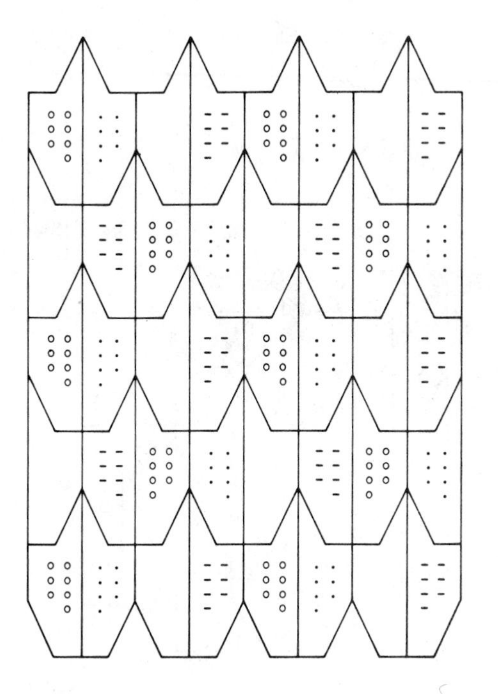

4.06.05

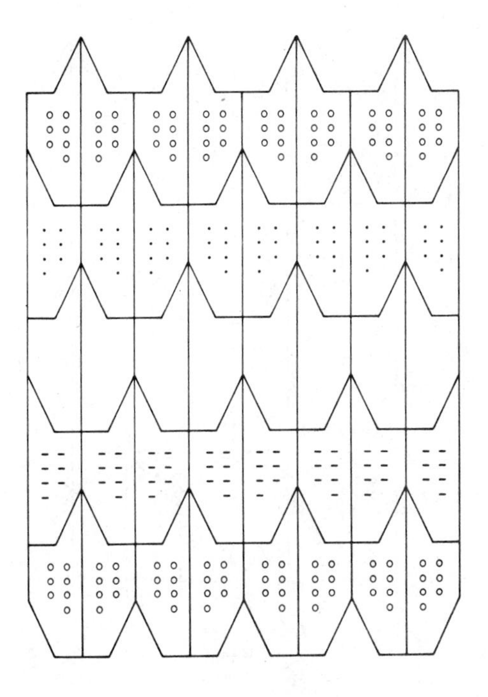

4.06.06

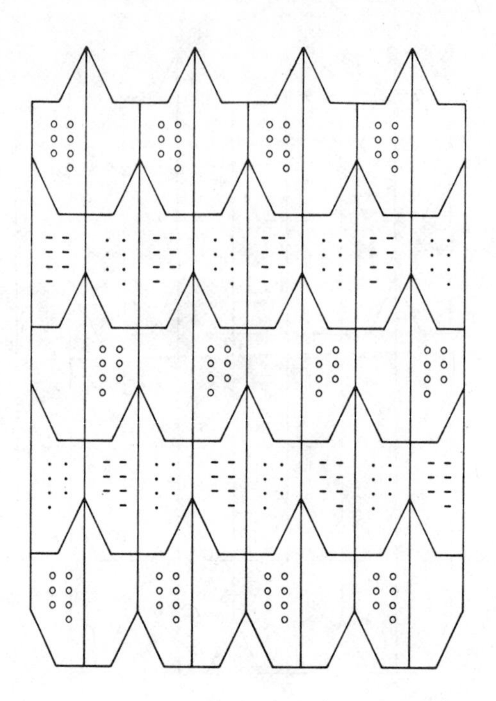

4.06.07

4.07.01

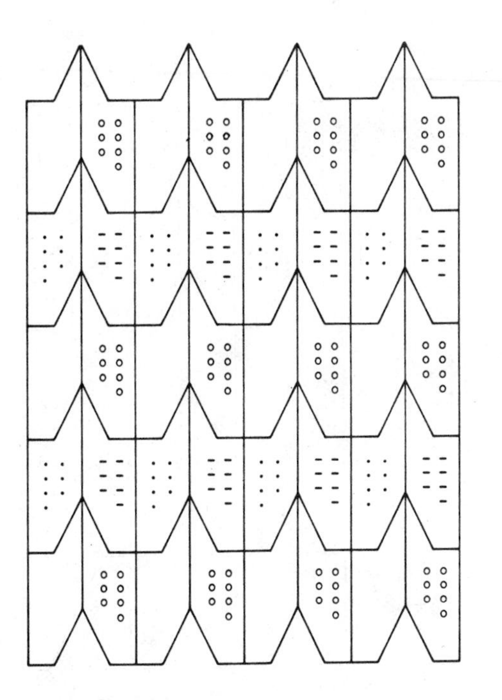

4.07.02

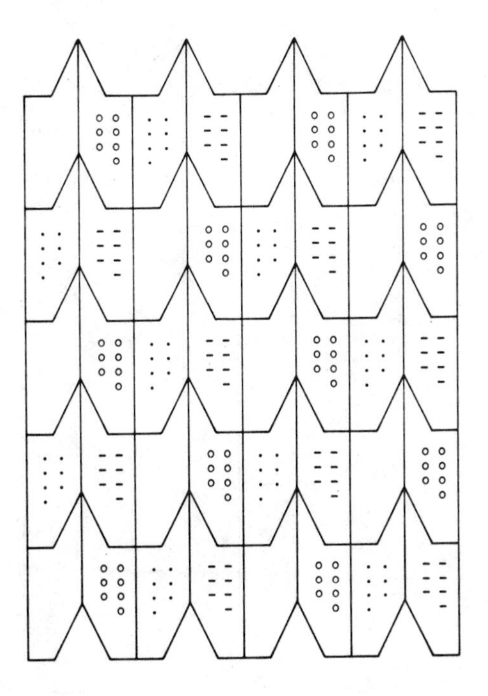

4.07.03

4.07.04

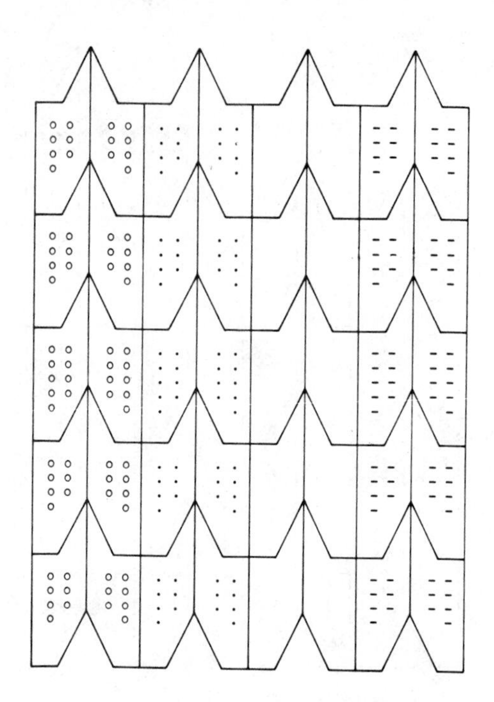

4.07.05

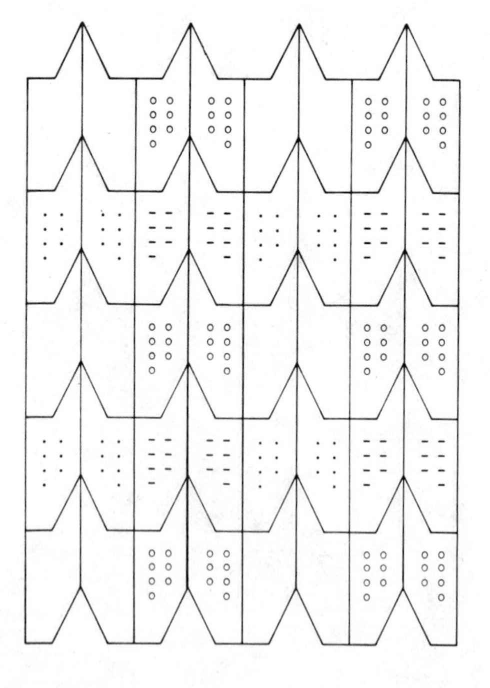

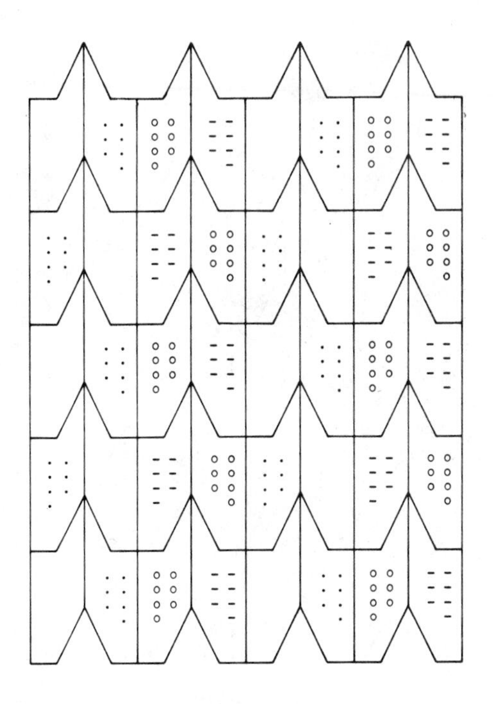

4.07.06

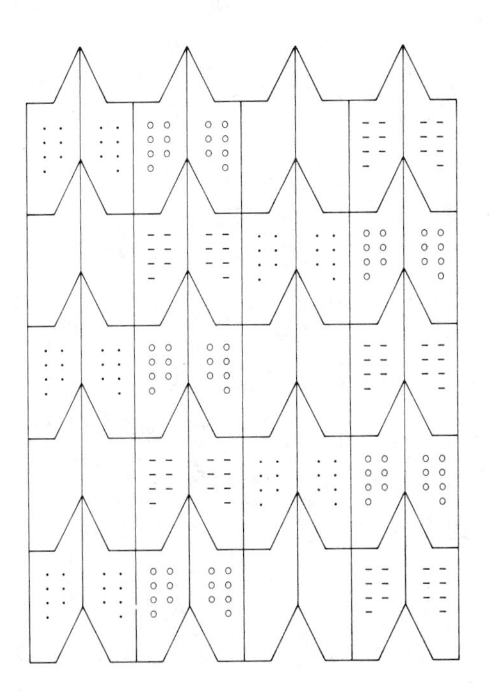

4.07.07

4.07.08

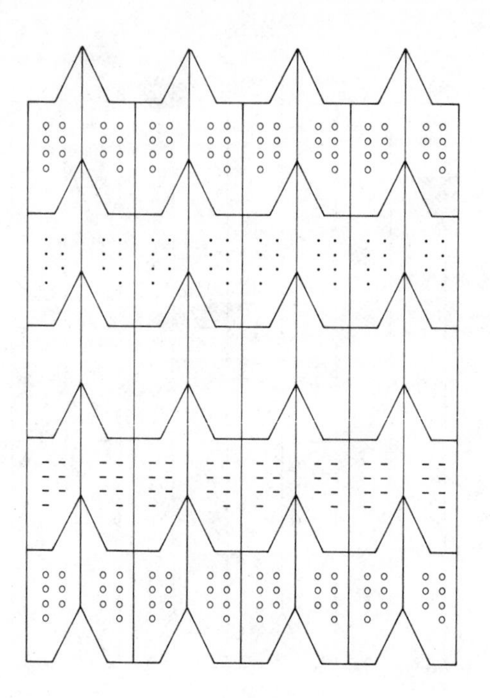

4.07.09

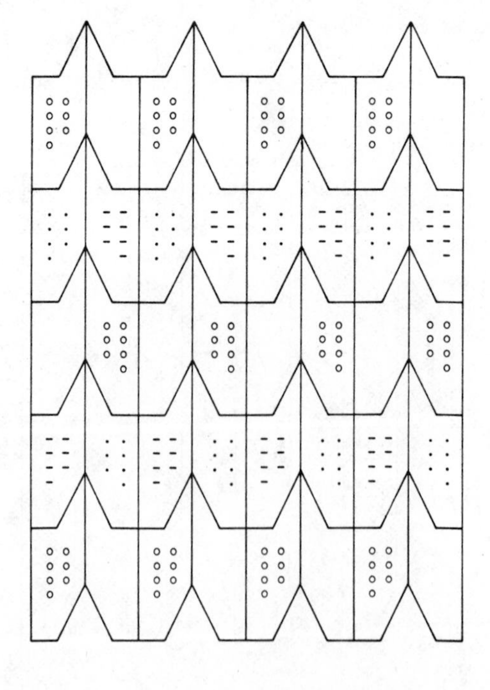

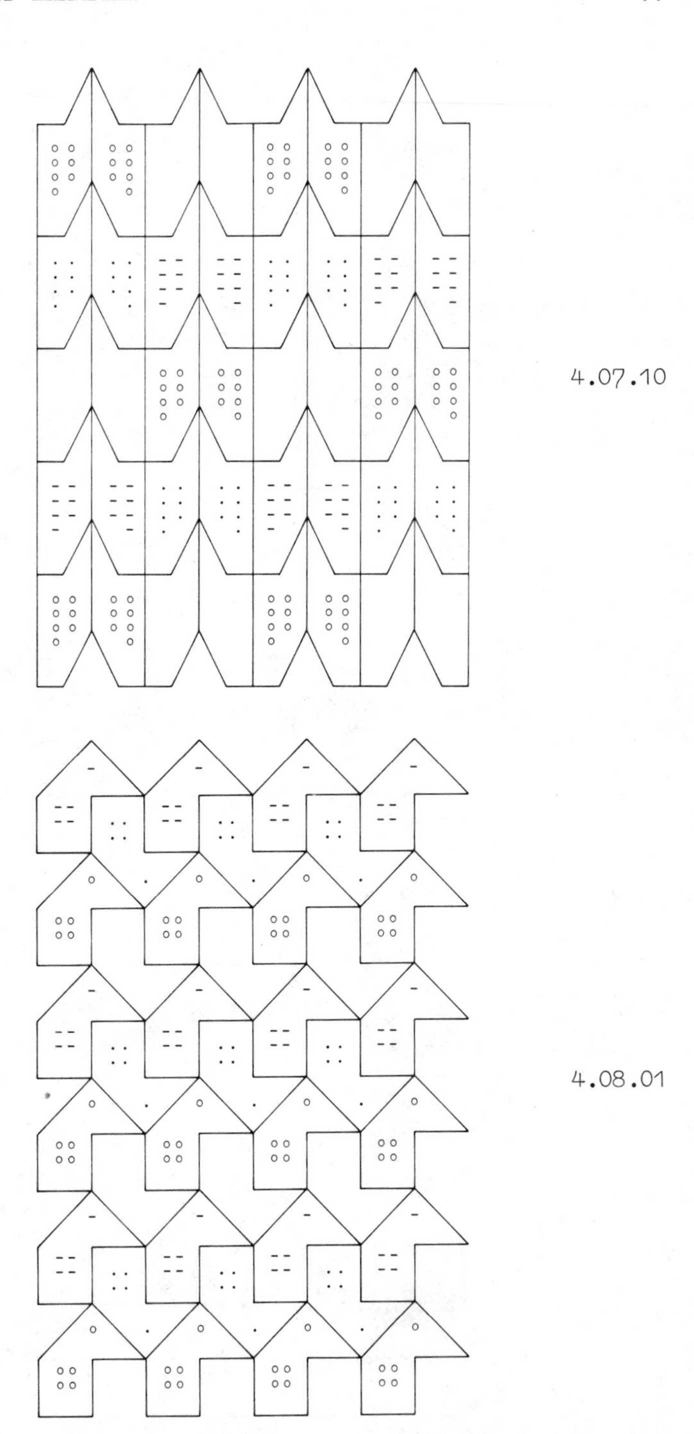

4.07.10

4.08.01

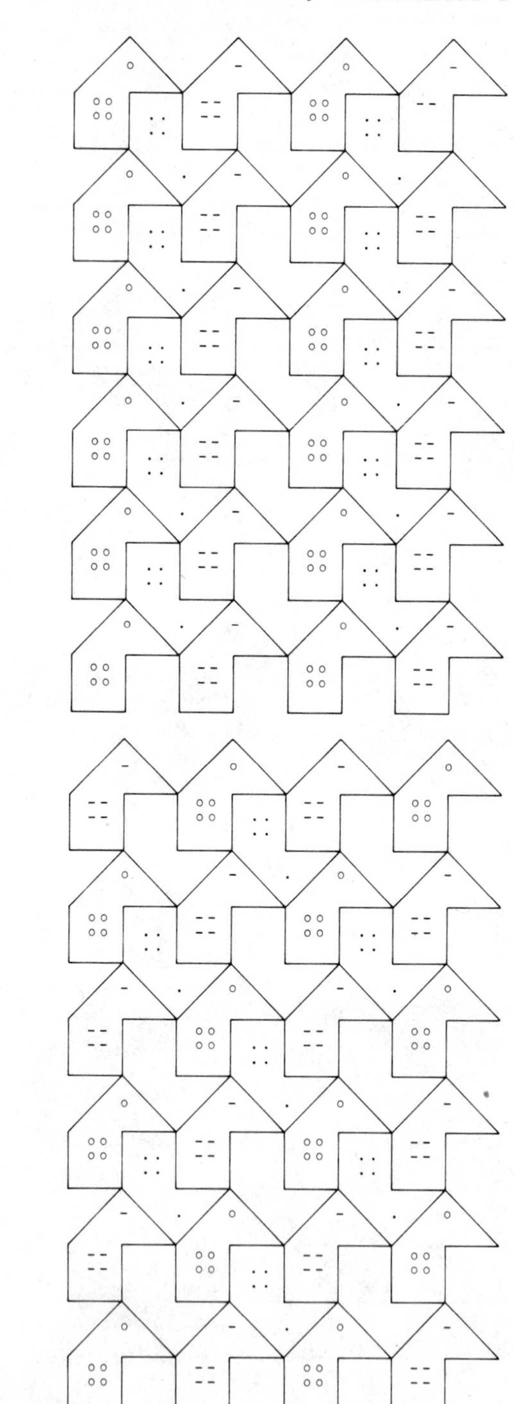

4.08.02

4.08.03

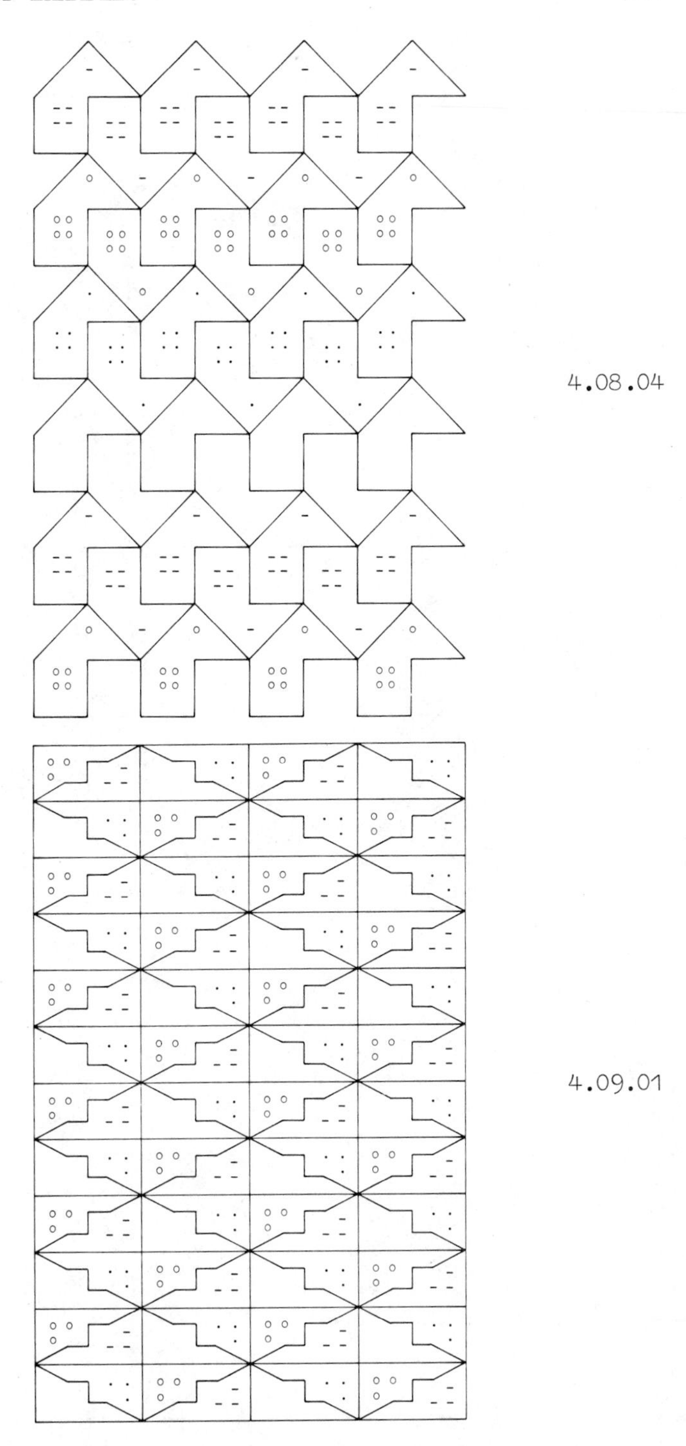

4.08.04

4.09.01

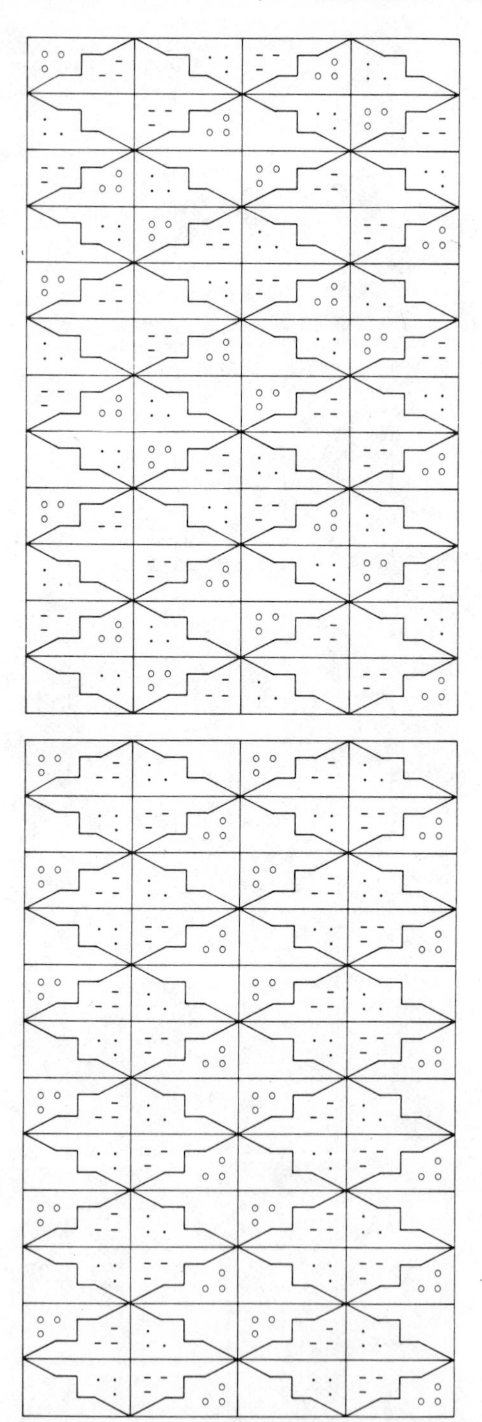

4.09.02

4.09.03

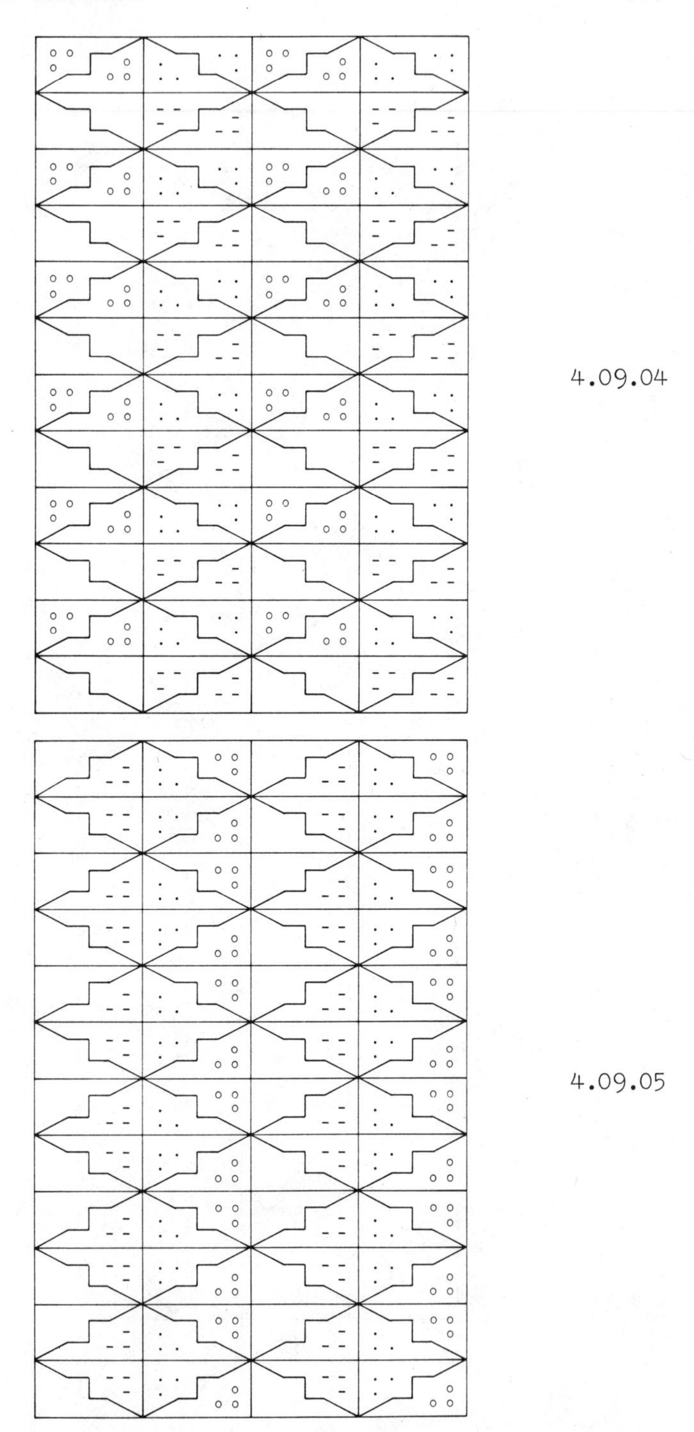

4.09.04

4.09.05

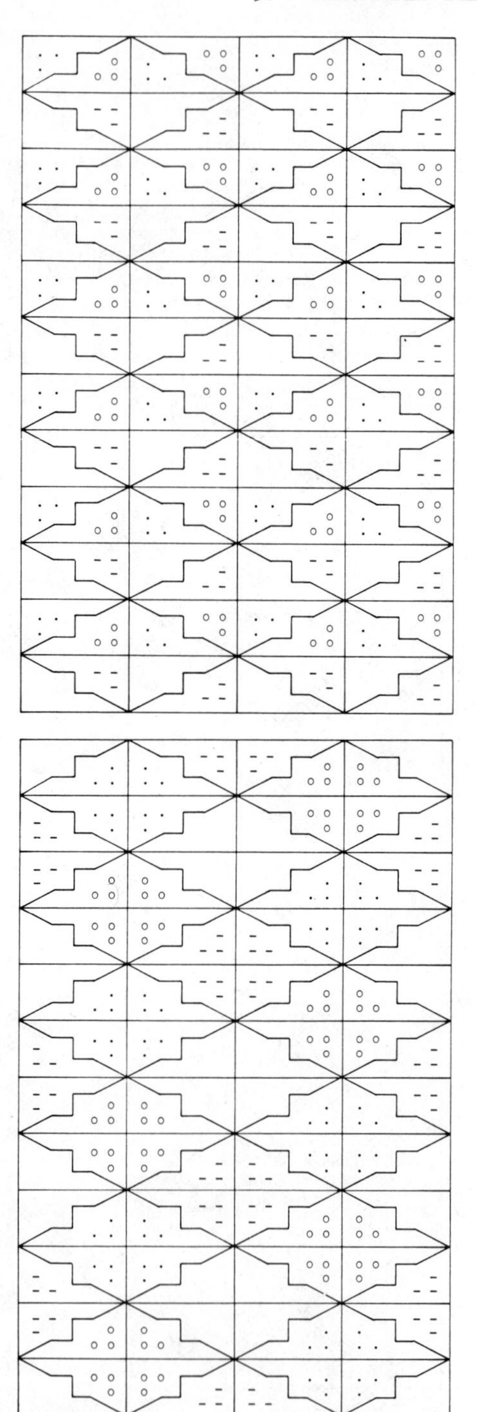

4.09.06

4.09.07

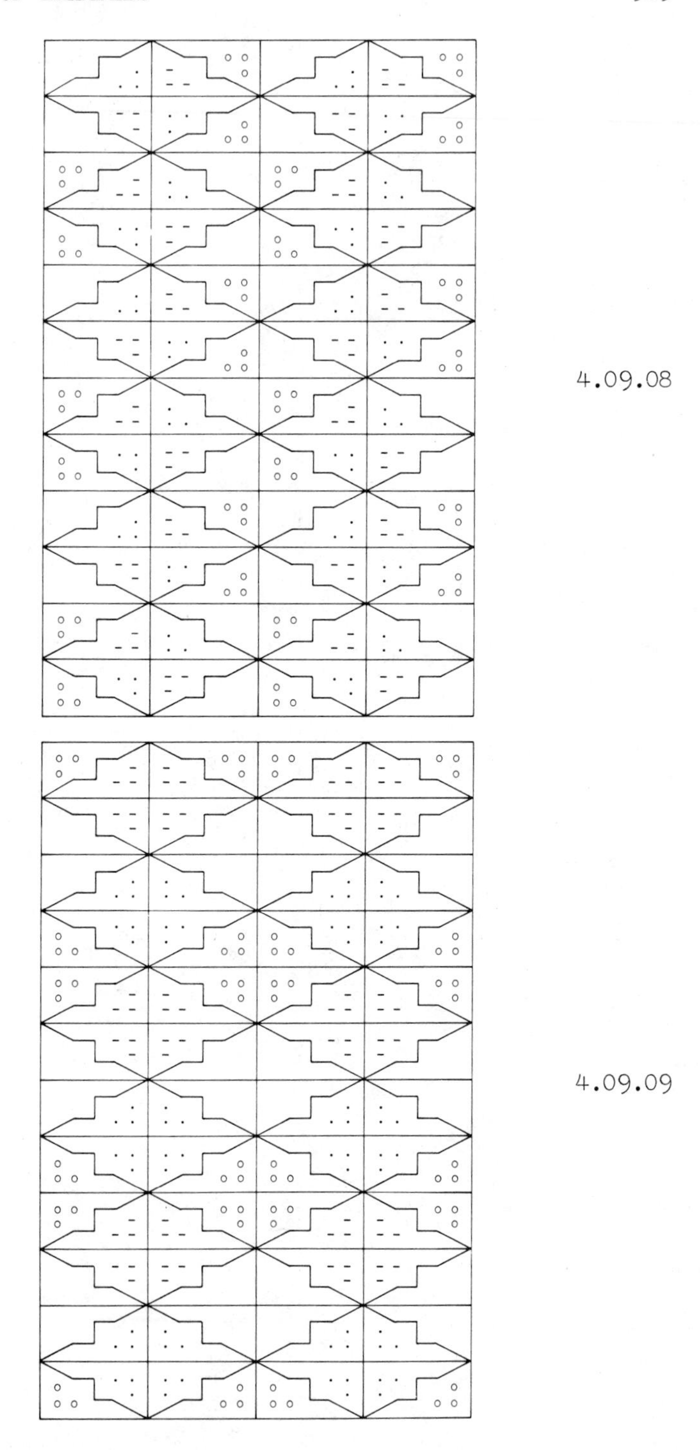

4.09.08

4.09.09

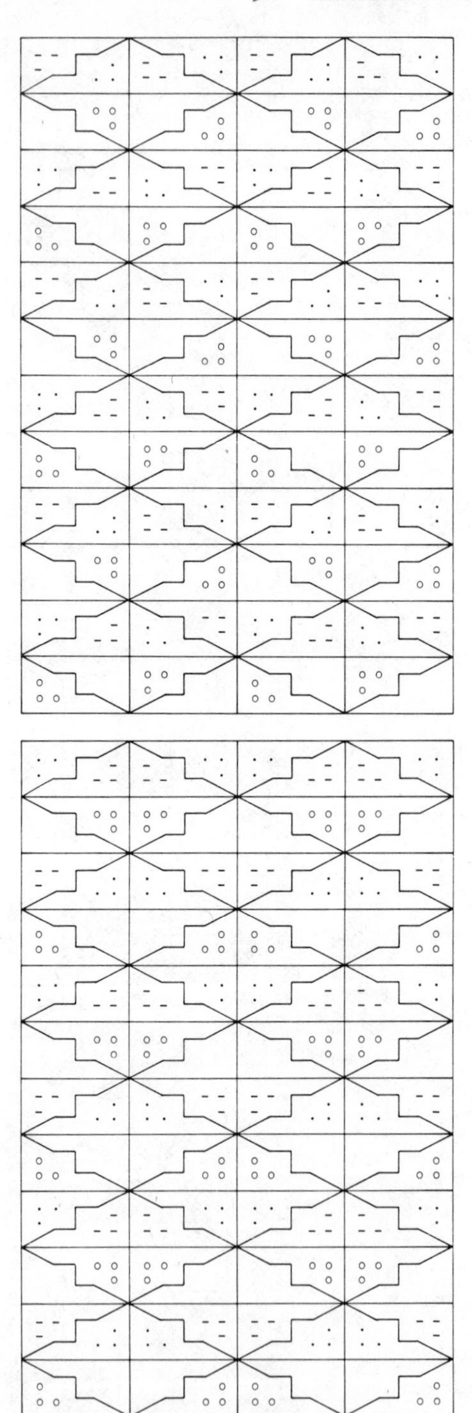

4.09.10

4.09.11

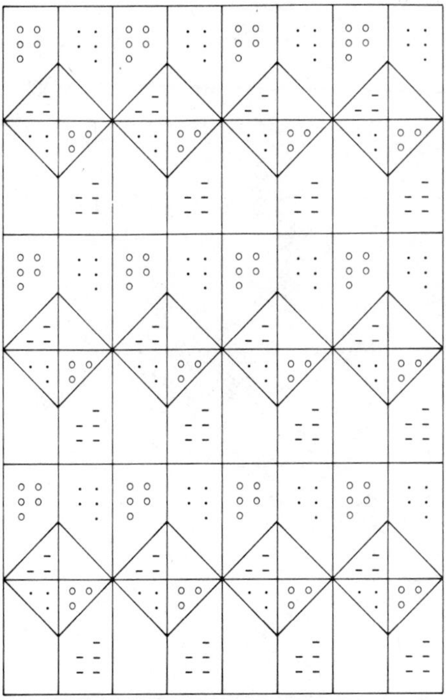

4.10.01

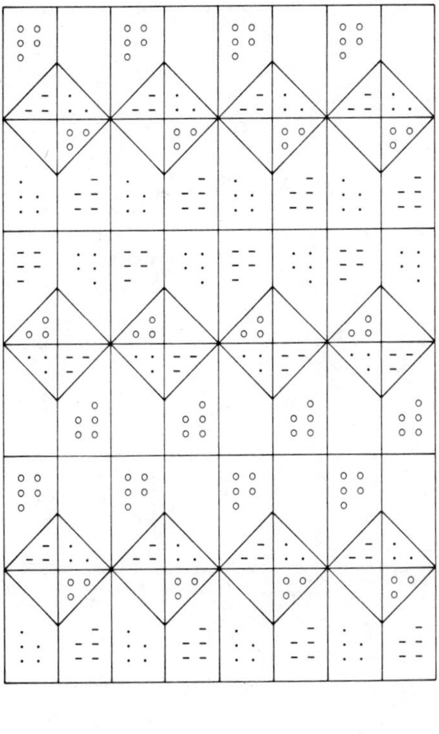

4.10.02

4.10.03

4.10.04

4.10.05

4.10.06

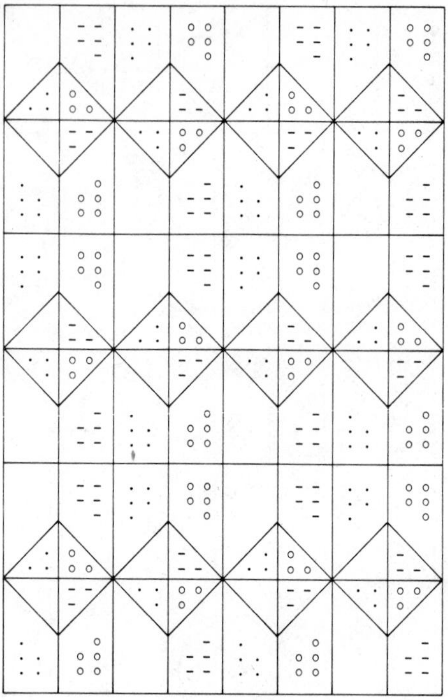

4.10.07

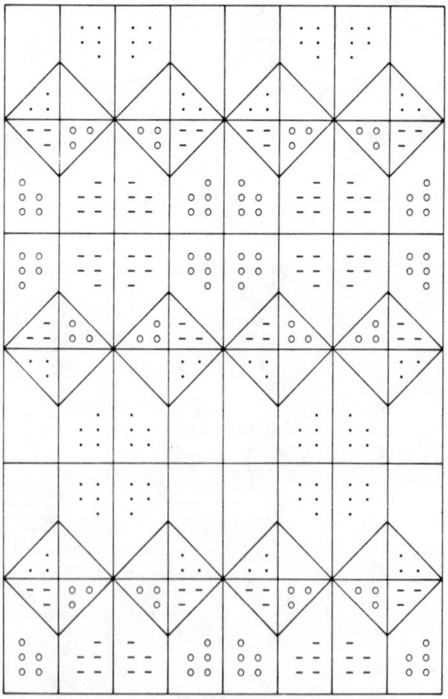

4.10.08

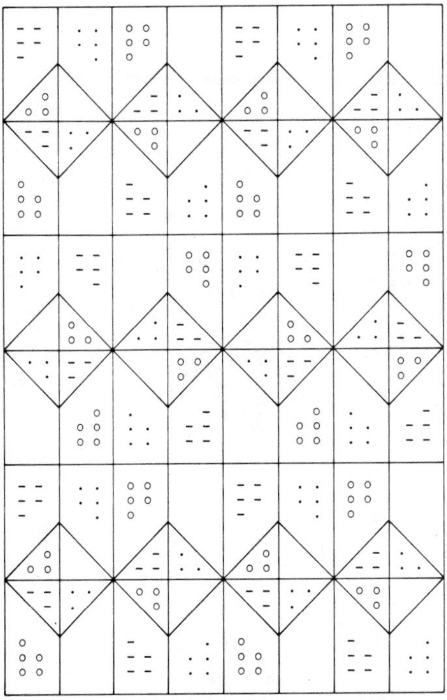

4.10.09

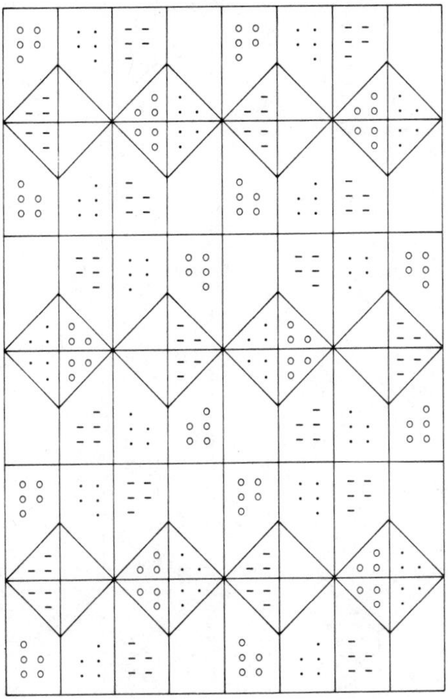

4.10.10

4.10.11

4.10.12

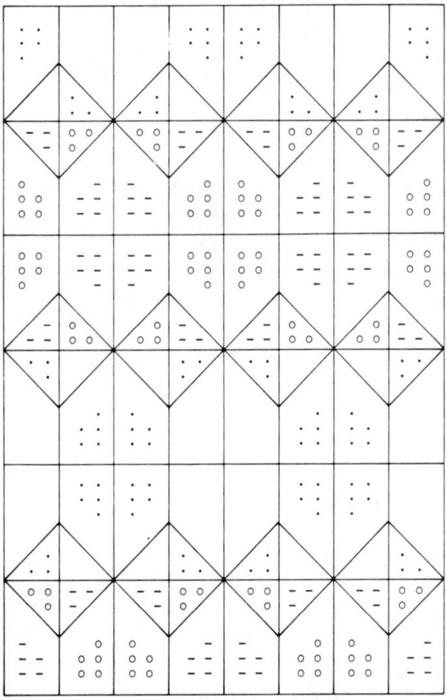

4.10.13

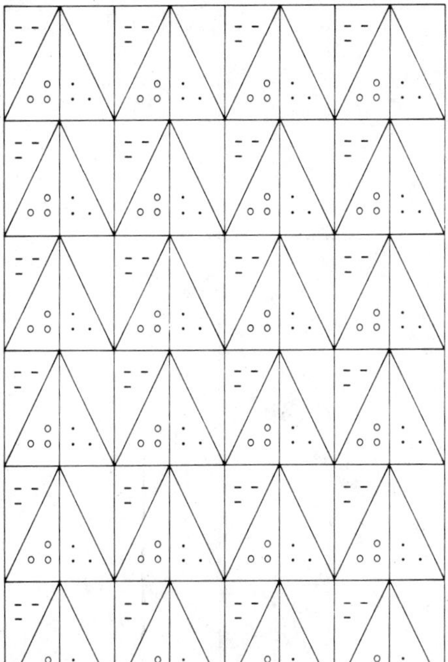

4.11.01

4.11.02

4.11.03

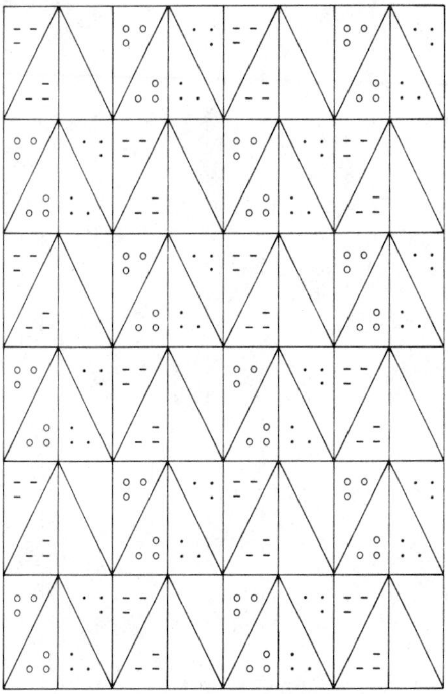

4.11.04

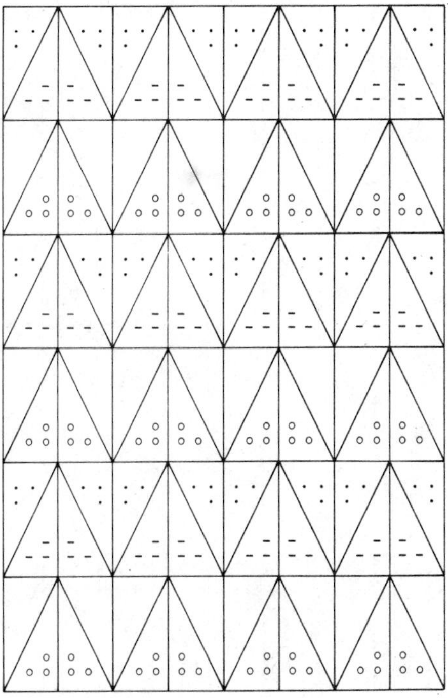

4.11.05

4.11.06

4.11.07

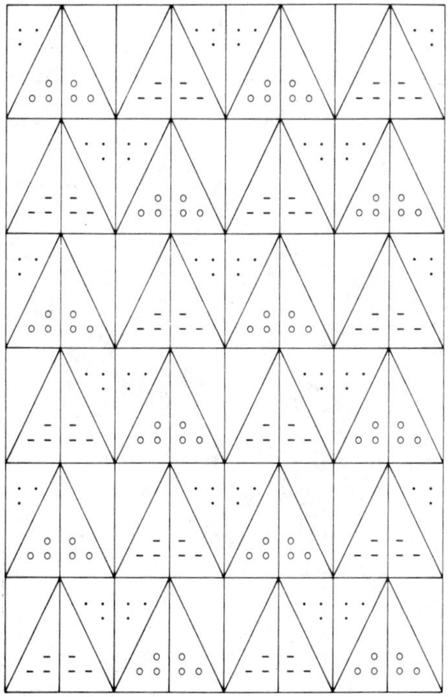

4.11.08

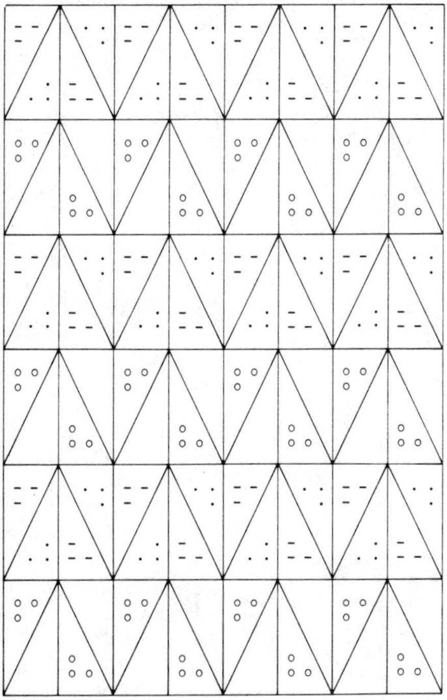

4.11.09

4.11.10

4.11.11

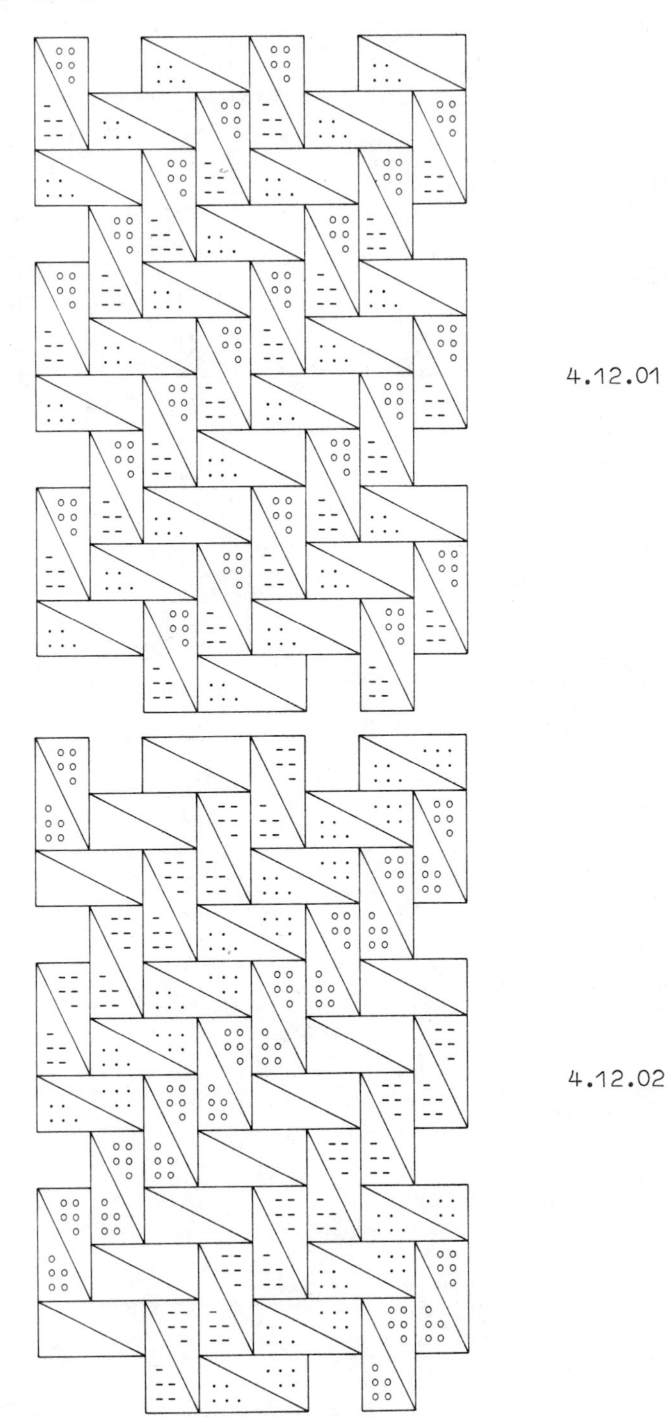

4.12.01

4.12.02

4.12.03

4.12.04

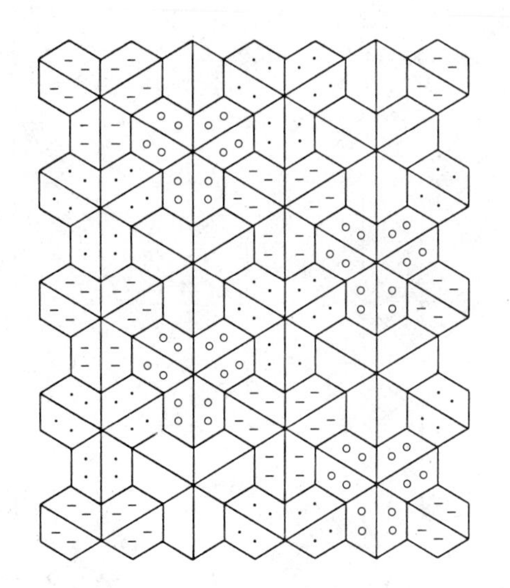

4.13.01

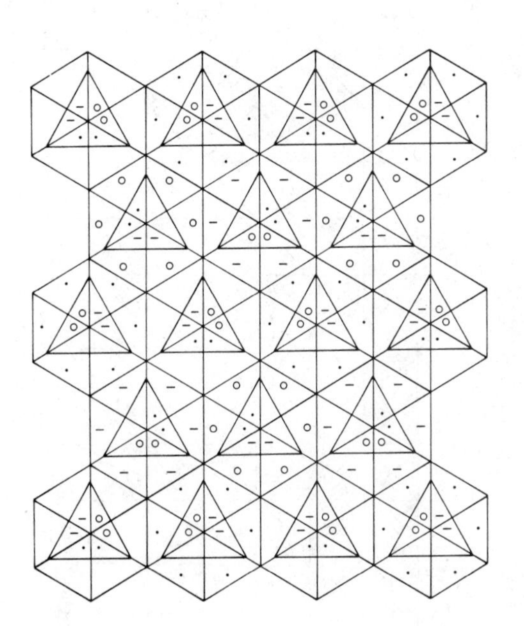

4.14.01

4.15.01

4.15.02

4.15.03

4.15.04

4.15.05

4.15.06

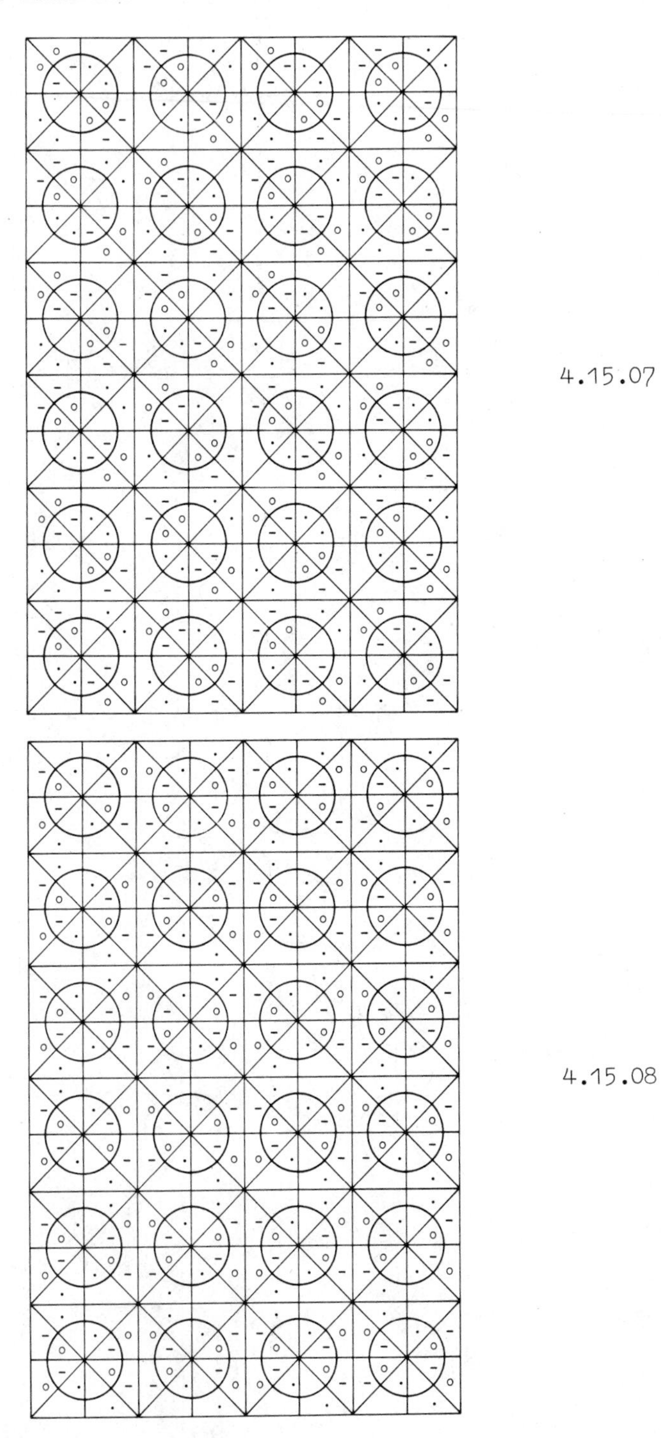

4.15.07

4.15.08

4.15.09

4.15.10

4.15.11

4.15.12

4.15.13

4.16.01

4.16.02

4.16.03

4.16.04

4.16.05

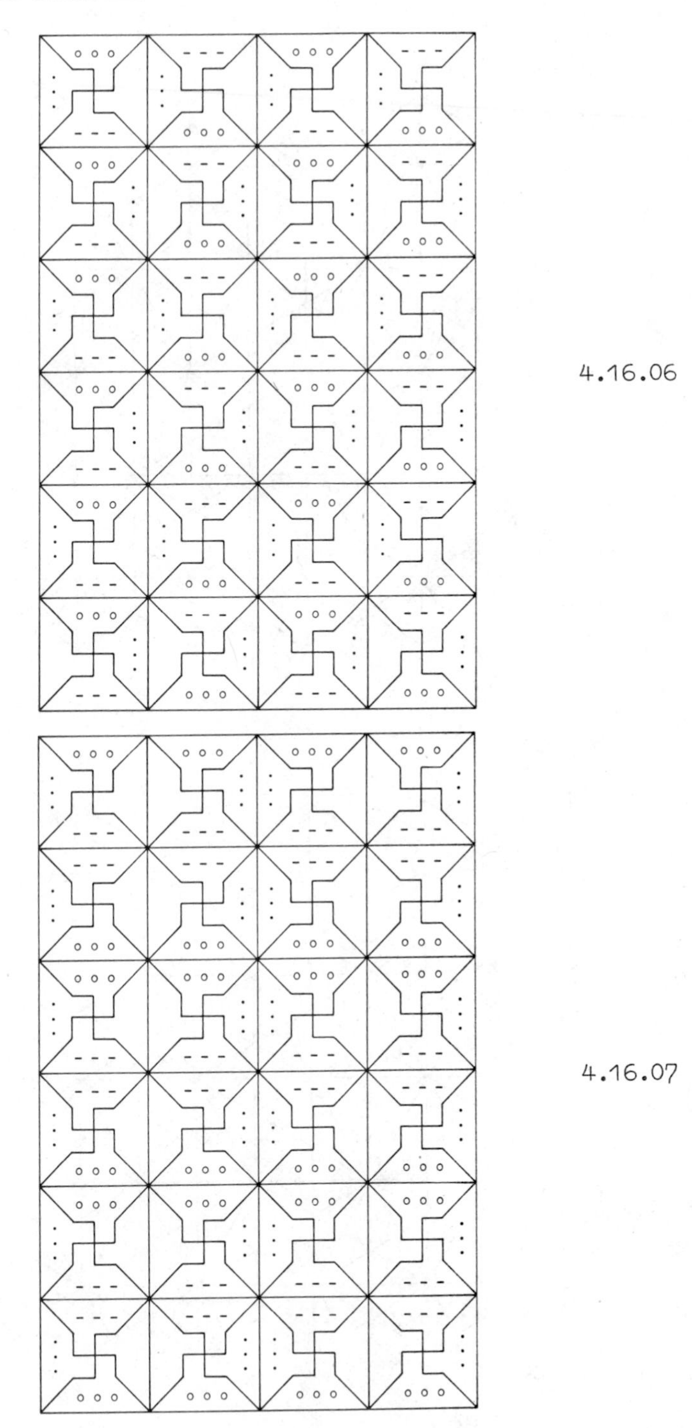

4.16.06

4.16.07

4.17.01

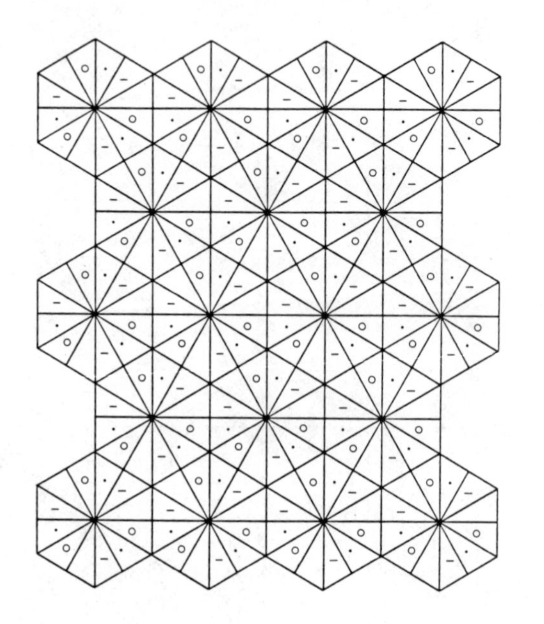

4.17.02

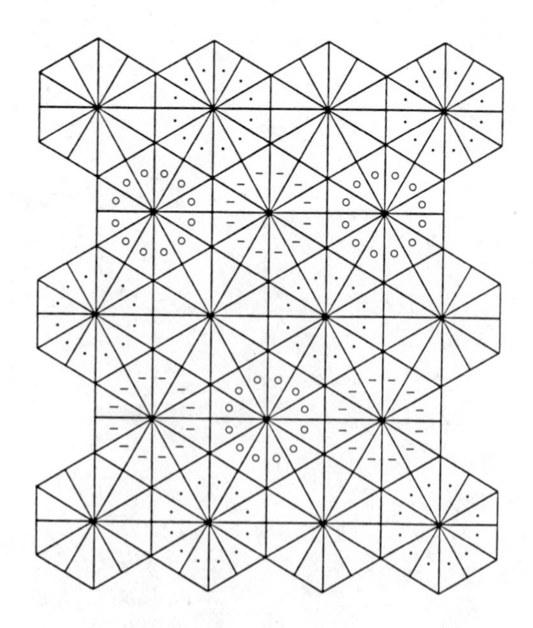

3.9 PROBLEMS

1. Let Ω be the standard quadratic mosaic in E. [See Figure 17.] Let N be $\{1,2\}$. Design a surjective mapping $\mathscr{k}$ carrying Ω to N, such that the (chromatic) symmetry group for the 2-chromatic mosaic $(\Omega,\mathscr{k})$ contains nothing but the identity, (ι,ε).

2. Complete the characterization of normalizers described in Proposition 25, by showing that conditions $(\Delta 1)$, $(\Delta 2)$, $(\Delta 3)$, and $(\Delta 4)$ are sufficient to guarantee that A lies in $\mathscr{F}^{\Delta}$.

3. Let T be any lattice in R^2, and let $\{u,v\}$ be a basis for T. By problem 8 in 2.8, the members of $\mathscr{L}_T$ can be presented as matrices (relative to $\{u,v\}$) of the form:

$$\begin{bmatrix} a & c \\ b & d \end{bmatrix},$$

where a, b, c, and d are any integers for which $ad - bc$ ± 1. Prove that $\mathscr{L}_T$ is generated by:

$$\begin{bmatrix} 1 & 1 \\ 0 & 1 \end{bmatrix}, \quad \begin{bmatrix} 0 & 1 \\ 1 & 0 \end{bmatrix}.$$

4. With reference to 4.4°, verify the description of the normalizers of the 17 ornamental groups, presented in Table 5.

5. With reference to 4.4°, verify that, for each of the 17 ornamental groups $\mathscr{F}$, the automorphisms of $\mathscr{F}$ implemented by $\underline{U}^{\Delta}$, $\underline{V}^{\Delta}$, $\underline{W}^{\Delta}$, and $\underline{X}^{\Delta}$ generate the automorphism group of $\mathscr{F}$. Moreover, verify the characterization of those generating automorphisms, presented in Table 6.

6. Let n be any positive integer for which $2 \leq n$. Show that the permutations

$$\mu \longleftrightarrow \begin{pmatrix} 1 & 2 & 3 & \ldots & n \\ 2 & 3 & 4 & \ldots & 1 \end{pmatrix}$$

and $$\nu \longleftrightarrow \begin{pmatrix} 1 & 2 & 3 & \ldots & n \\ 2 & 1 & 3 & \ldots & n \end{pmatrix}$$

on N generate S_N.

7. Show that, under application of ORBKT to H and $\mathcal{P}$, the number of "steps" in the computation cannot exceed $(p+4)(h-1)$. [See 5.7°.] In this context, the term "step" refers to any instance of an instruction which actually causes the ordered quadruple $(i^\circ, n^\circ, s^\circ, \ell^\circ)$ to change value.

8. Let T' and T'' be lattices in R^2, and let T' be included in T''. The object of this problem is to develop useful relations between bases for T' and T''.

(i) Prove that, for each basis $\{u'', v''\}$ for T'', there exists a basis $\{u', v'\}$ for T' of the form:

$$u' = p.u'' \quad ,$$
$$v' = r.u'' + s.v'';$$

where p, r, and s are integers for which $0 < p$, $0 \leq r < p$, and $0 < s$. Show that, relative to $\{u'', v''\}$, the integers p, r, and s are uniquely determined.
Show that ps is the index of T' in T''.
To establish these assertions, one may proceed as follows. Let $\{u^*, v^*\}$ be any basis for T', and let a, b, c, and d be the integers for which:

$$u^* = a.u'' + b.v'',$$
$$v^* = c.u'' + d.v''.$$

One should show that there exist integers a', b', c', and d' such that $a'd' - b'c' = 1$ and:

$$\begin{bmatrix} a & c \\ b & d \end{bmatrix} \begin{bmatrix} a' & c' \\ b' & d' \end{bmatrix} = \begin{bmatrix} p & r \\ 0 & s \end{bmatrix},$$

where p, r, and s are integers having the foregoing properties. The basis

$$u' = a'.u^* + b'.v^*,$$

$$v' = c'.u^* + d'.v^*,$$

for T' would be of the desired sort. One should then show that the members

$$i.u'' + j.v'' \qquad (0 \leq i < p, \quad 0 \leq j < s)$$

of T" represent the various cosets of T' in T". Incidentally, the index of T' in T" must also equal

$$\det \begin{bmatrix} a & c \\ b & d \end{bmatrix}.$$

(j) Prove that there exist bases {u',v'} and {u",v"} for T' and T", respectively, related as follows:

$$u' = p.u'',$$

$$v' = s.v'',$$

where p and s are integers for which $0 < p$, $0 < s$, and $p \mid s$ (that is, p divides s). Prove that, under the foregoing conditions, p and s are unique.
One may reformulate these assertions as follows. Let $\{u^-,v^-\}$ and $\{u^+,v^+\}$ be any bases for T' and T", respectively, and let a, b, c, and d be the integers for which:

$$u^- = a.u^+ + b.v^+,$$
$$v^- = c.u^+ + d.v^+.$$

One must show that there exist integers a', b', c', d', a'', b'', c'', and d'' such that $a'd' - b'c' = 1$, $a''d'' - b''c'' = 1$, and:

$$\begin{bmatrix} a & c \\ b & d \end{bmatrix}\begin{bmatrix} a' & c' \\ b' & d' \end{bmatrix} = \begin{bmatrix} a'' & c'' \\ b'' & d'' \end{bmatrix}\begin{bmatrix} \rho & 0 \\ 0 & \sigma \end{bmatrix},$$

where ρ and σ are integers meeting the conditions asserted earlier. The bases

$$u' = a'.u^- + b'.v^-,$$
$$v' = c'.u^- + d'.v^-,$$

for T' and

$$u'' = a''.u^+ + b''.v^+,$$
$$v'' = c''.u^+ + d''.v^+,$$

for T'', would have the desired properties.
To start the argument, one should note that ρ would have to be the greatest common divisor of a, b, c, and d, and that σ would have to be n/ρ, where n is the index of T' in T''.

9. Let T, T', and T'' be any lattices in R^2 for which T' and T'' are included in T. Let $\{u,v\}$, $\{u',v'\}$, and $\{u'',v''\}$ be bases for T, T', and T'', respectively:

$$u' = a'.u + b'.v,$$
$$v' = c'.u + d'.v,$$
$$u'' = a''.u + b''.v,$$
$$v'' = c''.u + d''.v,$$

where a', b', c', d', a'', b'', c'', and d'' are suitable integers. Prove that $T' = T''$ iff

$$\det\begin{bmatrix} a' & c' \\ b' & d' \end{bmatrix} = \det\begin{bmatrix} a'' & c'' \\ b'' & d'' \end{bmatrix}$$

and the entries in the matrix

$$\begin{bmatrix} a'' & c'' \\ b'' & d'' \end{bmatrix}^{-1}\begin{bmatrix} a' & c' \\ b' & d' \end{bmatrix}$$

are integral.

Now let $\{u',v'\}$ and $\{u'',v''\}$ be related to $\{u,v\}$ by the form described in part (i) of the preceding problem:

$$\begin{bmatrix} a' & c' \\ b' & d' \end{bmatrix} = \begin{bmatrix} \mathscr{p}' & \mathscr{r}' \\ 0 & \mathscr{s}' \end{bmatrix},$$

$$\begin{bmatrix} a'' & c'' \\ b'' & d'' \end{bmatrix} = \begin{bmatrix} \mathscr{p}'' & \mathscr{r}'' \\ 0 & \mathscr{s}'' \end{bmatrix},$$

where $0 < \mathscr{p}'$, $0 \leq \mathscr{r}' < \mathscr{p}'$, $0 < \mathscr{s}'$, $0 < \mathscr{p}''$, $0 \leq \mathscr{r}'' < \mathscr{p}''$, and $0 < \mathscr{s}''$. Prove that $T' = T''$ iff $\mathscr{p}' = \mathscr{p}''$, $\mathscr{r}' = \mathscr{r}''$, and $\mathscr{s}' = \mathscr{s}''$.

10. Let T and T^o be any lattices in R^2 for which T^o is included in T. Let $\{u,v\}$ and $\{u^o,v^o\}$ be bases for T and T^o, respectively, and let $\mathscr{p}$, $\mathscr{q}$, $\mathscr{r}$, and $\mathscr{s}$ be the integers for which

$$u^o = \mathscr{p}.u + \mathscr{q}.v,$$
$$v^o = \mathscr{r}.u + \mathscr{s}.v.$$

Let z be any member of R^2:

$$z = \mathscr{a}.u + \mathscr{b}.v,$$

where a and b are real numbers. Show that z is contained in T^o iff the entries in the matrix

$$\begin{bmatrix} p & r \\ q & s \end{bmatrix}^{-1} \begin{bmatrix} a \\ b \end{bmatrix}$$

are integral. Note that, when a and b are rational numbers, then there would exist a positive integer t such that ta and tb are integers. In that case, the foregoing condition for membership in T^o would be this, that the necessarily integral entries in the matrix

$$\begin{bmatrix} s & -r \\ -q & p \end{bmatrix} \begin{bmatrix} ta \\ tb \end{bmatrix}$$

be multiples of $t(ps - qr)$.

11. In this problem, we shall describe the basic properties of quadratic and hexagonal lattices in R^2. To develop the supporting arguments, one should use the results on prime factorization of (algebraic) integers in $Q(\sqrt{-1})$ and $Q(\sqrt{-3})$. These results can be found in the standard text by G. H. Hardy and E. M. Wright.
Let T be any lattice in R^2, and let T' and T" be any sublattices of T. We shall regard T' and T" as <u>equivalent</u> iff there exists a member L of the symmetry group $\mathcal{O}_T$ for T such that L(T') = T".
Let T be quadratic, and let n be any positive integer. Prove that there exists a quadratic sublattice of T having index n in T iff n has the form:

$$n = x^2 + y^2, \qquad (q)$$

where x and y are any integers.
Let T be hexagonal, and let n be any positive integer. Prove that there exists a hexagonal sublattice of T having index n in T iff n has the form:

$$n = x^2 - xy + y^2, \qquad \text{(h)}$$

where x and y are any integers.
Now let the prime factorization of n (in Z) be presented in "quadratic form," as follows:

$$n = 2^a p_1^{b_1} \dots p_\lambda^{b_\lambda} q_1^{c_1} \dots q_\mu^{c_\mu},$$

where p_1, ..., and p_λ are the prime factors of n congruent to 1 modulo 4, and q_1, ..., and q_μ are the prime factors of n congruent to 3 modulo 4. Show that n has the form (q) iff, for each j $(1 \leq j \leq \mu)$, c_j is even.
Given that T is quadratic, show that the number of equivalence classes of quadratic sublattices of T having index n in T (if there be any) equals

$$\tfrac{1}{2}[(b_1+1)(b_2+1)\dots.(b_\lambda+1)]$$

when, for some i $(1 \leq i \leq \lambda)$, b_i is odd, and equals

$$\tfrac{1}{2}[(b_1+1)(b_2+1)\dots.(b_\lambda+1) + 1]$$

when, for each i $(1 \leq i \leq \lambda)$, b_i is even. In the latter case, there is precisely one quadratic sublattice of T having index n in T, which is invariant under $\mathcal{O}_T$. Show that this sublattice must fall into one of two types, discriminated by the parity of a. [Consider, for example, the cases: $n = 2$, $n = 4$.]
Finally, let the prime factorization of n (in Z) be presented in "hexagonal form," as follows:

$$n = 3^a p_1^{b_1} \dots p_\lambda^{b_\lambda} q_1^{c_1} \dots q_\mu^{c_\mu},$$

where p_1, ..., and p_λ are the prime factors of n

congruent to 1 modulo 3, and q_1, ..., and q_μ are the prime factors of n congruent to 2 modulo 3. Show that n has the form (h) iff, for each j ($1 \leq j \leq \mu$), c_j is even.
Given that T is hexagonal, show that the number of equivalence classes of hexagonal sublattices of T having index n in T (if there be any) equals

$$\tfrac{1}{2}[(b_1+1)(b_2+1)....(b_\lambda+1)]$$

when, for some i ($1 \leq i \leq \lambda$), b_i is odd, and equals

$$\tfrac{1}{2}[(b_1+1)(b_2+1)....(b_\lambda+1) + 1]$$

when, for each i ($1 \leq i \leq \lambda$), b_i is even. In the latter case, there is precisely one hexagonal sublattice of T having index n in T, which is invariant under $\mathcal{O}_T$. Show that this sublattice must fall into one of two types, discriminated by the parity of a. [Consider, for example, the cases: n = 3, n = 9.]

12. Examine the data in Table 11, under:

$$m = 6, 7, 8, 9, 10, 11, 12,$$

for the cases:

$$n = p, p^2, pq,$$

where p and q are distinct odd primes. Explain the recurrent form of these data.

13. The object of this problem is to present a method by which one can, in a certain sense, describe all subgroups of the affine group $\mathscr{A}$: $\mathscr{A} = R^2 \rtimes \mathscr{L}$. The method depends upon the concept of "crossed-homomorphism."

Let T be any subgroup of R^2, and let $\mathscr{G}$ be any subgroup of $\mathscr{L}$ under which T is invariant. In this context, one may introduce an action of $\mathscr{G}$ by automorphisms on the quotient group R^2/T:

$$L.(z + T) = L(z) + T,$$

where L is any member of $\mathscr{G}$ and where z is any member of R^2. In turn, one may define the group

$$R^2/T \rtimes \mathscr{G},$$

consisting of all ordered pairs of the form $(z + T, L)$ (where z is any member of R^2 and where L is any member of $\mathscr{G}$) and carrying the following operation of multiplication:

$$(z' + T, L')(z'' + T, L'') = (z' + L'(z'') + T, L'L'').$$

One refers to $R^2/T \rtimes \mathscr{G}$ as the <u>semi-direct product</u> of R^2/T and $\mathscr{G}$, relative to the given action of $\mathscr{G}$ by automorphisms on R^2/T.

Let ζ be any mapping carrying $\mathscr{G}$ to R^2/T, satisfying the condition that, for any members L' and L'' of $\mathscr{G}$:

$$\zeta(L'L'') = \zeta(L') + L'.\zeta(L'').$$

One refers to such a mapping as a <u>crossed-homomorphism</u>, relative to the action of $\mathscr{G}$ on R^2/T.

Show that a mapping ζ carrying $\mathscr{G}$ to R^2/T is a crossed-homomorphism iff the mapping

$$\zeta^*(L \longmapsto (\zeta(L), L))$$

carrying $\mathscr{G}$ to $R^2/T \rtimes \mathscr{G}$ is a homomorphism.

Let ζ be a crossed-homomorphism carrying $\mathcal{G}$ to R^2/T, and let $\mathcal{B}$ denote the subset

$$\{[t,L] : \quad L \in \mathcal{G}, \quad t \in \zeta(L)\}$$

of $\mathcal{A}$. Show that $\mathcal{B}$ is a subgroup of $\mathcal{A}$, having translational part T and linear part $\mathcal{G}$. Conversely, show that every subgroup $\mathcal{B}$ of $\mathcal{A}$ having translational part T and linear part $\mathcal{G}$, must be of the foregoing form, and that the underlying crossed-homomorphism ζ is unique.

14. Let k be any positive integer, and let W be the rotation on R^2 having principal measure $2\pi/k$. [When $k = 1$, one should interpret W to be I.] Let X be any reflection on R^2. In this context, W and X generate the group $\mathcal{D}_k(X)$, and satisfy the relations:

$$W^k = I, \quad X^2 = I, \quad XWXW = I.$$

Now let $\mathcal{H}$ be an arbitrary group, with identity element e, and let w and x be any members of $\mathcal{H}$. Prove that the following statements are equivalent:

(1) there exists a homomorphism θ carrying $\mathcal{D}_k(X)$ to $\mathcal{H}$ for which $\theta(W) = w$ and $\theta(X) = x$;

(2) $w^k = e$, $x^2 = e$, and $xwxw = e$.

Using the results of this problem and of the foregoing, develop a smooth proof of Proposition 28 in 6.8°.

15. Let n be any positive integer, and let $(\Omega, \mathcal{k})$ be any n-chromatic mosaic in E. With reference to 2, let us generalize the definition of n-chromatic plane ornament, by substituting for conditions (o), (g), and (c) the following:

(γ) $\mathbf{E}_{(\Omega,\mathcal{k})}$ is an ornamental subgroup of $\mathbf{E}$.

Show that condition (γ) is equivalent to the operationally simpler condition:

(τ) for some (and hence for any) cartesian coordinate mapping K, there exist linearly independent members u and v of R^2 such that $(K^{-1}[u,I]K, \varepsilon)$ and $(K^{-1}[v,I]K, \varepsilon)$ are (chromatic) symmetries of (Ω, k).

As usual, ε stands for the identity permutation on N. Note that conditions (o) and (g) together imply condition (γ), and that condition (γ) implies condition (o). The effect of the new definition of n-chromatic plane ornament is to weaken condition (g), and to drop condition (c) altogether.
Now one can imitate the presentation in 3, to develop the more general problem of classification of chromatic plane ornaments. Describe the modifications of the algorithms GENMT and SUBMT necessary to treat the general problem. For the case of SUBMT, show that the new form of the problem would be the following:

for any positive integers n and m $(2 \le n \le n^*,$ $1 \le m \le 17)$, find all finite sequences

$$\mathcal{F}_m^1, \mathcal{F}_m^2, \ldots, \mathcal{F}_m^\nu$$

of subgroups of $\mathcal{F}_m$ such that

$$e_1 + e_2 + \cdots + e_\nu = n,$$

where, for each integer j $(1 \le j \le \nu)$, e_j is the index of $\mathcal{F}_m^j$ in $\mathcal{F}_m$; enumerate the equivalence classes of such sequences, where two such sequences

$$\mathcal{F}_m'^1, \mathcal{F}_m'^2, \ldots, \mathcal{F}_m'^\lambda$$

and

$\mathscr{F}''^{1}_{m}, \mathscr{F}''^{2}_{m}, \ldots, \mathscr{F}''^{\mu}_{m}$

are taken to be equivalent iff

$\lambda = \mu$

and there exist members A_1, A_2, ..., and A_λ of $\mathscr{F}_m^{\Delta}$ and a bijective mapping b carrying $\{1,2,\ldots,\lambda\}$ to itself such that, for any integers j' and j" $(1 \leq j',j'' \leq \lambda)$, $A_{j''}A_{j'}^{-1}$ is contained in $\mathscr{F}_m$, and such that, for any integer j $(1 \leq j \leq \lambda)$, $\mathscr{F}''^{b(j)}_{m} = A_j \mathscr{F}'^{j}_{m} A_j^{-1}$.

16. Building from problem 20 in 2.8, formulate and solve the problem of classifying all <u>chromatic border ornaments</u>. The results should be presented in "closed form," without recourse to machine computation.

CATALOGUE

The following tilings of the plane are those which have been employed in the text to illustrate the 17 types of plane ornament. They also served as a base for the folio of examples illustrating the 96 types of 4-chromatic plane ornament. One might make use of them to experiment with the coloring instructions given in the text for n-chromatic plane ornaments ($2 \leq n \leq 8$).

p1

p2

p3

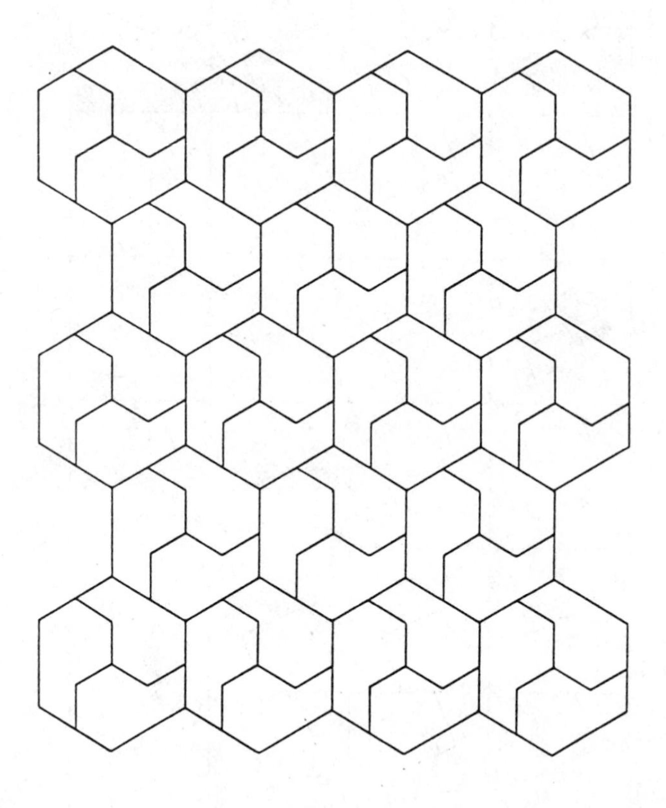

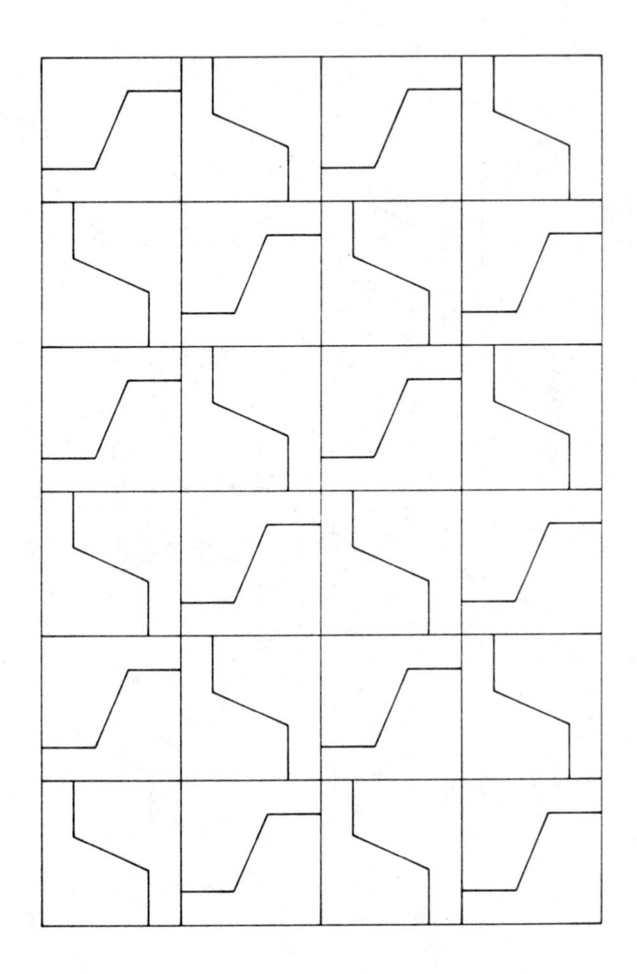

p4

p6

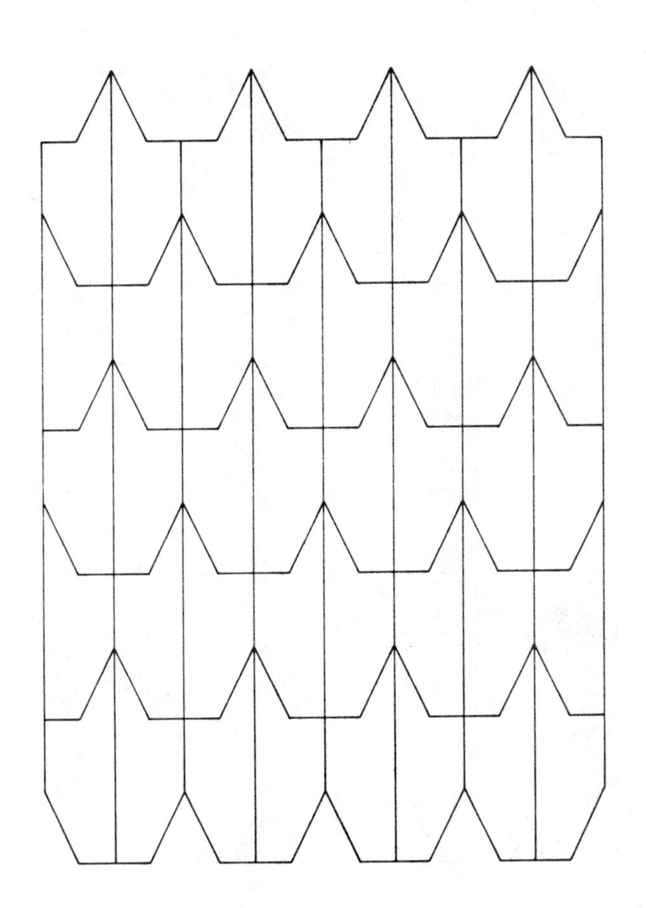

cm

pm

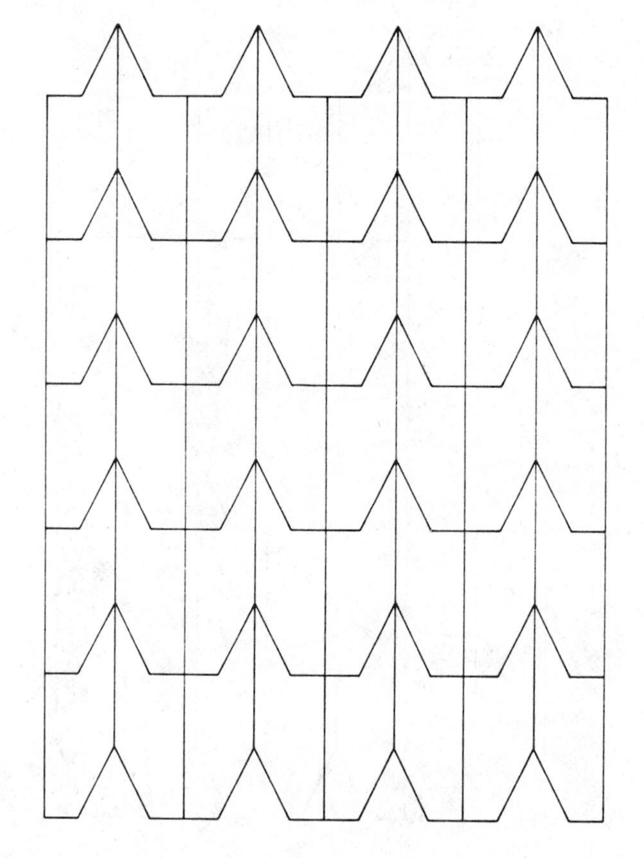

pg

cmm

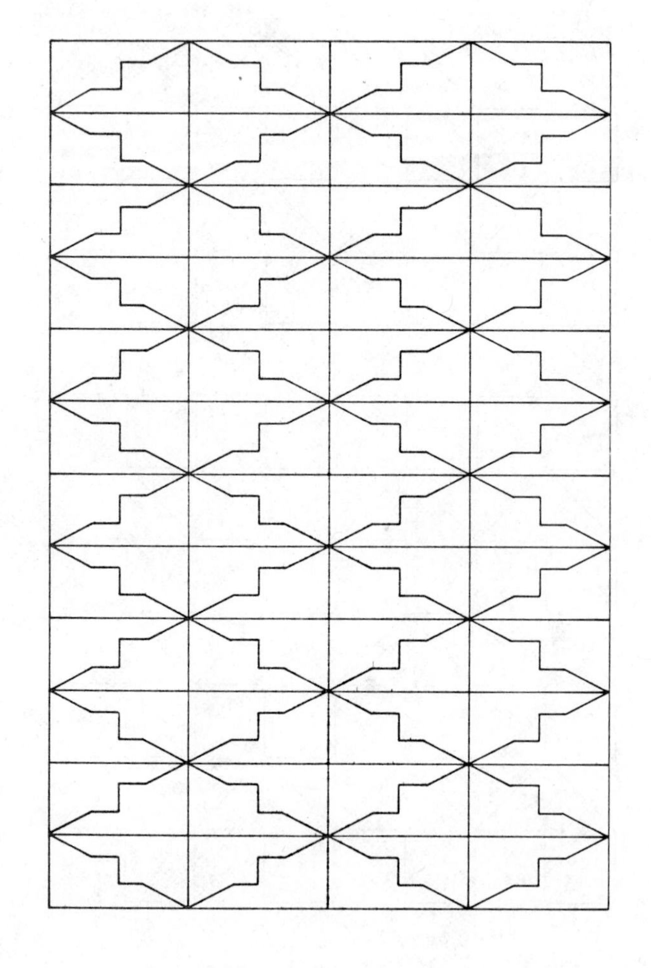

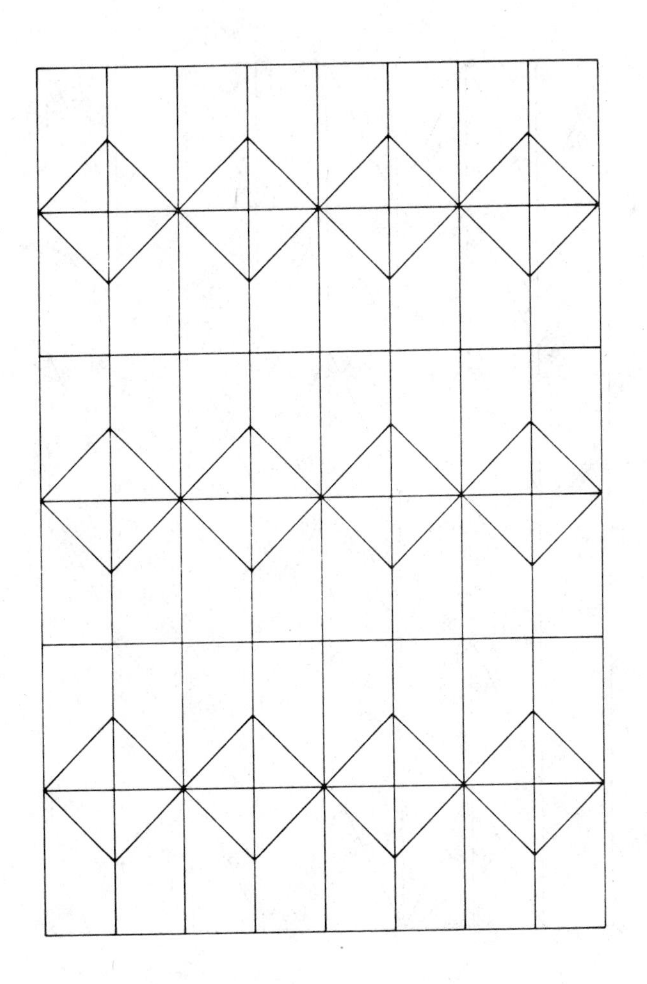

pmm

pmg

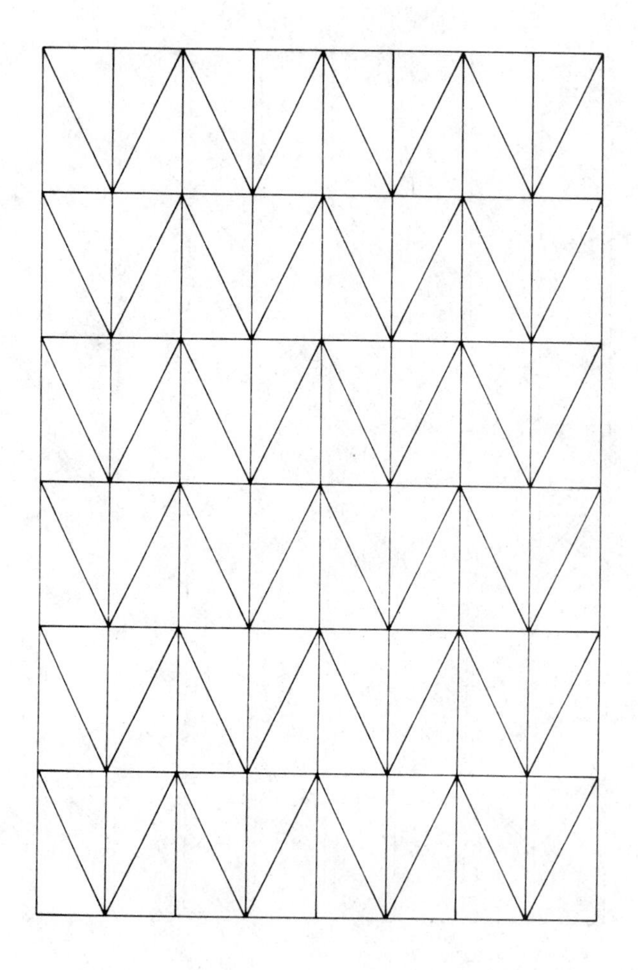

pgg

p31m

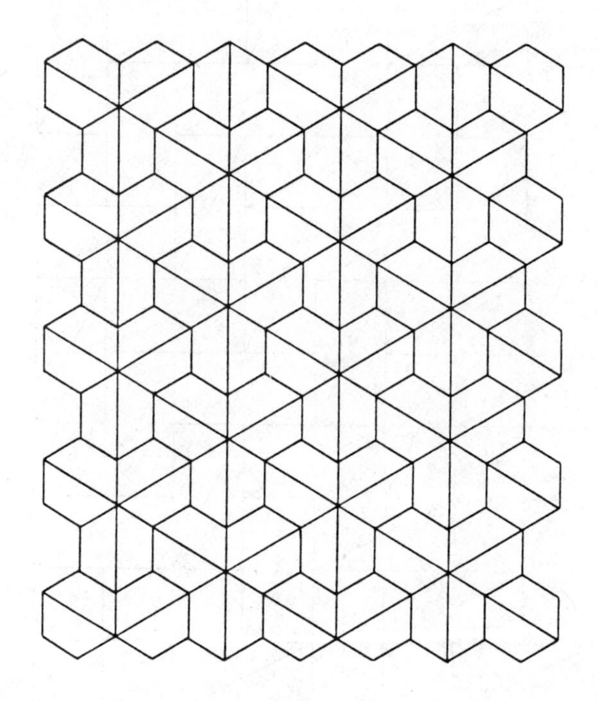

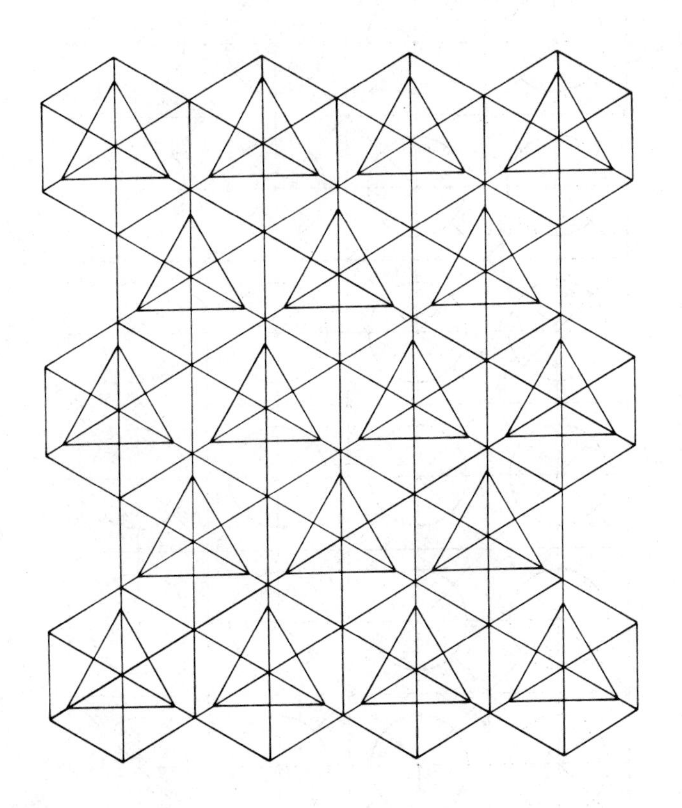

p3m1

p4m

p4g

p6m

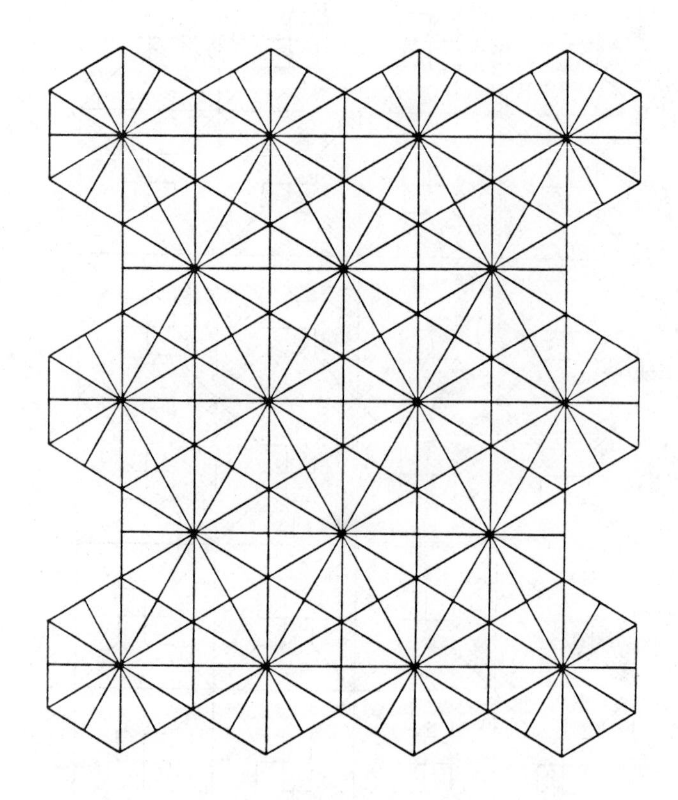

BIBLIOGRAPHY

Ascher, E. and Janner, A., "Algebraic aspects of crystallography: Space groups as extensions," Helv. Phys. Acta 38 (1965), 551-572.

Auslander, L., "An account of the theory of crystallographic groups," Proc. AMS 16 (1956), 1230-1236.

Bieberbach, L., "Über die Bewegungsgruppen des n-dimensionalen euklidischen Raumes mit einem endlichen Fundamental= bereich," Gött. Nachr. (1910), 75-84.

Bieberbach, L., "Über die Bewegungsgruppen der euklidischen Räume," Math. Ann. 70 (1910), 297-336 and 72 (1912), 400-412.

Bourgoin, J., Arabic Geometrical Pattern and Design, Dover, New York, 1973.

Brown, H., "An algorithm for the determination of space groups, Math. Comput. 23 (1969), 499-514.

Brown, H., Bülow, R., Neubüser, J., Wondratschek, H. and Zassenhaus, H., Crystallographic Groups of Four-Dimensional Space, John Wiley and Sons, New York, 1978.

Burckhardt, J. J., Die Bewegungsgruppen der Kristallographie, Birkhäuser Verlag, Basel, 1966.

Coxeter, H. S. M., Introduction to Geometry, John Wiley and Sons, New York, 1969.

Coxeter, H. S. M. and Moser, W. O. J., Generators and Relations for Discrete Groups, Springer-Verlag, New York, 1957.

Dye, D. S., Chinese Lattice Designs, Dover, New York, 1974.

Escher, M. C., The Graphic Work of M. C. Escher, Ballantine, New York, 1971.

Fejes Tóth, L., Regular Figures, Macmillan, New York, 1964.

Fořtová-Šámalová, P., Egyptian Ornament, Allan Wingate, London, 1963.

Gardner, M., "Mathematical Games," Scientific American 233 (July 1975), 112-117, 233 (August 1975), 112-115 and 236 (January 1977), 110-121.

Gombrich, E. H., The Sense of Order, Cornell U. Press, Ithaca, New York, 1979.

Grünbaum, B. and Shephard, G. C., "Perfect colorings of transitive tilings and patterns in the plane," Discrete Math. 20 (1977), 235-247.

Grünbaum, B. and Shephard, G. C., "Satins and twills: An introduction to the geometry of fabrics," Math. Mag. 53 (1980), 139-161.

Grünbaum, B. and Shephard, G. C., "Tilings with congruent tiles," Bull. Amer. Math. Soc. 3 (1980), 951-973.

Grünbaum, B. and Shephard, G. C., "A hierarchy of classification methods for patterns," Z. Kristallogr. 154 (1981), 163-187.

Grünbaum, B. and Shephard, G. C., Tilings and Patterns, Freeman and Co., San Francisco, 1980 (to appear).

Guggenheimer, H. W., Plane Geometry and Its Groups, Holden-Day, San Francisco, 1967.

Hardy, G. H. and Wright, E. M., The Theory of Numbers, Oxford U. Press, Oxford, 1960.

Heesch, H., "Aufbau der Ebene aus kongruenten Bereichen," Gött. Nachr. (1935), 115-117.

Jarratt, J. D. and Schwarzenberger, R. L. E., "Coloured plane groups," Acta Cryst. A36 (1980), 884-888.

Jones, O., The Grammar of Ornament, Van Nostrand, New York, 1972.

Loeb, A. L., Color and Symmetry, John Wiley and Sons, New York, 1971.

Loeb, A. L., Space Structures, Addison-Wesley, New York, 1976.

Macdonald, S. O. and Street, A. P., "The seven friezes and how to color them," Utilitas Math. 13 (1978), 271-292.

MacGillavry, C., Symmetry Aspects of M. C. Escher's Periodic Drawings, A. Oosthoek's Ultgeversmaatschappij, Utrecht, 1965.

MacLane, S. and Birkhoff, G., Algebra, Macmillan, 1967.

Milnor, J., "Hilbert's problem 18," Proc. Symp. Pure Math. 28 (AMS, 1976), 491-506.

Niven, I., "Convex polygons that cannot tile the plane," MAA Monthly 85 (1978), 785-792.

Pólya, G., "Über die Analogie der Kristallsymmetrie in der Ebene," Z. Kristallogr. 60 (1924), 278-282.

Reingold, E. M., Nievergelt, J. and Deo, N., Combinatorial Algorithms, Prentice-Hall, Englewood Cliffs, N. J., 1977.

Schattschneider, D., "The plane symmetry groups," MAA Monthly 85 (1978), 439-450.

Schattschneider, D., "Tiling the plane with congruent pentagons," Math. Mag. 51 (1978), 29-44.

Schwarzenberger, R. L. E., N-dimensional Crystallography, Pitman, San Francisco, 1980.

Senechal, M., "Color groups," Discrete Appl. Math. 1 (1979), 51-73.

Senechal, M. and Fleck, G., Patterns of Symmetry, U. of Massachusetts Press, Amherst, 1977.

Shubnikov, A. V. and Belov, N. V., Colored Symmetry, Pergamon Press, Oxford, 1964.

Shubnikov, A. V. and Koptsik, V. A., Symmetry in Science and Art, Plenum Press, New York, 1974.

Speltz, A., The Styles of Ornament, Dover, New York, 1959.

Stevens, P. S., Handbook of Regular Patterns, The MIT Press, Cambridge, Massachusetts, 1980.

van der Waerden, B. L. and Burckhardt, J. J., "Farbgruppen," Z. Kristallogr. 115 (1961), 231-234.

Warner, S., Classical Modern Algebra, Prentice-Hall, Englewood Cliffs, N. J., 1971.

Weyl, H., Symmetry, Princeton U. Press, Princeton, N. J., 1952.

Yale, P. B., Geometry and Symmetry, Holden-Day, San Francisco, 1968.

Zaslavsky, C., Africa Counts, Prindle, Weber and Schmidt, Boston, 1973.

Zassenhaus, H., "Neuer Beweis der Endlichkeit der Klassenzahl bei unimodularer Äquivalenz endlicher ganzzahliger Substitutionsgruppen," Abh. Math. Sem. Hamburg 12 (1938), 276-288.

INDEX